GRASS SYSTEMATICS

Grass Systematics

SECOND EDITION

By **FRANK W. GOULD** *and* **ROBERT B. SHAW**

TEXAS A&M UNIVERSITY PRESS
College Station

Manufactured in the United States of America

Library of Congress Cataloging in Publication Data

Gould, Frank W.
  Grass systematics.

  Includes bibliographies and index.
  1. Grasses—Classification.   I. Shaw, Robert Blaine,
1949–   .   II. Title.
QK495.G74G725   1983   584′.9   82-45894
ISBN 0-89096-145-X
ISBN 0-89096-153-0 (pbk.)

To the memory of Frank Gould.
His life was a demonstration of enthusiastic, careful,
honest research—"putting the pieces of the puzzle together."

Contents

Preface

The appearance in 1931 of the publication by N. P. Avdulov entitled "Cyto-taxonomic Investigation in the Family Gramineae" marked the beginning of a new era in the science of grass classification. In the ensuing years, information from the fields of cytology, anatomy, genetics, and physiology has provided a broad base for interpretation of variation patterns in grass structure and new concepts of phylogenetic relationships. There has been a significant lag, however, in the communication of research findings as presented in scientific journals to the teacher and student of agrostology.

This book has been designed to serve as a guide and source of information for an undergraduate course in agrostology. The documentation of facts, theories, and hypotheses by literature citations also makes it suitable for graduate studies and systematic grass research. Principal emphasis has been placed on the structure and growth of the grass plant and on the characteristics of the United States grass genera. Discussion of tribes and subfamilies relates particularly to grasses of subtropical and temperate North America, but reference is also made to the representation of these groups on a worldwide basis.

The listings of genera and higher taxa of United States grasses in Chaps. 4 and 5 are in phylogenetic sequence. The key to the United States genera, however, is mainly artificial and is based on the most readily observable characters of the inflorescence. Of the six major divisions of the key, two are tribes (Paniceae and Andropogoneae), and the remainder are artificial groups of genera with similar inflorescence features. It is hoped that the key will serve as a pattern for the development of state and other regional grass keys.

Thanks are due to numerous persons for their assistance and cooperation in manuscript preparation and review, including Peter H. Raven, Arthur H. Holmgren, Charlotte G. Reeder, Thomas R. Soderstrom, David E. Fairbrothers, Jimmie D. Dodd, Aly H. Mohamed, Robert I. Lonard, Girija P. Roy, M. Arshad Ali, and my wife, Lucile J. Gould. I am especially indebted to Dr. Clarence Cottam and Caleb Glazener of the Rob and Bessie Welder Wildlife Foundation for the use of the Welder Refuge facilities and for permission to use line drawings from *Grasses of the Texas Coastal Bend* (F. W. Gould and T. W. Box, Welder Wildlife Found. Contrib. 34, ser. C, revised, 1965) and unpublished line drawings prepared under Welder Foundation sponsorship. Several illustrations have been used from *Grasses of Southwestern United States* (F. W. Gould, *University of Arizona Biology Science Bulletin* No. 7, 1951) with permission from the University of Arizona Press. Most of the excellent line drawings were done by Mrs. Valloo Kapadia. Those from *Grasses of Southwestern United States* were prepared by Mrs. Lucretia Breazeale Hamilton.

FRANK W. GOULD

NOTE TO THE SECOND EDITION

Several months before Dr. Frank W. Gould's death on March 11, 1981, he asked if I would undertake a minor revision of his grass systematics textbook, which had recently gone out of print. He supplied a list of additions and corrections that he wished to appear in the revision. The list included correcting subfamily names; incorporating his most recent concept of the Paniceae, following the classification scheme in *The Grasses of Texas*; and including information on the importance of the photosynthetic pathway and C₄ subtype in grass systematics. Also, he requested that I update the literature cited sections in the systematics portion of the book. This I have tried to do with the aid of students and colleagues. However, no attempt has been made to update Chaps. 3 and 6 or to resolve the controversy concerning the generic relationships within the Triticeae (most recently addressed in *Grasses and Grasslands: Systematics and Ecology*, edited by James R. Estes, Ronald J. Tyrl, and Jere N. Brunken).

When I began the revision, seeking any corrections or additions to the book, I contacted several of Dr. Gould's former students and colleagues: Dr. Robert D. Webster (Dr. Gould's last graduate student and currently a Post-Doctoral Fellow at the Australian National University), Dr. Kelly W. Allred (Department of Range and Animal Science, New Mexico State University), and Dr. Stephan L. Hatch (who replaced Dr. Gould as curator of the Tracy Herbarium, Department of Range Science, Texas A&M University). Their many helpful comments were incorporated into this revision. Special thanks go also to Dr. Arnett C. Mace, Jr., Director, School of Forest Resources and Conservation, University of Florida, for the time and support he provided to allow me to complete much of this revision and to Wendy Zomlefer, Florida State Museum, Gainesville, Florida, for the excellent new illustrations. Finally, without Lucile Gould's tireless efforts to find support for this and all of Dr. Gould's unfinished projects, this revision would not have been possible.

ROBERT B. SHAW

GRASS SYSTEMATICS

1
Introduction

GRASSES AND MAN

Grasses inhabit the earth in greater abundance than any comparable group of plants. Some are adapted to warm, humid, tropical climates. Others are established in polar regions where the growing season is two months or less and direct sunlight is absent for many months of the year. Some are important elements of marsh and swamp vegetation, and others inhabit desert regions where the annual precipitation is 5 inches or less.

The Gramineae is one of the largest of the families of flowering plants. With an estimated 600 genera and 7,500 species,* it ranks third in number of genera, behind the Compositae and Orchidaceae, and fifth in number of species, behind Compositae, Orchidaceae, Leguminosae, and Rubiaceae (Good, 1953). With respect to completeness of representation in all regions of the world and to percentage of the total world's vegetation, it far surpasses all others.

The extent to which man's evolution and present position of domination in the biological world have been affected by grasses is an interesting point of speculation. Most civilizations have developed in grassland regions, and it is probable that were it not for the abundance and widespread distribution of grasses, the human population of the world would not have attained its present level.

Food for human consumption

Even before the time of recorded history, the grains of grasses undoubtedly provided a staple food supply for the human race. Rice, wheat, corn, barley, rye, oats, sorghum, and the various millets are at present the principal cultivated grain grasses. Sugar cane, *Saccharum officinarum*, is the main world source of sugar. The origins of these crop plants are buried in antiquity, with little or no direct evidence remaining of their wild ancestors.

Rice feeds more human beings than any other plant product. The rice plant, *Oryza sativa*, is native to tropical Asia, Africa, and possibly South America, but its cultivation has spread to all the warmer regions of the world. Rice was introduced into China three thousand years before Christ. In the United States, it was first grown in Virginia in 1647, and the culture of rice along the Louisiana-Texas Gulf Coast dates from 1718 (Hedrick, 1919).

Wheat, *Triticum aestivum*, is grown extensively throughout the cooler regions of the world, and the acreage of agricultural lands devoted to wheat exceeds that of any other crop plant. Wheat had its origin in Europe, where a number of wild species occur. Its culture spread to other areas at an early date, and it also was introduced into China about 3000 B.C. The first wheat grown in the New World came from a few grains preserved by a Negro slave of Cortez in 1530. In 1542 it was reported as being grown in the vicinity of

---

* Hubbard (1954) estimated 620 genera and 10,000 species.

Montreal, Canada. The first United States plantings were made on the Elizabeth Islands, off the coast of Massachusetts (Hedrick, 1919). Wheat not only is grown extensively in the cooler regions of the United States but also is planted in the southern states as a cool-season crop.

Corn (maize), *Zea mays*, is a New World plant that probably originated in Mexico from ancestral grasses that also gave rise to the closely related genus *Tripsacum*. It was not known in the Old World before the explorations of Columbus, but now is widespread on all continents. Mangelsdorf and Reeves summarized the present distribution by stating (1939, p. 7):

Today maize or Indian corn is grown in every state of the United States, in every suitable agricultural region on the globe; and a crop of corn is maturing somewhere in the world every month of the year. It grows from North Latitude 58° in Canada and Russia to South Latitude 40° in the Southern Hemisphere. Fields of maize are growing below sea-level in the Caspian Plain and at altitudes of more than 12,000 feet in the Peruvian Andes. Corn is cultivated in regions of less than ten inches of annual rainfall in the semi-arid plains of Russia and in regions with more than 200 inches of rain in the tropics of Hindustan. It thrives almost equally well in the short summer of Canada and the perpetual summer of tropical Colombia. No other crop is distributed over so large an area, and only one, wheat, occupies a larger acreage.

Many thousands of varieties of corn have been developed, more than six thousand in Russia alone. All varieties fall under the general classifications of *dent*, *flint*, *flour*, *sweet*, and *pop*, based on differences in the nature of the storage material in the grain. Most of the corn grown in the United States today is hybrid, combining the characteristics of two or three carefully inbred lines.

Forage for domesticated animals

It is an accepted fact that direct human consumption of grain foods is a more efficient means of utilization with respect to energy relationships than is indirect consumption through animal products. Meat and dairy foods, however, are basic constituents of the American diet and are necessary if high nutritional levels are to be attained. In the United States the principal sources of these products are cattle, sheep, and swine. Dairy cattle, for the most part, are maintained in grazing areas termed *tame pastures* or *improved pastures*. Many species of native and introduced grasses are utilized in improved pastures, often in combination with leguminous plants such as clover and vetch. Fertilization and irrigation are part of high-production management procedures. Beef cattle and sheep are raised mainly on western and southern ranges or pastures and then moved to livestock centers for fattening and "finishing" in feedlots.

Range forage. During the early stages of man's civilization, grass vegetation covered between one-fourth and one-fifth of the world's total land surface. Salter (1952) noted that the original grasslands of the seventeen states comprising the American "West" totaled some 700 million acres. Over this vast expanse roamed the bison, antelope, and deer. When settlers from the East came to the Mississippi Valley in about 1830 and merged with livestock men moving northward from Texas, the livestock industry had its start (Stoddard and Smith, 1955). There were tremendous increases in the numbers of cattle and sheep being grazed on the open ranges up to 1885, when an extremely severe winter, followed by a long period of summer drought, drastically reduced the numbers. With settling, fencing, and cultivation of the fertile middle western plains, the range livestock industry was pushed westward into the mountains, valleys, and arid plains.

The number of highly important native forage grasses in the United States is impressive. Included are the dominants of the grassland associations discussed in Chap. 6. Perhaps best known are the four widespread Mid-western "tall grasses": big bluestem (*Andropogon gerardii*), little bluestem (*Schizachyrium scoparium* or *A. scoparius*), Indiangrass (*Sorghastrum nutans*), and switchgrass (*Panicum virgatum*). Sideoats grama (*Bouteloua curtipendula*) and blue grama (*B. gracilis*) are among the most valuable of the mid-grass forage species. Buffalograss (*Buchloë dactyloides*) is a common element of semiarid plains, especially those subjected to heavy grazing by livestock.

Soil conservation

A high proportion of the world's most fertile and productive soils was developed under a vegetational cover of grass. Roots, stolons, rhizomes, and litter from the annual replacement of leafy culms not only are soil builders but also are effective soil stabilizers. In both agricultural and range forage areas, over-utilization and abuse have resulted in the loss of vast quantities of topsoil by the action of wind and water. Through experience man has learned that a perennial grass cover provides the best means of checking surface soil loss and rebuilding depleted soils.

Bamboo use

The place of bamboo in the lives of millions of human dwellers of tropical regions is an intriguing subject that can only be touched upon here. In tropical Asia, bamboo shoots are an important part of man's diet. Watt (1908) noted that in India alone the annual consumption of shoots for food was 150 million culms. Crops of bamboo grain, though irregular in production, provide a nutritious food reported to have prevented famine on more than one occasion during the past century. Church (1889), in an analysis of the food value of the grain of *Dendrocalamus strictus*, reported it to contain 13.9 percent water, 66.3 percent starch, 11.5 percent albuminoids, and small amounts of fiber, ash, and oil. Non-food uses of bamboo and bamboo products are almost endless. Some of these are as follows: poles and posts for building construction, fence posts, bridges, boat masts, ladders, and animal cages; shafts and rods, spears, bows and arrows, and fishing poles; handles for tools, whips, and knives; furniture of all types; window shades; woven articles such as baskets, trays, and mats; rope and cordage; water conduits and drainpipes; musical instruments, toys and ornaments, and cooking utensils; miscellaneous items such as chopsticks, pipestems, whetstones, sieves, writing paper, facial tissue, and cigarette papers.

The two species of bamboo native to the continental United States, *Arundinaria gigantea* and *A. tecta*, provide valuable livestock forage on the humid, forested ranges of the Southeast. Many exotic species of bamboo have been introduced into the United States as garden and lawn ornamentals. McClure (1951) listed twenty-one bamboo genera represented in cultivation. He noted that *Bambusa multiplex*, *B. bambos*, *B. vulgaris*, and *Sinocalamus oldhami* are the most commonly introduced bamboos in southern Florida and that *A. simonii*, *Phyllostachys aurea*, *P. bambusoides*, and *Pseudosasa japonica* are most frequently grown in cooler regions of the country.

Wildlife management

Many types of wildlife are dependent upon grass and grassland habitats for food, shelter, and normal completion of their life cycle. Nearly all North American large game animals utilize grass as a major or minor part of their food intake. Antelope and the now nearly extinct bison are products of a prairie environment. Such animals cannot exist except in regions of extensive grasslands. Small mammals such as the prairie dog, rabbit, fox, and coyote are restricted to, or frequently associated with, prairie habitats. Similarly, the biota of grasslands includes large birds such as the prairie chicken and certain species of hawks and owls, as well as numerous smaller birds including quail and the ever-present meadowlark.

Marsh and swamp grasses, together with the associated species of *Scirpus* (rushes), *Typha* (cattails), and *Carex* and *Cyperus* (sedges), provide food and cover for numerous species of birds and small mammals. Important in this respect are species of *Glyceria* (mannagrasses), *Zizania aquatica* (annual wildrice), *Zizaniopsis miliacea* (southern wildrice), and several species of the genus *Leersia*, including the widely distributed rice cutgrass, *L. oryzoides*. Migratory birds, especially ducks and geese, during their annual southerly treks feed not only in aquatic and marsh habitats but also in cultivated grain fields. Many of these birds overwinter in the coastal rice and pasture regions of Louisiana and Texas. Geese feed extensively on rice and the weed seedlings of rice fields and also take significant amounts of forage from temporary and permanent pastures of the Gulf Coastal Prairie. It is estimated that 1½ to 2 million geese spend four to six months in this region, and their forage consumption frequently poses a serious problem to the rancher and farmer.

Turf

Numerous benefits are derived from nonagricultural uses of sod-forming grasses, collectively referred to as *turf* or *turfgrasses*. These benefits are mainly aesthetic in nature or associated with sports and recreation. Most apparent of the numerous uses of turf in our highly favored "American way of life" is the grass cover of lawns, parks, highway rights-of-way, cemeteries, and the playing fields of numerous sports including golf, baseball, football, lawn tennis, and soccer. Probably of no less value to urban dwellers is the reduction of dust in the atmosphere and dirt in the streets. Lawn maintenance not only provides a common focus of neighborhood interest and community pride but also is the basis of a multimillion-dollar business in seed, fertilizers, insecticides, herbicides, fungicides, mowers, rollers, sprinklers, and other equipment.

Most turfgrasses are strains or hybrids of a relatively few native and introduced species. Southern lawns and other turfs are mainly species of *Cynodon* (bermudagrass), *Zoysia* (zoysia grasses), *Stenotaphrum secundatum* (St. Augustine grass), *Eremochloa ophuriodes* (centipedegrass), *Paspalum notatum* (bahiagrass), and *Buchloë dactyloides* (buffalograss). Turfgrasses of the cooler regions of the United States include species of *Poa* (bluegrasses), *Agrostis* (bentgrasses), *Festuca* (fescues), and *Lolium* (ryegrasses). These are also grown to a limited extent in the South during the cool months and are here referred to as *cool-season* grasses.

CHARACTERISTICS
OF MONOCOTYLEDONS

Grasses are typical members of the Monocotyledonae, the morphology and anatomy of which have been treated with insight and inspiration by Agnes Arber in her book *Monocotyledons* (1925). Except in the seedling stage, roots of most monocotyledons are adventitious, following the early deterioration of the primary root system. In contrast, the primary root of dicotyledonous plants usually develops into a long-lived taproot.

Correlated with the relatively limited and unspecialized root systems of monocotyledons is the frequent development of highly modified stem and leaf structures such as rhizomes, stolons, corms, and bulbs. These function in mechanical support of the plant, food storage, and asexual reproduction. Although rhizomes generally grow parallel to the surface of the ground, some are positively geotropic and grow straight downward. A species of *Cordyline* of the Liliaceae has a stout vertical rhizome. In some orchids such as *Epipogium aphyllum*, rhizomes with absorbing hairs have taken over the root function of absorption. In *E. nutans*, the surface of the rhizome is smooth and "the intake of water seems to be due to long hairs produced by the scale leaves" (Arber, 1925). Vascular tissue of monocotyledonous stems is in bundles surrounded by parenchyma cells. The vascular bundles may be scattered or arranged in rings. Typically, all cells at the same level of the stem mature at the same time, and secondary growth usually does not take place. Arborescent, woody plants such as *Aloe* and *Yucca* do have a cambium in the cortex which produces cells on the inner side that develop into closed vascular bundles and, on the outer side, parenchyma cells.

Monocotyledonous foliage leaves usually have a sheathing base and a simple, linear, ovate, or oblong blade with parallel nervation and entire margins. Stipules and petioles are usually lacking. The genus *Trillium*, with trifoliate, petiolate, reticulate-veined leaves, is an exception to the general rule. Frequently, as in the Cyperaceae and Commelinaceae, the leaf sheath is tubular. Arber (1925) noted correlations between the characteristic monocotyledonous inflorescence spathe and the leaf sheath. Of interest in this respect is the enlarged and inflated sheath that subtends and partially encloses the inflorescence of many grasses. In *Sporobolus cryptandrus*, the inflorescence occasionally remains completely enclosed.

The floral plan of three or six perianth segments in one or two whorls, three or six stamens in one or two whorls, and a gynoecium of three united carpels is characteristic throughout the Monocotyledonae. This is in striking contrast to the typical dicotyledonous flower, with parts usually in fours or fives. Various reductions in floral structures have resulted in the loss of parts and the development of unisexual flowers in many monocotyledons. *Typha*, the common cattail, with numerous male and female flowers densely clustered in separate portions of the same inflorescence, represents one line of specialization. Very little remains of the flower except the essential structures (stamens and pistils). In the Orchidaceae, the flower has become highly elaborated and specialized for insect pollination. The perianth is zygomorphic and generally ornate. Grasses, which are characteristically wind-pollinated, being insect-pollinated in only a few cases, have small, inconspicuous flowers in which the perianth is reduced to lobes of tissue, the lodicules.

PHYLOGENETIC RELATIONSHIPS OF THE GRAMINEAE

The fossil record

The fossil record indicates that during successive geologic eras, the vegetation of the world has been dominated first by one major plant group and then by another. Ferns, pteridosperms, and lycopods were the dominant plant forms during the Carboniferous periods of the Paleozoic. By the early Mesozoic, the gymnosperms had reached their peak of development. At some time during the late Paleozoic or Mesozoic, flowering plants arose, to become the most successful form of plant life in the world today.

As is true of angiosperms in general, the fossil record of the Gramineae is entirely too incomplete to be of much help in determining phylogenetic relationships. Grasses probably came into being in the Mesozoic, after flowering plants were well diversified. Circumstantial evidence suggests that they arose in a tropical climate. Thomasson (1980a), in an excellent historical review of paleoagrostology, reports that the earliest supposed grass fossils are of doubtful validity. In 1820 the name Poacites was applied to grasslike leaves from late Paleozoic formations (Schlotheim, 1820). Later, on the basis of more complete specimens, these fossils were shown to belong in part to the now extinct gymnosperm order, Cordaitales, and in part to Paleozoic Equisetales (Potonie, 1893). The latter group is represented in modern floras by the single genus *Equisetum*.

The relative position in time of the first definitely known grass fossils is shown in Table 1-1. Well-preserved fossils of vegetative grass structures have been obtained from late Tertiary rocks of Europe. The specimens consist of rhizomes and culms with roots, buds, and basal parts of leaves. They have been identified as reed grasses of the Arundo-Phragmites group (Heer, 1859). Vegetative shoots supposedly of the bamboo *Chusquea* have been reported from late Tertiary rocks of Argentina by Frenguelli and Parodi (1941). Carbonized grass fruits from Tertiary (probably lower Miocene) deposits of the Florissant Beds of Colorado have been placed in various extant genera by Cockerell (1908), Brues and Beirne (1908), Knowlton (1916), MacGinitie (1953), and Beetle (1958).

More recently, well-preserved fossil grass anthoecia (lemma and palea of the upper or fertile floret) have been found and described from the southwestern United States (Frye and Leonard, 1959; Leonard, 1958; Leonard and Frye, 1978) and the central Great Plains (Thomasson 1977, 1978, 1979, 1980b). Thomasson (1978, 1980a) has shown the taxonomic importance of epidermal characteristics of the anthoecia, such as bicellular microhairs, silica-bodies, papillae and cellular arrangement visible with the scanning electron microscope, in relating fossil grasses to extant genera.

Relationships of the Gramineae

Plants of the grass, sedge (Cyperaceae), and rush (Juncaceae) families are often popularly grouped together as "grasses and grasslike plants." Although similar to grasses and sedges in general habit, the rushes differ strikingly in flower and fruit characters. The rush flower regularly has a well-developed

Table 1-1. Grass fossils and the geological time chart

Era	Period	Epoch	Estimated age of time boundaries (in millions of years)	Organisms
	Quarternary	Recent Pleistocene	1	First record of man
Cenozoic		Pliocene	10	
	Tertiary	Miocene	25	First fossil grass fruits
		Oligocene	40	
		Eocene	70	First horse
Mesozoic	Cretaceous (upper and lower)		125	First Angiosperm fossils
	Jurassic		150	Mammals, dinosaurs, reptiles
	Triassic		180	
	Permian		205	Conifers, tree ferns Insects, arachnoids
Paleozoic	Carboniferous (Pennsylvanian and Mississippian)		255	Equisetales
	Devonian		315	Amphibians, scorpions
	Silurian		350	
	Ordovician		430	Marine life only
	Cambrian		510	
Precambrian			3000	

Source: Time chart modified from Andrews, 1961.

perianth of six scarious segments and a dehiscent capsular fruit with numerous seeds. In grasses and sedges, the perianth is absent or represented by highly modified scales or (in the Cyperaceae) bristles, and the fruit is one-seeded and indehiscent.

Grasses and sedges are similar in general habit, in having small flowers borne in bracteate clusters, in the lack of or the extreme modification of perianth segments, in being wind-pollinated, and in the dry, indehiscent, single-seeded fruit. The most apparent morphological differences between the Gramineae and the Cyperaceae are shown in Table 1-2.

Table 1-2. Morphological differences between the Gramineae and the Cyperaceae

	Gramineae	Cyperaceae
Stems	Terete or flattened	Triangular (at least just below the inflorescence)
Leaves	2-ranked	3-ranked
Leaf sheath	Mostly open	Mostly closed
Bract immediately subtending the flower	Usually 2-nerved	With odd number of nerves
Fruit	Usually a caryopsis	Usually an achene, never a caryopsis
Embryo	Lateral to the endosperm	Embedded in the endosperm

The families Gramineae and Cyperaceae have been placed in the same order (Glumiflorae, Graminales, or Poales) by many systematists. Not only has the close relationship of grasses and sedges been assumed, but these families have also been considered close to the Liliaceae and Juncaceae. Recently doubt has been expressed as to the validity of these assumptions. Concerning Gramineae-Cyperaceae relationships, Lawrence (1951) noted that grasses have terminal flowers whose ovaries probably evolved from ancestral types with parietal placentation, whereas sedges have axillary flowers whose ovaries evolved from types with free-central placentation.

Stebbins (1956) suggested close relationships of grasses to "primitive Liliaceae" and Old World, mostly Southern Hemisphere monocotyledons of the families Flagellariaceae and Restionaceae. The Flagellariaceae is a small group of tropical herbs with usually grasslike leaves and small flowers in terminal panicles. The Restionaceae, comprising several genera and many species, is widely distributed in the Southern Hemisphere, but north of the Equator is represented by a single species, this in the Indo-Malaya region

(Hutchinson, 1959). Hutchinson noted that in certain regions of South Africa and Australia, these plants take the place of grasses in the vegetation.

Dr. Robert Thorne (in personal communication) commented that to understand the origin of the grasses one should look closely at the Commelinaceae and particularly the members of the Flagellariineae, whose development from commelinoid ancestors has probably paralleled that of the grasses. He noted that the grasses, restiads, centrolepids (Centrolepidaceae), and flagellariads are all wind-pollinated and have many of the same syndrome of characteristics found in grasses, presumably the result partly of convergent or parallel evolution and partly of common ancestry in the not too distant past, perhaps in the late Cretaceous.

HISTORICAL DEVELOPMENT
OF GRASS CLASSIFICATION

Plant classification is the natural result of man's need and desire to differentiate between the kinds of plants with which he comes in contact. The first plants to be named and grouped in special categories were those with food or medicinal values. The naming of agricultural grasses must have antedated the first historical records by many thousands of years. As man's scientific knowledge increased, the classification of plants developed into the science of taxonomy. From the early efforts of Theophrastus, three hundred years before the Christian era, to the middle of the eighteenth century, plant taxonomy was relatively disorganized. Early taxonomic publications were mainly annotated lists of names, with the names consisting of short descriptive phrases. One of the first papers to deal specifically with grasses was that of Johann Scheuchzer, published in 1708 under the title *Agrostographiae Helvetica Prodromus*. This could be considered the starting point of *agrostology*, the science of grass classification.

Systems of classification

The publications of Linnaeus did much to promote and systematize plant classification. In the first edition of *Species Plantarum* (1753), the designated starting point for the binomial nomenclature of flowering plants, Linnaeus keyed and listed a total of forty grass genera, including such familiar names as *Andropogon, Cenchrus, Panicum, Hordeum, Triticum,* and *Phalaris.* The so-called sexual system of classification developed by Linnaeus was based mainly on the number and arrangement of flower parts. It was highly mechanical and artificial and in some cases resulted in the grouping together of very dissimilar plants. During the nineteenth century, there was a general shift in the objectives of systematics to the grouping together of morphologically similar plants; this is termed *natural classification.*

Robert Brown (1810) was the first to understand the true nature of the spikelet and to recognize it as a reduced inflorescence branch. Linnaeus had interpreted it as a single flower. Brown clearly recognized the two great subdivisions of the Gramineae, the Panicoideae and Pooideae, which he referred to as the Paniceae and Poaceae. He accurately described the spikelet characteristics of these groups and noted the tropical-subtropical distribution of the former, as contrasted with the cool-climate adaptation of the latter.

In 1812 Palisot de Beauvois authored a significant treatise, *Essai d'une nouvelle Agrostographie*, in which he stated that the grass family is undoubtedly the least well known of the higher plant groups. Perhaps the greatest contribution of Beauvois to grass systematics was in the naming and describing of a large number of genera. The classification system of Kunth (1833) received much recognition and was adopted by Endlicher, Palatore, and Steudel. Kunth distinguished thirteen tribes, but recognized no subfamilies. Outstanding among the natural systems was that of Bentham (1881), which was based mainly on morphological characters of the inflorescence and flower. The arrangement of Bentham was again presented in *Genera Plantarum*, by Bentham and Hooker (1883), and was used with modifications by Hackel (1887, 1889), Stapf (1917–1934), Hitchcock (1920, 1935), and Bews (1929). In the system of Bentham, thirteen tribes were grouped in two subfamilies, as shown in Table 1.3.

Table 1-3. The natural classification system of Bentham (1881)

Subfamily PANICOIDEAE	Subfamily FESTUCOIDEAE
Paniceae	Bambuseae
Andropogoneae	Festuceae
Maydeae (Tripsaceae)	Hordeae
Trisetegineae (Melinideae)	Aveneae
Zoysieae	Agrostideae
Oryzeae	Chlorideae
	Phalarideae

A. S. Hitchcock, in *The Genera of Grasses of the United States, with Special Reference to the Economic Species* (1920) and *Manual of the Grasses of the United States* (1935, revised 1951), shifted the tribes Oryzeae and Zoysieae from the Panicoideae to the Festucoideae (Pooideae) and added another small tribe by splitting off the Zizanieae from the Oryzeae. Since its publication, Hitchcock's manual has served as a basis for most ecological and taxonomic treatments of North American grasses and grasslands.

At about the same time that Hitchcock was completing his manual, systems of grass classification were published in Russian by N. P. Avdulov and in French by H. Prat that were to profoundly affect the course of grass systematics. Both were based to a large extent on microscopic characters. Avdulov (1931) reported the results of chromosome studies of some 232 grasses and correlated these data with leaf anatomy, the first seedling leaf, the organization of the resting nucleus, the nature of the starch grains in the fruit, and geographical distribution. He grouped the grasses in two subfamilies, the Poatae (Pooideae) and Sacchariferae (Panicoideae), with the Poatae subdivided into the Festuciformes and Phragmitiformes. In 1932 Prat reported on the significance of the grass leaf epidermis in classification and four years later (1936) published a ninety-three-page treatise entitled *La Systematique des Graminées*. He recognized three subfamilies—the Festucoideae, Panicoideae, and Bambusoideae—and correlated characters of leaf epidermis and anatomy; cytology; morphology of seedlings, embryos, fruits, and the inflorescence; the nature of starch grains; physiology; ecology; and serology in grouping tribes and genera. The systems of both Avdulov and Prat are phylogenetic, with the grasses arranged according to genetic and evolutionary relationships.

One of the first to take up the "new taxonomy" was Dr. C. E. Hubbard of the Royal Botanic Gardens, Kew, England, whose publications of 1948 and 1954 were oriented along the lines established by Avdulov and Prat. The decade 1950–1960 was one of great activity in the field of grass systematics, especially in respect to evidence for the phylogenetic arrangement of the major grass groups. New subfamily groupings were proposed by Pilger (1954), Jacques-Felix (1955), Beetle (1955), Stebbins (1956), and Tateoka (1957). Outstanding among the other numerous contributions were those of de Wet (1954, 1956) concerning the position of *Danthonia* and related genera, Reeder (1957) on the taxonomic significance of the grass embryo, and W. V. Brown (1958) on leaf anatomy in respect to classification. As a fitting climax to this period of progress in grass systematics, a symposium entitled Natural Classification of the Gramineae was held in conjunction with the Ninth International Botanical Congress (Montreal, Canada, 1959). The numerous papers presented were grouped under headings of Anatomy, Embryology, Cytology, Histology, and Physiology. Many of the contributions were subsequently published in *Recent Advances in Botany* (University of Toronto Press). Recent papers concerned with grass phylogeny include those of Prat (1960), Tateoka (1960), Stebbins and Crampton (1961), Parodi (1961), Jacques-Felix (1962), and Reeder (1962). Although total harmony in respect to the arrangement of genera in tribes and tribes in subfamilies has not been achieved, much progress has been made. There is general agreement that the Bambusoideae, Pooideae (Festucoideae), Chloridoideae (Eragrostoideae), and Panicoideae are subfamilies. Prat, Stebbins and Crampton, and Parodi recognized two additional smaller subfamily groups, the Oryzoideae (Pharoideae) and Arundinoideae (Phragmitiformes, Phragmitoideae).

Agrostology in the United States

Outstanding among the early American agrostologists were George Vasey, botanist for the U.S. Department of Agriculture; F. L. Scribner, chief of the Division of Agrostology of the U.S. Department of Agriculture; W. C. Beal, professor of botany in the Michigan Agricultural College (now Michigan State University); and George V. Nash of the New York Botanical Garden. Included in Vasey's publications were two volumes in a series entitled *Illustrations of North American Grasses*: volume I (1890–1891), *Grasses of the Southwest*, and volume II (1892–1893), *Grasses of the Pacific Slope*. These include large, attractive line drawings of 200 species, with short descriptions of each. Scribner published a series of three "pocket-sized" books entitled *American Grasses* (1897, 1899, 1900) containing line drawings and short descriptions of 846 species. Beal published a two-volume book series entitled *Grasses of North America for Farmers and Students* (1887, 1896). Volume I is concerned mainly with general botanical, agricultural, and economic aspects of agrostology, and volume II with the taxonomy of North American genera and species. Nash contributed the section on grasses for Small's *Flora of the Southeastern United States* (1903) and also did the first grass treatments of *North American Flora* (volume 17).

The productive career of A. S. Hitchcock in grass systematics started with his appointment in 1901 as assistant chief of the Division of Agrostology of the U.S. Department of Agriculture. In the following thirty-four years, Hitchcock produced over 250 publications and established himself as one of the truly great American botanists. Most significant of Hitchcock's publications were *The Genera of Grasses of the United States with Special Reference to the Economic Species* (1920, revised 1936), and *Manual of Grasses of the United States* (1935, revised 1951). Throughout most of his career, he was ably assisted by Agnes Chase, who came to Washington as botanical illustrator for the U.S. Department of Agriculture and in 1907 was made assistant in agrostology. Following the death of Hitchcock in 1935, the research activities of the Washington grass herbarium were continued by Mrs. Chase and by Jason R. Swallen. When, after numerous reprintings, Hitchcock's manual became unavailable, Mrs. Chase was called upon to prepare a new edition with nomenclatural corrections and additional information from current research publications. The revision became available for distribution in 1951.

LITERATURE CITED

Andrews, H. N., Jr. 1961. *Studies in paleobotany*. John Wiley & Sons, Inc., New York.

Arber, A. 1925. *Monocotyledons*. Cambridge University Press, London.

Avdulov, N. P. 1931. Karyo-systematische Untersuchungen der Familie Gramineen. (Russian with German summary.) *Bull. Appl. Bot. Suppl.*, 44.

Beal, W. J. 1887. *Grasses of North America for farmers and students*, vol. I. Thorp & Godfrey, Lansing, Mich.

———. 1896. *Grasses of North America for farmers and students*, vol. II. Henry Holt and Company, Inc., New York.

Beauvois, A. M. F. J. Palisot de. 1812. *Essai d'une nouvelle agrostographie; ou nouveaux genres des Graminées, avec figures représentant les caractères de tous les genres*. Paris, typ. Fain.

Beetle, A. A. 1955. The four subfamilies of the Gramineae. *Bull. Torrey Bot. Club*, 82:196–197.

———. 1958. *Piptochaetium* and *Phalaris* in the fossil record. *Bull. Torrey Bot. Club*, 85:179–181.

Bentham, G. 1881. Notes on Gramineae. *J. Linn. Soc. Bot.*, 19:14–134.

———, and J. D. Hooker. 1883. *Genera plantarum*. 3(2):1074–1215.

Bews, J. G. 1929. *The world's grasses: Their differentiation, distribution, economics, and ecology*. Longmans, Green & Co. Ltd., London.

Brown, R. 1810. *Prodomus florae novae Hollandiae*, vol. I. R. Taylor, London, pp. i–viii, 145–590.

Brown, W. V. 1958. Leaf anatomy in grass systematics. *Bot. Gaz.*, 119:170–178.

Brues, C. T., and B. Beirne. 1908. A new fossil grass from the Miocene of Florissant, Colorado. *Bull. Wis. Natur. Histor. Soc.*, 6:170–171.

Church, A. H. 1889. Food grains of India (continued). *Kew Bull.*, 1889:283–284.

Cockerell, T. D. A. 1908. The fossil flora of Florissant, Colorado. *Amer. Mus. Natur. Histor. Bull.*, 24:71–110.

de Wet, J. M. J. 1954. The genus *Danthonia* in grass phylogeny. *Amer. J. Bot.*, 41:204–211.

———. 1956. Leaf anatomy and phylogeny in the tribe Danthonieae. *Amer. J. Bot.*, 43:175–182.

Frenguelli, J., and L. R. Parodi. 1941. Una *Chusquea* fosil de el Mirador. *Notas Del Museo de la Plata: Paleontologia*, 6:235–238.

Frye, J. C., and A. B. Leonard. 1959. Correlation of the Ogallala Formation in western Texas with type localities in Nebraska. *Univ. Texas Bureau Econ. Geol. Report. Invest.*, 32:1–62.

Good, R. 1953. *The geography of the flowering plants*, 2d ed. Longmans, Green & Co., Ltd., London.

Hackel, E. 1887. Gramineae. In A. Engler and K. Prantl, *Die Naturlichen Pflanzenfamilien*, vol. II, pp. 1–97.

———. 1889. Andropogoneae. In A. DeCandolle and C. DeCandolle, *Monogr. Phan.*, vol. VI, pp. 1–716.

Hedrick, U. P. 1919. Sturdivant's notes on edible plants. *Rep. N.Y. Exp. Sta. 1919*.

Heer, O. 1859. *Flora Teriaria helvetiae*, vol. III. Winterthur, Paris.

Hitchcock, A. S. 1920. *The genera of grasses of the United States, with special reference to the economic species.* U.S. Dep. Agr. Bull. 772.

———. 1935. *Manual of the grasses of the United States.* U.S. Dep. Agr. Misc. Publ. 200.

———. 1951. *Manual of the grasses of the United States,* 2d ed. (Revised by Agnes Chase.) U.S. Dep. Agr. Misc. Publ. 200.

Hubbard, C. E. 1948. The genera of British grasses. In Hutchinson, *British flowering plants,* pp. 284–348.

———. 1954. *Grasses: A guide to their structure, identification, uses, and distribution in the British Isles.* Penguin Books, Bungay, Suffolk.

Hutchinson, J. 1959. *The families of flowering plants,* vol. II, *Monocotyledons,* 2d ed. Oxford University Press, London.

Jacques-Felix, H. 1955. Notes sur les Graminées d'Afrique Tropicale. VIII. Les tribus de la serie Oryzoide. *J. Agr. Trop. Bot. Appl.,* 2:600–619.

———. 1962. Les Graminées d'Afrique Tropicale. I. Généralités, classification description des genres. *Bull. Sci. No. 8. Instit. de Recherches Agron. Trop. et des Cult. Vivrieres.*

Knowlton, F. H. 1916. *Mesozoic and Cenozoic plants of North America.* U.S. Geol. Surv. Bull. 696.

Kunth, C. S. 1833. *Enumeration plantarum.* I. *Agrostographie enumeratio Graminearum.*

Lawrence, G. H. M. 1951. *Taxonomy of vascular plants.* The Macmillan Company, New York.

Leonard, A. B. 1958. Two new fossil plants from the Pliocene of Northwestern Texas. *Univ. Kans. Sci. Bull.,* 38:1393–1403.

———, and J. C. Frye. 1978. Paleontology of Ogallala Formation, northeastern New Mexico. *New Mexico Bureau Mines, Mineral Resources Circular,* 161:1–21.

Linnaeus, C. 1753. *Species plantarum.* Holmiae, impensis Laurentii Salvii. 2 vols.

McClure, F. A. 1951. In Hitchcock, *Manual of the grasses of the United States,* 2d ed., p. 29.

MacGinitie, H. D. 1953. *Fossil plants of the Florissant Beds, Colorado.* Carnegie Inst. Wash. Publ. 599.

Mangelsdorf, P. C., and R. G. Reeves. 1939. *The origin of Indian corn and its relatives.* Tex. Agr. Exp. Sta. Bull. 574.

Nash, G. V. 1903. Poaceae. In J. K. Small, *Flora of the southeastern United States,* pp. 48–161.

Parodi, L. R. 1961. La taxonomia de las Gramineae Argentinas a la luz de las investigaciones más recentes. *Recent Advances in Bot.,* 1:125–129.

Pilger, R. 1954. Das system der Gramineae. *Bot. Jahrb.* 76:281–384.

Potonie, R. 1893. Die Flora des Rothliegenden von Thuringen. *Jahrbuch Kaiser Preuseen Geologie Landesant,* 9:1–298.

Prat, H. 1932. L'épiderme des Graminées: étude anatomique et systématique. *Ann. Sci. Nat. Bot.,* 14:117–324.

———. 1936. La systématique des Graminées. *Ann. Sci. Nat. Bot.,* 18:165–258.

———. 1960. Vers une classification naturelle des Graminées. *Bull. Soc. Bot. Fr.,* 107:32–79.

Reeder, J. R. 1957. The embryo in grass systematics. *Amer. J. Bot.*, 44:756–768.

———. 1962. The bambusoid embryo: A reappraisal. *Amer. J. Bot.*, 49:639–641.

Salter, R. M. 1952. Conservation and improvement of soil and water resources with grasses. In *Proc. 6th Int. Grassland Congr.*, vol. 2, pp. 124–133.

Scheuchzer, J. 1708. *Agrostographiae Helvetica Prodromus.* (Tiguri.) Sumtibus auctoris.

Schlotheim, E. F. 1820. *Die Petrefacten-kunde auf ihrem jetzig Standpunkte durch die Beschreibrung seiner Sammlung versteinerter und fossiler Uberreste des thur Pflanzenreichs der Vorwelt erlauter Gotha.*

Scribner, F. L. 1897. *American grasses.* U.S. Dep. Agr. Divis. Agrost. Bull. 7.

———. 1899. *American grasses.* II. U.S. Dep. Agr. Divis. Agrost. Bull. 17.

———. 1900. *American grasses.* U.S. Dep. Agr. Divis. Agrost. Bull. 20.

Stapf, O. 1917–1934. Gramineae. In D. Prain (ed.), *Flora of tropical Africa*, vol. 9. (Published periodically in parts.) L. Reeve and Co., Ltd., Kent, pp. 1–1100.

Stebbins, G. L. 1956. Cytogenetics and the evolution of the grass family. *Amer. J. Bot.*, 43:890–905.

———, and B. Crampton. 1961. A suggested revision of the grass genera of temperate North America. *Recent Advances in Bot.*, 1:133–145.

Stoddard, L. A., and A. D. Smith. 1955. *Range management.* McGraw-Hill Book Company, New York.

Tateoka, T. 1957. Miscellaneous papers on the phylogeny of Poaceae (10). Proposition of a new phylogenetic system of Poaceae. *J. Jap. Bot.*, 32:275–287.

———. 1960. Cytology in grass systematics: A critical review. *Nucleus*, 3:81–110.

Thomasson, J. R. 1977. Fossil grasses, borages and hackberries from southwestern Nebraska. *Univ. Wyoming Contr. Geol.*, 16:34–39.

———. 1978. Observations on the characteristics of the lemma and palea of the late Cenozoic grass *Panicum elegans* Elias. *Amer. J. Bot.*, 65:34–39.

———. 1979. Tertiary grasses and other angiosperms from Kansas, Nebraska and Colorado: Biostratigraphy and relationships to living taxa. *Kans. Geol. Surv. Bull.*, 218:1–68.

———. 1980a. Paleoagrostology: A historical review. *Iowa State J. Res.*, 54:301–317.

———. 1980b. *Archaeoleersia nebraskensis* gen. et. sp. nov. (Gramineae-Oryzeae), a new fossil grass from the late Tertiary of Nebraska. *Amer. J. Bot.*, 67:876–882.

Vasey, G. 1890–1891. *Illustrations of North American grasses*, vol. I, *Grasses of the Southwest.* U.S. Dep. Agr. Bot. Bull. 12. Part I (1890), plates 1–50; part II (1891), plates 1–50.

———. 1892–1893. *Illustrations of North American grasses*, vol. II, *Grasses of the Pacific slope.* U.S. Dep. Agr. Bot. Bull. 13. Part I (1892), plates 1–50; part II (1893), plates 51–100.

Watt, Sir G. 1908. *The commercial products of India.* John Murray, Ltd., London.

2
The Grass Plant

THE ROOT

Enlargement of the coleorhiza and the coleoptile, followed by elongation of the primary root and the mesocotyl, are the first manifestations of grass seed germination (Fig. 2-1). Additional roots, *transitionary node roots*, develop from the embryo within a few hours to several days after germination (Ho-

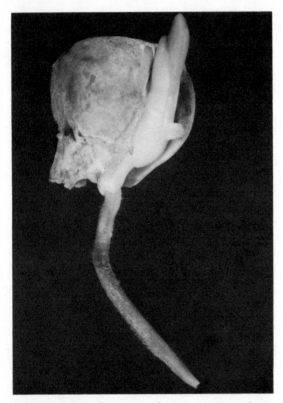

Fig. 2-1. Corn embryo soon after germination, showing the primary root extending downward from the coleorhiza, the elongating primary shoot partially enclosed by the coleoptile, and a seminal root developing between the primary root and shoot.

shikawa, 1969). The primary root and the transitionary node roots, produced from the embryo below the *scutellar* or *transitionary node*, constitute the *seminal* or *primary root system*. The total number of roots in this system is usually one to seven, differing with the species, seedling vigor, and environmental conditions.

Frequently roots arise adventitiously from the mesocotyl (Fig. 2-2) and

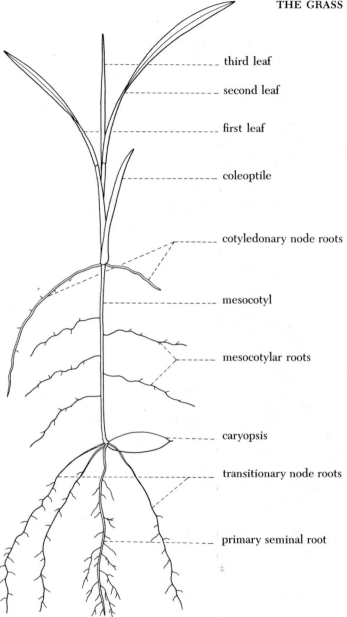

third leaf

second leaf

first leaf

coleoptile

cotyledonary node roots

mesocotyl

mesocotylar roots

caryopsis

transitionary node roots

primary seminal root

Fig. 2-2. Hypothetical grass seedling showing development of adventitious roots (cotyledonary node roots and mesocotylar roots). The primary root system of the seedling consists of the primary seminal root and the transitionary node roots (redrawn from Hoshikawa, 1969).

are referred to as *mesocotyl roots* (Hoshikawa, 1969). Also, during the emergence of the third leaf stage in seedling development, *crown roots* emerge from the cotyledonary node or the basal node of the coleoptile, except in the subfamily Bambusoideae (Hoshikawa, 1969). The mesocotyl roots and the crown roots are the first indication of the *adventitious root system* that eventually will develop into the only root system of most mature grass plants. The significance of the underground organs of seedlings to grass taxonomy will be examined in more detail in Chap. 4.

Knobloch (1944) found that in *Bromus inermis* adventitious roots began developing between the fifth and fourteenth day after germination and constituted over 50 percent of the total root system at about the fortieth day (greenhouse conditions). Weaver and Zink (1945), working with a number of prairie grasses, found one to four adventitious roots twenty-one days after germination and a fairly complete adventitious system at fifty-three to seventy-four days. These investigators also found that the seminal root system remained alive and active as an absorbing organ up to four months. When, by artificial means, the development of adventitious roots was prevented, the seminal root system gave full support to top growth for about three months, after which the plant died.

As has been noted for monocotyledonous plants in general, grass roots are considered to be fibrous. Enlargement of root structures for support or food storage is exceedingly rare. The distribution and quantity of roots depend on both genetic and environmental factors.

In a study of twelve grasses of eastern Nebraska, Weaver (1919) found that six, including *Bouteloua gracilis*, *Buchloë dactyloides*, and *Koeleria pyramidata*, had roots confined to the surface 2 feet of soil, four had root systems reaching 4 to 5 feet in depth, and two, *Panicum virgatum* and *Andropogon gerardii*, had roots reaching the 7- to 9-foot levels.

Weaver (1926) showed that the size of the root system is correlated with the amount of top growth. In general, species producing the greatest top growth also produced the greatest root growth. On a ratio, basis, however, the poorer top-producing species often have a higher ratio of roots to tops than the better top-producing species. Ratios of roots to tops by weight vary from about 0.7:1 to 4:1, and the majority of range grasses have ratios between 0.8:1 and 1.5:1. Cook (1943) found that strains of *Bromus inermis* having a ratio of more than 1:1 roots to tops were more drought-resistant than those with a lower proportion of roots.

Range-management recommendations are often based on the fact that much of the perennial grass root system dies and is replaced each year. Sprague (1933) found this replacement to be about 50 percent in *Poa pratensis* and *Agrostis tenuis*. Weaver and Zink (1945) followed the mortality of the main adventitious root axis in a number of prairie grasses. In one test of grasses growing in their natural habitats, the percentages of main root axes surviving for two growing seasons were as follows: *Koeleria pyramidata*, 30; *Stipa spartea*, 57; *Schizachyrium scoparium*, 23; *Andropogon gerardii*, 45; *Bouteloua curtipendula*, 36; and *Sorghastrum nutans*, 37.

Soil factors influencing root growth are *moisture, temperature, structure, depth, fertility*, and *chemical reaction*. Soil moisture is generally the most important in respect to extension of root systems. In regions of shallow soil moisture, the roots are correspondingly shallow. With deeper moisture, as in well-developed prairie soils, the root systems reach to a much greater depth.

Grasses vary widely in their ability to grow in saturated, poorly aerated soils. Excessive soil wetness inhibits the root growth of most grasses, but some, such as species of *Oryza, Leersia, Zizania, Zizaniopsis*, and *Spartina*, are adapted to and require soil saturation. *Luziola caroliniensis* is an aquatic with floating leaves.

Soil temperature is one of the most important factors in periodicity of grass root growth. Working with grasses of the Midwest which replace most of their roots each year, Stuckey (1941) noted the following general growth pattern: active root growth from October until the advent of freezing air temperatures; slow growth maintained through the cold weather of winter; active growth resumed in the spring and continued until the appearance of flower primordia; slow growth from time of flowering until late June; and cessation of growth in the summer during periods of high soil temperatures.

THE SHOOT SYSTEM

The grass shoot may arise directly from the embryo, or it may originate from a vegetative bud in the axil of a leaf of an older shoot (Evans, 1946). The simple shoot stage in the development of a seedling has been termed the *primary shoot*. This together with one or more orders of *lateral shoots* makes up the *compound shoot system* (Fig. 2-3). In the vegetative stage, the shoot consists of a leafy stem axis with leaves and buds or branches borne alternately at successive nodes. Adventitious roots are developed at the lower nodes of shoots in contact with the ground.

Stems

The erect grass stem is termed a *culm*. Branching at the base of the culm may produce erect lateral shoots; horizontal, above-ground *stolons*; or subterranean *rhizomes*. Lateral shoots, including stolons, rhizomes, and vegetative branches at the upper culm nodes, differ from primary shoots only in their origin from pre-existing shoots rather than from the embryo. In grasses with a cespitose habit, the culms are usually branched only at or near the base, but there are many exceptions. In *Muhlenbergia porteri, Andropogon glomeratus*, and *A. virginicus*, there is profuse branching and rebranching at the upper nodes, with the ultimate branches all floriferous and the entire plant having the appearance of a dense bouquet. In creeping, sod-forming grasses such as *Cynodon dactylon* and *Stenotaphrum secundatum*, the branching continues indefinitely, with only a portion of the leafy stems ter-

Fig. 2-3. Compound shoot system of *Sorghum hale-pense*.

minating in inflorescences. Lateral shoots may be *intravaginal* or *extra-vaginal*. In intravaginal branching the shoot tip emerges from the apex of the enclosing leaf sheath, whereas in extravaginal branching the shoot tip breaks through the sheath. Intravaginal branching is much more common than extravaginal.

With the exception of bamboos, most perennial grasses have herbaceous culms that die back to the base annually and are replaced by shoots from the basal buds. The perennial grass "plant" resulting from several seasons of growth is thus made up of lateral shoots. In time, the plant may assume the aspect of a clone of relatively independent units. As the diameter of such a clone increases, there is often a corresponding dying out in the center, resulting in a ring formation. A striking example of this is provided by ring muhly, *Muhlenbergia torreyi*, in which the relatively narrow band of living grass reaches a diameter of several feet. Harberd (1961; 1962) reported on clonal development in *Festuca rubra* and *F. ovina*. One clone of *F. rubra* was observed to have a spread of over 240 yards. It was estimated that this clone had a possible age of more than one thousand years.

Lateral shoots of cultivated crop and pasture plants are frequently referred to as *tillers* or *suckers*. The former term is used for wheat and other cereals, and the latter for corn and sorghum. In Hitchcock's manual (1951), the basal lateral shoots are referred to as *innovations*. The term *stooling* is applied to cespitose grasses in which the outer culms of a clump are decumbent and spreading at the base.

Culms of bamboos differ from those of herbaceous grasses not only in their perennial nature but also in respect to variability of size and structure. Bamboo culm heights vary from 10 to 15 cm in pygmy bamboos of Japan and the Kuriles to over 40 m in species of *Dendrocalamus* (Arber, 1934).

Stolons may be similar to the erect culms in leafiness and general appearance, as in *Cynodon dactylon* (Fig. 5-94), or highly modified and distinctly different, as in *Buchloë dactyloides* (Fig. 5-101) and *Hilaria belangeri* (Fig. 5-105). The ability of stolons to root at the nodes not only serves to establish and spread the plant but also is a form of asexual reproduction when the stolons are broken.

Rhizomes are recognizable as stems by the regular nodes and internodes and by the production of scale leaves and adventitious roots at the nodes and thus at regular intervals. Rhizomes may be similar to, and intergrading with, stolons, as in *Cynodon dactylon* (Fig. 5-94), or slender, highly modified, and deeply subterranean, as in *Poa arachnifera*, *Agropyron smithii*, and *Chloris chloridea* (Fig. 5-97). Short, stout rhizomes and/or hard, "knotty" culm bases are sometimes incorrectly termed "rootstocks." The most specialized development of the grass rhizome is in the bamboo, where, as noted by McClure (1966), the rhizome system constitutes the structural foundation of the plant. Erect culms arise from nodes or jointed, segmented rhizomes. The length, manner of branching, and orientation of the rhizome system determine the number and spacing of erect culms and thus the general plant habit. McClure recognizes two basic forms of bamboo rhizome systems, pachymorph and leptomorph. In the former (Fig. 2-4A), the rhizome is short and thick, with lateral buds giving rise only to rhizomes and with culm axes arising only from the apex of the usually pear-shaped rhizome segment. In the latter (Fig. 2-4B), the rhizome is long and slender, and every node bears a shoot bud and a verticil of roots. The volume of rhizomes developed by some bamboos is tremendous. In old clumps, with rhizomes doubled and redoubled on themselves, the mass may reach as much as a meter in height above the surface of the ground.

Corms or bulbs may be developed at the base of culms. These structures function in food storage and also as organs of vegetative reproduction. The enlarged food storage cells are usually in short internodes or in the basal portion of scale leaves. *Panicum bulbosum*, a species of mountainous regions of the northwestern United States and Mexico, has hard corms as much as 1

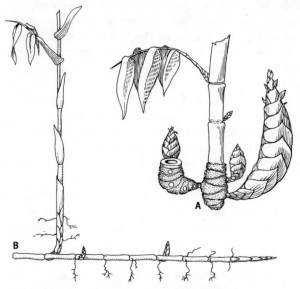

Fig. 2-4. Pachymorph rhizome system of (A) *Arundinaria ambilis* and (B) leptomorph system of *Bambusa beecheyana*. (Reprinted by permission of the publishers from F. A. McClure, *The Bamboos: A Fresh Perspective*, 1966.)

cm in diameter. Several North American species of *Melica* have culms with swollen, scaly bases and thus have acquired the name oniongrass. European species with corms or bulbs that have been introduced into the United States include *Poa bulbosa, Arrhenatherum elatius* var. *bulbosum, Phleum pratense*, and *Molinia caerulea*.

Culm anatomy and histology

The culm in transverse section exhibits typical monocotyledonous stem structure. Vascular bundles of the internode may be scattered in persistent pithy ground tissue, as in *Zea mays*; arranged in two to several rings around a persistent central region of ground tissue, as in *Bouteloua hirsuta* (Fig. 2-5); or arranged in two to several rings around a central ground tissue that breaks down to form a hollow stem, as in typical *Triticum aestivum* (Fig. 2-6). Solid or semisolid internodes are characteristic of grasses of the Pan-

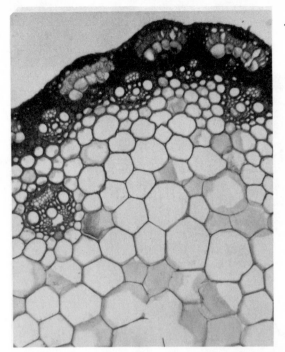

Fig. 2-5. Transverse section of *Bouteloua hirsuta* culm internode showing vascular bundles arranged around the margin of a persistent central ground tissue.

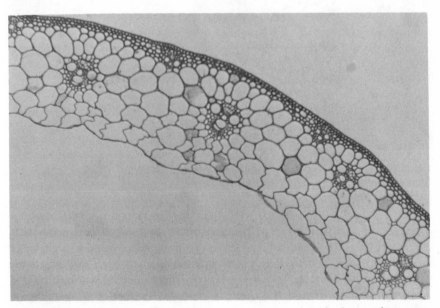

Fig. 2-6. Transverse section of *Triticum aestivum* culm internode showing large central cavity. From Turtox microscope slide preparation.

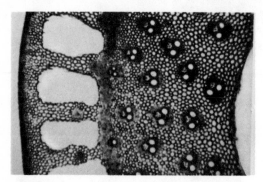

Fig. 2-7. Transverse section of rhizome of *Arundinaria tecta* showing large air canals. (From McClure, 1963.)

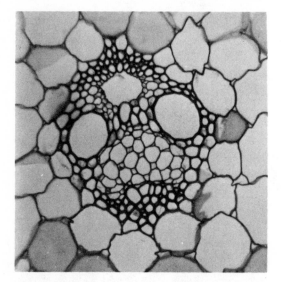

Fig. 2-8. Transverse section of vascular bundle from culm of *Zea mays*. From Turtox microscope slide preparation.

icoideae, especially of the tribe Andropogoneae, and hollow internodes are characteristic of the Pooideae. Variations do occur, even within a single species. Strains of *T. aestivum* with solid internodes have been developed to combat infestation by the wheat-stem sawfly (Booth, 1964; McNeal, 1961).

In addition to the central cavity, culms of a number of grasses, especially those adapted to aquatic or marshy habitats, have other cavities and air chambers. Metcalfe (1960) noted the presence of intercellular air spaces in *Oryza sativa, Leersia hexandra,* and species of *Stipa* and *Sacciolepis.* Mc-Clure (1963) has used the air canals of the rhizomes of *Arundinaria tecta* (Fig. 2-7) as a taxonomic character in the separation of this species from *A. gigantea.*

The epidermis of the culm is similar to that of the leaf blade, consisting mainly of rows of long- and short-cells, with stomata developed at intervals in some of the columns of long-cells. Hair cells of various types may also be present. Beneath the epidermis is a *cortex* of variable width. In the early stages of culm development, the cortex is made up of both parenchyma and sclerenchyma (fiber) cells. In young, green stems the parenchyma cells contain chloroplasts and thus are referred to as *chlorenchyma cells.* As the culm matures, lignification of cortical cells often continues, and one to several layers of sclerenchyma are developed immediately beneath the epidermis.

The number of xylem and phloem elements and of the associated fiber and parenchyma cells of the vascular bundle varies with the size of the bundle (Fig. 2-8). Often there are two *protoxylem* and two *metaxylem vessels* and a number of *tracheids.* At maturity one or both of the protoxylem vessels usually break down, leaving irregular cavities or lacunae. Frequently the outer vascular bundles of the stem are in the cortical region and are partially or completely embedded in mechanical tissue.

Developmental morphology of the grass shoot

In grass seed germination, the enlargement of the coleorhiza and the extension of the primary root are soon followed by rapid elongation of the embryonic shoot and emergence of the coleoptile. Although growth and development of the seedling result from both cell division and cell enlargement, the initial growth is due mainly to elongation of cells differentiated in the embryo. Emergence of the coleoptile is accomplished by elongation of the internode immediately below the coleoptilar node of the embryonic shoot and by growth of the coleoptile itself. Continued growth brings the shoot apex above ground, and the function of protection of the terminal meristem is shifted from the coleoptile to the leaf sheaths and basal portion of the blades.

The *apical* or *terminal meristem* of the shoot is termed the *growing point.* At the extreme tip there is a one- or two-layered *tunica* covering the irregularly arranged cells of the *corpus* (Fig. 2-9). Barnard (1964) stated that two layers are always present and that both the outer layer (*dermatogen*) and the inner layer (*hypodermis*) divide only by anticlinal walls except where leaf primordia are being initiated. He disagreed with the conclusions of Brown et al. (1957) that in a number of grasses, especially those of the Panicoideae and Chloridoideae, there is but one tunica layer.

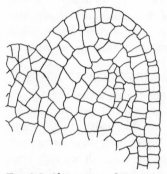

Fig. 2-9. Shoot apex of *Zizaniop-sis miliacea* showing two tunica layers covering the corpus. (Redrawn from Brown et al., 1957.)

The phytomere concept

Just back of the growing-point tip, two rows of ridges are developed alternately on either side of the axis. These constitute the first external indication of a segmental differentiation of the shoot. The ridges make rapid lateral growth until they completely surround the axis and become somewhat disk-shaped (Fig. 2-10). As interpreted by Weatherwax (1923), Evans and Grover

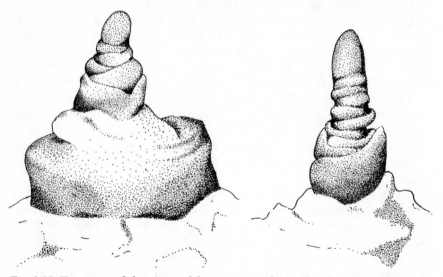

Fig. 2-10. Two stages of elongation of the vegetative shoot apex of *Triticum aestivum*. (Redrawn from Barnard, 1955.)

(1940), and Sharman (1942), these are rudimentary *phytomeres*. The phytomere is considered to be the basic unit of structure of the grass shoot, defined by Weatherwax as "an internode together with the leaf at its upper end and the bud at its lower end." The close relationship of the axillary bud to the leaf above has been noted by Evans and Grover (1940) and by Sharman (1945). The bud originates close to the leaf above it and is later separated from this leaf by the interpolation of an internode between it and the leaf (Esau, 1953). Etter (1951) in studies of *Poa pratensis* found that phytomeres characteristically tend to organize into three main types of shoots: rhizomes, tillers, and flowering shoots. He noted, however, that young phytomeres are quite plastic and may develop into any one of a number of structures. They may become an internode with leaf sheath and blade, as in the lower portion of the culm, or they may become an internode, sheath, bud, and root, as in a rhizome. The terminal phytomere of the flowering shoot is leafless except for the bracts of the spikelet (glumes, lemmas, and paleas).

At first, growth of the phytomere is general throughout, but later it is restricted to the basal portion of the internode and the true node of the culm. The sequence of maturation of the phytomere is from the apex to the base. Cells in the basal portion of the internode usually retain the capability of division for some time, and this region is an *intercalary meristem*. Intercalary meristems make it possible for a shoot that has been blown or knocked down (lodged) to become erect again. Through the action of hormones, the meristem on the lower side of a lodged culm is stimulated to renewed growth, forcing the shoot into an erect position.

Growing-point characteristics

To return to consideration of the growing point, it has been noted that development of rudimentary phytomeres may be more rapid than stem maturation, with the result that there is an accumulation of rudimentry phytomeres on an elongated meristematic core. The length of the entire growing point, however, rarely exceeds 2.0 cm and is usually from 0.5 to 1.0 cm. Sharman (1942) divided the grasses of several genera into three groups based on the number of primordia (rudimentary phytomeres) maintained in the growing point:

1. *The type with twelve to twenty or more primordia*. Species of *Lolium*, *Anthoxanthum*, and *Melica*. In a later paper, Sharman (1947) noted as many as thirty primordia in grasses of this group.
2. *An intermediate type with five to ten primordia*. This was the most common situation in the forage grasses examined (all pooid), including species of *Agropyron*, *Agrostis*, *Festuca*, *Phleum*, and *Phalaris*.

3. *A type with a very short axis and only one to three primordia*. Present in *Avena*, *Secale*, *Triticum*, *Oryza*, *Sorghum*, *Saccharum*, and *Zea*.

In the study, Sharman found that no special significance was associated with the number of primordia maintained at the growing point.

Two factors of great importance in respect to the suitability of specific grasses for livestock ranges and pastures are (1) the time during the growing season at which the apical meristem is elevated to within reach of the grazing animal and (2) the number of basal nodes from which regrowth may occur if the terminal meristem is removed (Rechenthin, 1956). Branson (1953) determined the average heights of shoot growing points during the summer grazing season for several perennial grasses of the central Great Plains region. In *Panicum virgatum* and *Agropyron smithii*, he found that the growing points were elevated early in the growing season to well above the minimum height to which cattle can graze. On the other hand, the growing points of *Schizachyrium scoparium*, *Poa pratensis*, *Bouteloua gracilis*, *Bouteloua curtipendula*, and *Buchloë dactyloides* remained essentially at ground level throughout the summer.

LEAVES
The grass leaf consists of a basal *sheath*, which tightly enfolds the culm, and a flattened *blade* or *lamina*. A membranous or hairy *ligule* is commonly present on the adaxial surface at the apex of the sheath. (Fig. 2-11). Projections of tissue termed *auricles* (Fig. 2-12) may be developed laterally at the apex of the sheath or at the base of the blade. Both sheath and blade carry on the normal leaf functions of photosynthesis and respiration. They also play an important role in the support and protection of the developing shoot. The apical meristem is usually protected by several leaves of the "telescoped" young shoot, and the intercalary meristems of the stem likewise are protected and held erect by the more mature leaf structures. Lateral shoots and branches also develop under the protection of leaf sheaths on the main axis.

Prophyll
The first leaf of a culm branch or lateral shoot characteristically lacks a blade. This modified sheath structure is the *prophyll* or prophyllum, the function of which is to protect the immature lateral stem axis. The prophyll is borne between the branch and the main axis (Fig. 2-13). In the early stages of development, it is tightly appressed to the latter. The central portion is usually curved to fit the contour of the culm, and the margins are folded back over the developing lateral bud or stem. The prophyll is typically membra-

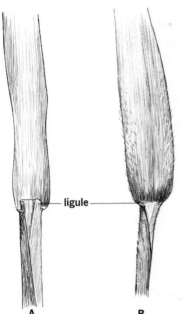

ligule

A   **B**

Fig. 2-11. Grass ligules. (A) Membranous ligule of *Avena fatua*; (B) hairy ligule of *Dichanthelium lanuginosum*.

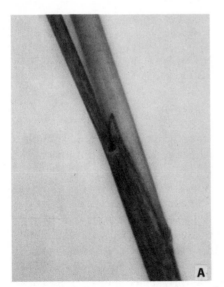

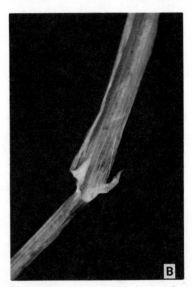

Fig. 2-12. Leaf auricles. (A) Sheath auricle of *Sorghastrum nutans*; (B) blade auricles of *Hordeum vulgare*.

Fig. 2-13. Prophyll (p) at base of lateral shoot (ls) of
Sorghum halepense.

nous and has numerous fine, uniformly spaced nerves, as in the typical leaf
sheath. The nerves on the two lateral folds or "keels," however, are greatly
enlarged.

Sheath

The sheath has been interpreted by some investigators as a flattened petiole.
Typically it has the general shape of a hollow cylinder split down one side.
The margins of the sheath commonly overlap, both at the point of attachment and for all or most of the length of the sheath. In *Glyceria*, species of
Bromus, *Festuca*, and a few other genera, the margins of the sheath are completely or incompletely united (connate) from the base upward. The nerves
of the sheath are usually numerous and relatively uniform in development,
but frequently there is a distinct midrib. Grasses such as *Andropogon virginicus*, *Muhlenbergia emersleyi*, *Poa compressa*, and many species of *Chloris* that have flattened culms also have sharply keeled and laterally compressed leaf sheaths.

Ligule

The ligule is usually a thin, white or brownish membrane, but in some grasses, especially those of the Chloridoideae, it consists of a fringe of hairs or is absent. The Mexican grass *Muhlenbergia macroura* has a broad, firm ligule 2 to 4 cm long. *Sorghastrum nutans* has a stiff, brownish ligule that is usually divided into a rounded central lobe and two stiff, pointed lateral projections. The latter have been referred to as *sheath auricles* (Hitchcock, 1951). In *Leptoloma cognatum* there is a gradual transition from sheath to ligule on either side of the blade attachment. The green lateral nerves of the sheath extend upward into the marginal portions of the ligule. The type of ligule is usually consistent for all species of a genus, but in *Panicum* there are both membranous and hairy ligules, and in some species the ligule is absent.

The junction of sheath and blade is typically marked by a band of mechanical tissue that is different in appearance and texture from either the sheath or the blade. The abaxial surface of this area is termed the *collar*. In most bamboos, there is a petiole-like contraction between the blade and sheath. This is also well developed in species of the tropical genera *Pharus* (Fig. 2-14) and *Zeugites* and in *Setaria palmifolia*, the leaves of which have broad, often ovate or oblong blades.

Fig. 2-14. Leaves of *Pharus parvifolius* showing the petiole-like construction between sheath and blade.

Blade

The leaf blade, similar to that of many other monocotyledonous plants, is typically linear or lanceolate, with parallel nerves and entire, smooth or scabrous margins. Blade size and shape vary within wide limits. The blades of *Monanthochloë littoralis* are infrequently over 1 cm long. At the other extreme, the blades of the bamboo *Neurolepis nobilis* reach 4.5 meters in length and 30 cm in width (Arber, 1934). Grasses of the humid tropics tend to have large, often ovate or oblong blades. In contrast, grasses of semiarid regions commonly have narrow, linear blades that are often involute (inrolled) under drought conditions. *Aciculate* blades are extremely narrow and permanently inrolled to the extent that the internal structure is altered. Metcalfe (1960) noted that in *Miscanthidium teretifolium* the blade has become almost cylindrical in transverse section, and the adaxial surface can be recognized only by a minute groove on the adaxial side of the cylinder.

Variation in nervation of the blade is somewhat greater than in the sheath. Broad, flat blades, such as in *Sorghum halepense* and *Zea mays*, have a large, strongly developed midnerve or "midrib." Narrow blades that become involute usually have uniformly developed nerves and lack a prominent midnerve.

An exception to the typical leaf structure is seen in *Neostapfia colusiana*, a low, tufted annual that is endemic to the dry beds of vernally moist pools in northern California. In this grass the sheath loosely envelops the culm, and there is little or no differentiation of sheath and blade. Species of the closely related genus *Orcuttia* have similar leaves.

Epidermis (Fig. 2-15). The arrangement of epidermal cells on the adaxial surface of the leaf blade is generally quite different from that of the abaxial surface. Epidermal cells are arranged in longitudinal columns or zones, with the zones over the vascular bundles (*costal zones*) usually different from those between the vascular bundles (*intercostal zones*). Most epidermal cells are classified as *long-cells* or *short-cells*, the former being much longer than wide, and the latter frequently as wide as, or wider than, long on the longitudinal axis of the blade.

Long-cells vary in length and width, wall thickness, and the extent to which the walls are sinuous, papillate, or pitted. Significant differences in long-cells exist among genera and even among species of the same genus.

Short-cells occur in rows, in pairs, or singly. They are classified as *silica-cells* when the lumen of each is filled with a *silica-body*, and as *cork-cells* when the walls have the characteristics of cork. The shape of silica-cells varies from round or oblong to linear, crescent-shaped, dumbbell-shaped, nodular (with three or more expanded areas), sinuous, saddle-shaped, or

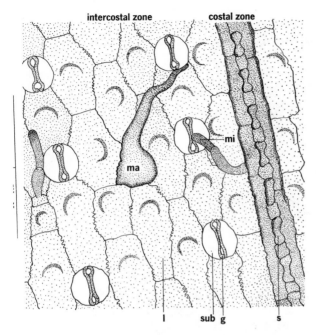

Fig. 2-15. Epidermis of *Echinochloa crusgalli* showing a
narrow costal zone and a portion of a broad intercostal
zone, silica-cells (s), long-cells with sinuous walls and
single large papilla (l), bicellular microhairs (mi) and a
macrohair (ma), and stomata with guard cells (g) and subsidiary cells (sub).

cross-shaped (Fig. 2-16). Saddle-shaped silica-cells are also referred to as
"double-ax-shaped."

Stomata are restricted to the intercostal zones and are usually arranged
in well-defined bands, the character of which varies from species to species
and even from plant to plant. The two *guard cells* of each *stoma* have a characteristic appearance, being somewhat enlarged at the ends and constricted
in the middle. A *subsidiary cell* is present on the side of the guard cell away
from the stoma. This is usually straight-sided, low and mound-shaped or triangular, and it appears to be functionally associated with the guard cell.

Bands of large, colorless *bulliform* cells are frequently present in the
intercostal zones of the blade. Most commonly these are present at the base
of furrows on the adaxial surface, but they may also occur in the abaxial epidermis and on nonfurrowed surfaces.

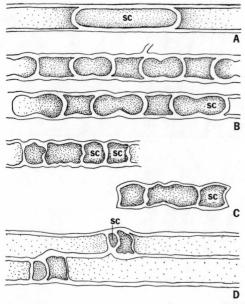

Fig. 2-16. Epidermal silica-cells (sc) of (A) *Bromus willdenowii*, (B) *Cenchrus myosuroides*, (C) *Cynodon dactylon*, and (D) *Sporobolus indicus*.

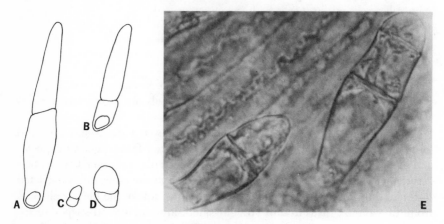

Fig. 2-17. Bicellular microhairs of the leaf epidermis of (A) *Cenchrus myosuroides*, (B) *Paspalum dilatatum*, (C) *Sporobolus indicus*, and (D) *Cynodon dactylon*. (E) Bicellular and tricellular microhairs of palea of *Echinochloa muricata*. (Photo by Arshad Ali.)

Dermal appendages. Metcalfe (1960) recognized four types of dermal appendages: microhairs, macrohairs, prickle-hairs, and papillae. *Microhairs* (Fig. 2-17), as the name implies, are microscopic and can be observed only with a compound microscope. The magnification necessary for study usually lies between 100× and 450×. Microhairs are usually absent from the leaves of typical pooid grasses but are characteristic of most other groups. They are generally two-celled, with a rounded or elongate, extremely thin-walled distal cell. One-celled microhairs have been reported in *Sporobolus* and a few other genera. Metcalfe (1960) illustrated multicellular microhairs for the bamboo *Guaduella oblonga*.

Macrohairs vary greatly in size, shape, and wall thickness. Usually they are much larger than microhairs, have thicker walls, and are one-celled. Short macrohairs grade into *prickle-hairs*, which are also referred to as *asperities*. Prickle-hairs are stout, thick-walled, sharply pointed cells with swollen bases. *Papillae* are outgrowths of various shapes on the epidermal cells. One or more may occur on a single cell, as in *Echinochloa crusgalli* (Fig. 2-15), and they may or may not be cutinized.

The Leaf Blade in Transverse Section. The ground tissue of the blade, the *mesophyll*, is composed of thin-walled chlorenchyma cells and associated colorless parenchyma cells. Rarely in grasses is there a differentiation of the mesophyll into palisade parenchyma and spongy tissue. The chlorenchyma tissue exhibits various cellular arrangements from irregular with large air spaces to tightly packed forms moderately or distinctly radiating around the vascular bundles.

Vascular bundles of the leaf blade contain the same characteristic xylem and phloem elements as those of the stem. Small vascular bundles may be reduced to one or a few tracheids and a similar number of sieve elements. Most veins or nerves of the blade are made up of a single vascular bundle, but large midnerves may contain several. In all cases, the vascular bundles are encircled by one or two *bundle sheaths*. A single sheath, or the outer sheath when there are two, is composed of large, usually thin-walled cells and is referred to as the *parenchyma sheath* (Brown 1958). The inner sheath usually consists of small cells with greatly thickened inner and radial walls. This has been called an *endodermis* or *mestome sheath* by Brown (1958), while Metcalfe (1960) and Barnard (1964) simply refer to it as the *inner sheath*.

Sclerenchyma fibers regularly are associated with all but the smallest vascular bundles. They may be present in clusters between the epidermis and the outer bundle sheath, or they may interrupt the bundle sheath on one or both sides to connect with sclerenchyma of the vascular bundle.

When the sclerenchyma is continuous from the vascular bundle to the epidermis on either side, this is termed *girder* or *I-beam construction.*

Brown (1958) recognized six types of leaf blade anatomy in grasses: pooid, bambusoid, arundinoid, panicoid, aristidoid and chloridoid. The pooid type (Fig. 2-18) is composed of a well-developed mestome sheath sur-

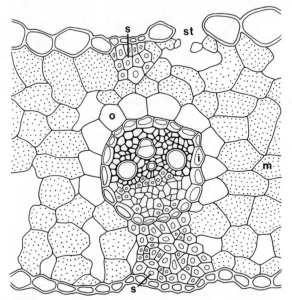

Fig. 2-18. Pooid-type leaf. Portion of transverse section of *Poa* sp. leaf showing vascular bundle with double sheath, the outer of large, thin-walled cells (*o*) and the inner of small cells with thickened inner and radial walls (*i*). The mesophyll (*m*) is of large, loosely and irregularly arranged chlorenchyma cells. Sclerenchyma strands (*s*) are present above and below the vascular bundle, that on the abaxial (lower) side large, interrupting the outer bundle sheath and forming a continuous "girder" of supporting tissue with thick-walled cells of the bundle. A stoma (*st*) is present on the adaxial surface. The abaxial (lower) epidermis is covered with a cuticle.

rounded by a poorly defined parenchyma sheath of small and thin-walled cells. Chloroplasts are present in the parenchyma sheath cells. The chlorenchyma cells of the mesophyll are loosely or irregularly arranged with large intercellular air spaces. The bambusoid type (Fig. 2-19) typically has both a mestome and a parenchyma sheath. The cells of the parenchyma sheath vary in size but are always thick-walled and round or elliptical and contain chlo-

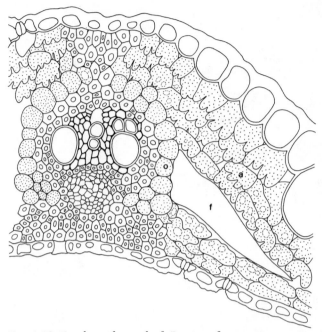

Fig. 2-19. Bambusoid-type leaf. Portion of transverse section of *Maclurolyra tecta* leaf showing vascular bundles with double sheath. The cells of the outer or parenchyma (*o*) sheath are thick-walled and contain chloroplasts. The mesophyll chlorenchyma is composed of arm cells (*a*) and large fusoid cells (*f*) (redrawn from Calderón and Soderstrom, 1973).

roplasts. Mesophyll chlorenchyma is composed of arm-cells with large fusoid-cells that are perpendicular to the vascular bundles. The arundinoid type (Fig. 2-20) usually has a poorly defined mestome sheath with thinner cell walls than the previous types. The parenchyma sheath is composed of large cells which lack chloroplasts. The chlorenchyma cells in this type are characteristically tightly or densely packed. Arm-cells are occasionally present, but fusoid-cells apparently are completely lacking. The panicoid type (Figs. 2-21–2-23) is a variable form. A mestome sheath is typically lacking (Fig. 2-21); however, some species in this group do have a mestome sheath (Fig. 2-22). The parenchyma sheath is composed of large cells with or without specialized starch plastids. The chlorenchyma cells of the mesophyll are irregular or radiate from the parenchyma sheath in various degrees. The aristidoid type (Fig. 2-24) is unique to the genus *Aristida*. A mestome sheath is completely lacking, but a double parenchyma sheath is present. The cells

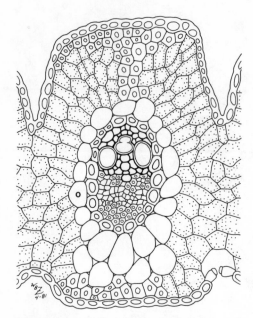

Fig. 2-20. Arundinoid-type leaf. Portion of transverse section of *Cortaderia selloana* showing one large vascular bundle with outer or parenchyma sheath (*o*) lacking chloroplasts (redrawn from Brown, 1958).

of the inner sheath are larger than those of the outer sheath. The cell-walls of both sheaths are thicker than in other types and contain specialized plastids. The chloridoid type (Figs. 2-25, 2-26) contains a mestome sheath at least around the larger bundles, while a single parenchyma sheath with specialized plastids is always present around all bundles. The chlorenchyma cells are characteristically long, narrow, and distinctly radiating from the bundles.

Specialized cells for starch storage in the parenchyma sheaths have been found in all genera with the chloridoid- and aristidoid-type leaf anatomy and in some genera with panicoid- and arundinoid-type anatomy. These specialized cells are characterized by walls thicker than walls of typical mesophyll cells and by numerous plasmodesmata and pits in the cell walls, and the chloroplasts present in these cells differ in some way from mesophyll plastids (Brown, 1975). The term *Kranz cells* (first used by Haberlandt [1884]) has been applied to these specialized cells (Brown, 1977). Two subtypes of Kranz cells have been described based on paradermal leaf sections (Brown, 1974). The M.S. subtype corresponds to Kranz cells that are elongated parallel to the vein and are twice as long as wide. The P.S. subtype correlates to Kranz cells lying perpendicular to the vein and typically twice as wide as long or square. A third subtype, D.S., has been used by Brown (1977) to distinguish the double sheath in most aristidoid grasses. Presence or absence of *grana stacks* within the chloroplasts of the Kranz cells has been found to correspond to the M.S. and P.S. subtypes (Gutierrez, Gracen, and Edwards, 1974). Species with the M.S. subtype lack grana stacks, while the P.S. subtype species contain grana stacks in the chloroplasts of the Kranz cells.

An additional feature of plants with Kranz-type cells is the number of cells containing chlorophyll which lie between adjacent parenchyma sheaths. Hattersly and Watson (1975, 1976) reported adjoining parenchyma sheaths in grasses with Kranz cells were segregated by two to four chlorophyll-containing cells. In contrast, grasses which lack Kranz cells (non-Kranz) always have more than four chlorophyll-containing cells between parenchyma sheaths. The presence of Kranz cells and the corresponding small lateral cell count are indicative of a specialized form of photosynthesis, the C_4 pathway, which is discussed in the next section.

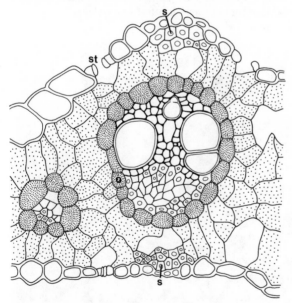

Fig. 2-21. Panicoid-type leaf. Portion of transverse section of leaf of *Echinochloa crusgalli* showing vascular bundle with single outer sheath (o) surrounded by somewhat radically arranged chlorenchyma cells of the mesophyll. Sclerenchyma strands (s) are present above and below the bundle. Stomata (st) are present on both surfaces.

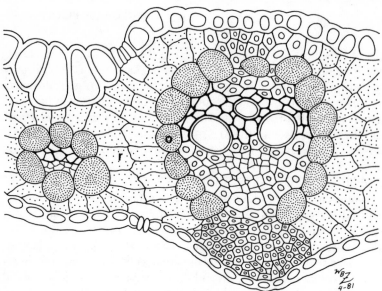

Fig. 2-22. Panicoid-type leaf. Portion of transverse section of *Eriochloa sericea* leaf showing vascular bundle with outer or parenchyma sheath (*o*) with numerous chloroplasts and inner or mestome sheath (*i*). Note the radiating chlorenchyma cells of the mesophyll.

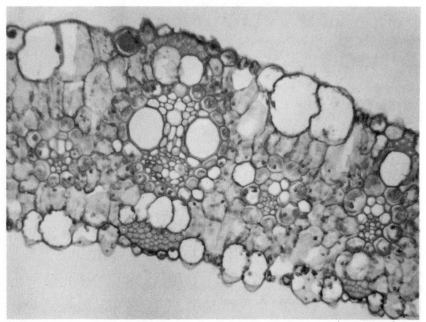

Fig. 2-23. Panicoid-type leaf. Portion of transverse section of *Echinochloa crusgalli* leaf showing one large and three small vascular bundles. Note the papillate cells of the lower (abaxial) surface and the large bulliform cells of the upper (adaxial) surface. (Photo by Arshad Ali.)

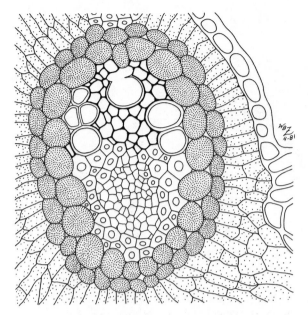

Fig. 2-24. Aristidoid-type leaf. Portion of transverse
section of *Aristida* sp. showing vascular bundle with
two well-developed parenchyma sheaths which contain specialized plastids (redrawn from Brown, 1958).

Photosynthetic pathways

The photosynthetic process in grasses (like in all higher plants) is divided
into two interrelated segments, the *light reaction* and the *dark reaction*.
The light reaction, so called because it only occurs in the presence of light, is
a photochemical reaction in which light energy is converted into chemically
bound energy by chlorophyll and cytochromes. This conversion of energy is
achieved through a process known as *photophosphorylation*. Two types of
photophosphorylation (cyclic and noncyclic) occur simultaneously in the
chloroplasts present in the chlorenchyma cells of the leaf mesophyll. During
cyclic photophosphorylation (Fig. 2-27), adenosine diphosphate (ADP) is
transformed to adenosine triphosphate (ATP). Noncyclic photophosphorylation (Fig. 2-28) also converts ADP to ATP, and in addition transforms nico-
tineamide adenine dinucleotide phosphate (NADP) to its reduced form
(NADPH). The hydrogen ions (H^+) necessary for the conversion of NADP to
NADPH come from the ionization of water. Hydroxyl ions (OH^-), also a
product of the ionization of water, react to yield oxygen (O_2) and water. It is
important to note that the oxygen produced during photosynthesis comes

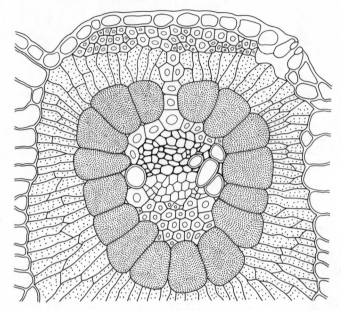

Fig. 2-25. Chloridoid-type leaf. Portion of transverse section of *Bouteloua pectinata* leaf showing large outer bundle sheath and irregular inner sheath of smaller cells with uniformly thickened walls. Note the small, tightly packed, radially arranged chlorenchyma cells surrounding the bundle. For a larger portion of the leaf section, see Fig. 2-26.

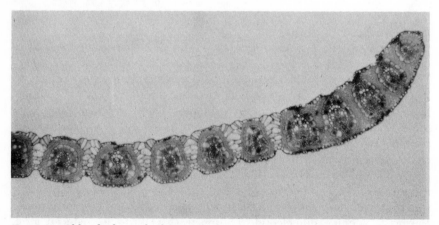

Fig. 2-26. Chloridoid-type leaf. Portion of transverse section of *Bouteloua pectinata* leaf showing separation of vascular bundles by bands of colorless cells continuous with the bulliform cells of one or both epidermises. The bands of colorless cells are absent from the marginal regions of the blade. (Photo from slide prepared by Girija Roy.)

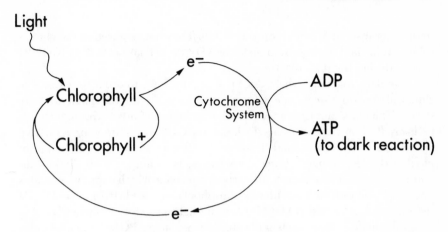

Fig. 2-27. Cyclic photophosphorylation showing conversion of light energy to chemical energy (ADP to ATP) which is used in the dark reaction of photosynthesis (redrawn from Rahn, 1980).

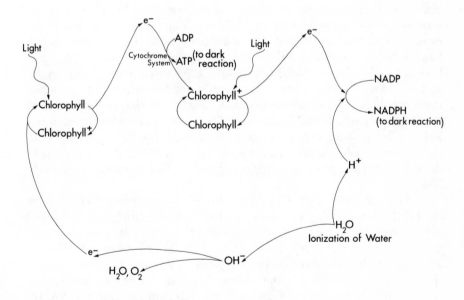

Fig. 2-28. Noncyclic photophosphorylation showing conversion of light energy to chemical energy (ADP to ATP and NADP to NADPH). Note that the H^+ ion necessary for the reduction of NADP to NADPH comes from the ionization of H_2O, which also is the source of O_2 released during photosynthesis (redrawn from Rahn, 1980).

from water instead of carbon dioxide (CO_2). The energy stored in the chemical bonds formed during the transfer of ADP to ATP and NADP to NADPH is used during the dark reaction.

The dark reaction, which does not require light to occur, is a series of chemical reactions in which CO_2 is incorporated into the plant to form starch. The classical form of CO_2 fixation in higher plants is through the C_3 pathway (Calvin-Benson cycle, RuDP carboxylation pathway, or reductive pentose pathway) (Calvin and Bassham, 1962). CO_2, which enters the mesophyll of the leaf through stomatal apertures, is combined with ribulose diphosphate (RuDP) to form a six-carbon molecule which rapidly separates into two three-carbon molecules (3-phosphoglyceric acid, PGA). The PGA is reduced with ATP and NADPH (from the light reaction) to form phosphoglyceraldehyde (a three-carbon phosphorylated sugar, PGH). Two molecules of PGH react to form a sugar, while other molecules of PGH enter into a complex series of reactions to regenerate RuDP to continue the cycle (Fig. 2-29). The entire dark reaction in C_3 plants generally takes place in chlorenchyma cells present in the leaf mesophyll.

An alternate form of the dark reaction (the C_4, β decarboxylation, dicarboxylic acid, Hatch-Slack, or Kranz-type pathway), has been found in numerous grasses (Kortschank et al., 1965; Hatch and Slack, 1966; Smith and Brown, 1973; Brown, 1977; Waller and Lewis, 1979). Grasses with the C_4 form of the dark reaction combine CO_2 and phosphoenolpyruvate (PEP) carboxylase in the chlorenchyma cells to form the four-carbon molecule oxaloacetate (OAA). This substance is converted to malate or aspartate, which is transported inward through the plasmodesmata of the chlorenchyma cells to the Kranz or parenchyma sheath cells. The acids are decarboxylated in the Kranz cells, releasing CO_2 and pyruvate or alanine. The CO_2 then enters the Calvin-Benson cycle as in C_3 plants while the pyruvate or alanine returns to the chlorenchyma cells and is converted back to PEP to continue the cycle (Fig. 2-30).

Three subtypes of the C_4 pathway have been identified (Gutierrez, et al., 1974). These subtypes are based on the decarboxylating enzymes found in the Kranz cells which cause the release of the CO_2 molecules. Plant species which form malate are decarboxylated by the substance NADP-malic enzyme, which is the most common type in C_4 grasses. Aspartate-forming species are decarboxylated by the NAD-malic enzyme or PEP-carboxykinase. Grasses with M.S. subtype Kranz cells characteristically have the NADP-malic enzyme decarboxylating substance, while P.S. subtype Kranz cells occur in grasses with PEP-carboxykinase and NAD-malic enzyme substances (Brown, 1977).

Significant differences in physiological and biochemical characteristics

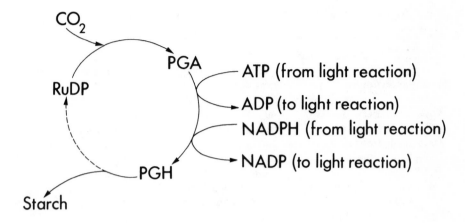

Fig. 2-29. The C_3, Calvin-Benson, or non-Kranz cycle showing the incorporation of carbon (from CO_2) and storage of energy (from the light reaction) in the form of starch (redrawn from Rahn, 1980).

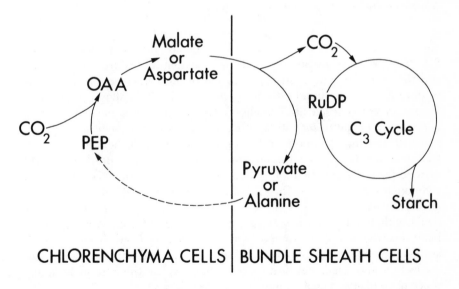

Fig. 2-30. The C_4, or Kranz cycle showing the incorporation of carbon and storage of energy in the form of starch. Note the spatial segregation of this cycle where CO_2 is assimilated in the chlorenchyma cells and incorporated through the C_3 cycle in the bundle sheath cells (adapted from Moore, 1977).

Table 2-1. Summary of physiological and biochemical differences between C_3 and C_4 plants

Characteristics	C_4	C_3
Initial products of photosynthesis	oxaloacetate, malate, or asparate	3-phosphoglyceric acid or hexose phosphates
Photorespiration	no	yes
CO_2 compensation point	0–10 ppm	50–320 ppm
Oxygen suppression of photosynthesis	no	yes
$^{13}C/^{12}C$ ratios	−9 to −19 0/00	−22 to −38 0/00
Light saturation point	1.5 to 1.8 langleys	0.2 to 0.4 langleys
CO_2 assimilation rate	50–80 mg/sq dm/hr	15–35 mg/sq dm/hr
Optimum temperature for photosynthesis	30°–40° C	15°–25° C
Accumulated nonstructural polysaccharides	usually starch	usually fructosans
Water requirements	low	high
Photoperiod	short-day or day-neutral	long-day

Source: Adapted from Waller and Lewis, 1979; Brown, 1977.

have been correlated with the C_3 and C_4 photosynthetic pathways (Table 2-1). The primary function of the distinct anatomical features of C_4 plants, and the corresponding spatial segregation of the dark reaction, is to concentrate CO_2 in the bundle-sheath or Kranz cells. This allows the Calvin-Benson cycle to operate at the highest possible concentration of CO_2, which is often one of the major rate-limiting substrates in photosynthesis (Bjorkman, 1976). Thus, C_4 photosynthesis is a much more efficient form of CO_2 fixation at low CO_2 concentrations than is the C_3 pathway. Bjorkman (1976) divided the characteristics of C_4 plants into three categories: (1) biochemical components necessary for the C_4 pathway, (2) functional components that are a response to increased efficient use of CO_2, and (3) correlated components with but not causally related to C_4 photosynthesis. Apparently C_4 plants evolved in arid environments where these correlated characters, such as tolerance of high light intensity and temperature and water stress, were of a selective advantage. This feature could help explain the infrequent occurrence of C_4 plants in cool, alpine, and arctic habitats.

As demonstrated by numerous researchers, determination of any of the many characteristics of C_4 plants predicts the presence of the others. Brown (1975) suggested that a single term, *Kranz syndrome*, should be used to refer to all of the anatomical and physiological characteristics common to C_4 plants. At present, Kranz anatomy, the C_4 pathway, and the Kranz syndrome have become synonymous. The term *Kranz* can be used in reference to cells, sheaths, plants, species, genera, and families. According to Brown (1975), it has the advantage of being short and unambiguous, and it implies the syndrome of characteristics. Conversely, all plants utilizing the C_3 photosynthetic pathway, and their cells, sheaths, and so on, are designated as *non-Kranz*.

THE INFLORESCENCE

The flowering portion of the shoot, the *inflorescence*, is delimited at the base by the culm node bearing the uppermost leaf. Normally no leaves or bracts except those of the spikelets are present in the grass inflorescence. The leaf at the base of the inflorescence stalk generally has a reduced blade,

Fig. 2-31. *Andropogon glomeratus*. (A) Flowering shoot with prophyll at base and spathe-like leaf at the middle node; (B) cluster of flowering shoots terminating a main culm axis.

but the sheath is often enlarged and spatheate. In grasses with highly compound shoot systems, such as *Andropogon glomeratus*, the ultimate floriferous branchlets represent reduced lateral shoots, each complete with a prophyll at the base, one bracteate leaf, and an inflorescence above (Fig. 2-31).

The *spikelet* (Fig. 2-32), containing one or more flowers, is often referred to as the basic unit of the grass inflorescence. The spikelet consists of a shortened stem axis, the *rachilla*, bearing one to several *florets* and usually

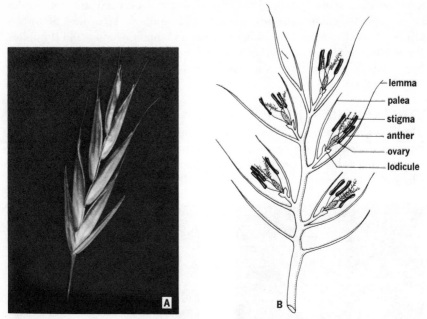

Fig. 2-32. (A) Spikelet of *Bromus carinatus*; (B) diagrammatic interpretation of same spikelet.

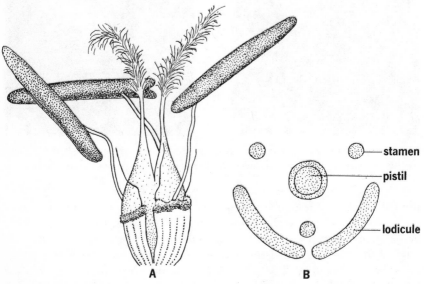

Fig. 2-33. The grass flower. (A) Flower of *Sorghum halepense*; (B) diagram of transverse section of same flower.

delimited at the base by two "empty" floral bracts, the *glumes*. Glumes are said to be empty because they do not have buds or flowers in their axils. The glumes and florets are alternately arranged on the rachilla except in a few highly modified spikelets where the glumes have become opposite by suppression of the internode separating them. Each floret consists of a flower enclosed by two bracts, the *lemma* (lower) and *palea* (upper). The lemma is borne on the rachilla, but the palea is developed at the base of the flower on a lateral axis. Most grass flowers have two lodicules, three stamens, and a pistil with a one-loculate, one-ovulate ovary (Fig. 2-33). The bamboo flower typically has three lodicules and six stamens. Grass flower structure will be taken up in more detail later.

Spikelet arrangement

Spikelets are sessile or pediceled on the main inflorescence axis or in simple or compound branching systems. In considering the type of inflorescence, the spikelet is treated as comparable to a flower. Growth and maturation of the inflorescence are *determinate*, with the terminal spikelets maturing first and the basal ones maturing last. In a few highly modified types, such as the familiar ear of corn, the pistillate inflorescence of *Zea mays*, maturation may be initiated in the middle and proceed both ways. Associated with the determinate sequence of spikelet development, growth in the intercalary meristem at the base of the main flower stalk continues after apical meristem activity has ceased. This is correlated with the well-developed system of protective leaf bases of the shoot and is in full harmony with the general pattern of grass shoot growth and the phytomere concept.

Classically, the terms *raceme*, *spike*, and *panicle* are used to describe the basic types of grass inflorescence. *Compound raceme*, *compound spike*, and *compound panicle* are terms that have been applied to specialized forms. The use of this terminology is unfortunate, as the determinate pattern of growth is cymose rather than racemose. Both spike and panicle are generally defined as variations of the raceme, and by exact definition these also should have the indeterminate flowering of the true raceme.

In the absence of satisfactory alternative terminology, grass inflorescences will be classified as panicle, raceme, and spike, defined as follows:

Panicle (Fig. 2-34)—all inflorescences where the spikelets are not sessile or individually pediceled on the main axis. Most grasses have paniculate inflorescences, and in genera such as *Bromus*, *Poa*, *Avena*, *Eragrostis*, and *Panicum* the branching pattern is relatively simple. Contracted, densely flowered, cylindrical panicles are characteristic of *Phleum*, *Alopecurus*, *Phalaris*, and *Lycurus* and occasional in many genera. Most striking of the panicle modifications are those in which the spikelets are sessile or

Fig. 2-34. Panicle-type grass inflorescences. (A) Open panicle of *Briza minor*; (B) tightly contracted panicle of *Phleum pratense*; (C) panicle of *Leptochloa filiformis* with numerous, scattered, unbranched primary branches; (D) panicle of *Dactyloctenium aegyptium* with digitately arranged unbranched primary branches; (E) panicle of *Bothriochloa ischaemum* with spicate primary branches, the spikelets in pairs of one sessile and perfect and one pediceled and staminate or sterile.

short-pediceled along the primary inflorescence branches. These branches may be digitate at the culm apex or distributed along the main inflorescence axis. All grasses of the tribe Chlorideae have inflorescences of this type, as do many genera of the Paniceae and the Andropogoneae.

Raceme (Fig. 2-35)—an inflorescence in which the spikelets are borne on individual flower stalks, *pedicels*, developed directly on the

Fig. 2-35. Raceme-type grass inflorescences. (A) *Pleuropogon californicus*; (B) *Hordeum pusillum* with a spicate raceme in which the spikelets are in threes at the nodes, the two outer ones short-pediceled and sterile, the central one sessile and fertile; (C) *Danthonia unispicata*, with the inflorescence commonly reduced to a single spikelet.

main inflorescence axis. Included in this category are the spicate racemes of grasses such as *Hordeum* which have both sessile and short-pediceled spikelets at the same node. Racemes with spikelets borne singly at the nodes are infrequent, but are characteristic of *Pleuropogon, Sclerochloa,* and *Brachypodium* and occasional in depauperate plants of *Bromus, Festuca,* and other genera that usually have panicles. Inflorescences of *Brachypodium mexicanum* and *Danthonia unispicata* are usually reduced to a single, terminal spikelet.

Spike (Fig. 2-36)—an inflorescence in which all spikelets are sessile on the main axis. The spikelets may be solitary at the nodes, as in *Lolium Triticum* and *Agropyron,* or two or more per node, as in *Hilaria* and typical plants of *Elymus* and *Sitanion.*

Fig. 2-36. Spike-type grass inflorescence. (A) *Agropyron smithii,* with one spikelet at each node; (B) *Hilaria mutica,* with three spikelets at each node.

Spikelet and floret sexuality

The grass flower may be perfect, staminate, pistillate, or sterile. These terms are also applied to the floret or spikelet as a whole. A *fertile floret* is one with a functional pistil, regardless of the presence or absence of stamens.

Unisexual florets and spikelets are occasional throughout the grass family, but are much less frequent in subfamily Pooideae than in the other two major subfamilies, the Panicoideae and Chloridoideae. In the two-flowered spikelets of the tribe Paniceae, the lower floret with rare exception is staminate or sterile. Both florets are perfect in the tropical genus *Isachne*.

In the tribe Andropogoneae, the lower floret of the two-flowered spikelet is usually sterile and rudimentary, but one of the pair of spikelets borne at each node of the rachis is often staminate. *Hackelochloa*, *Tripsacum*, *Coix*, and *Zea* are monoecious, with strikingly different male and female spikelets. In the spicate raceme of *Heteropogon*, both spikelets of the lower pairs are staminate, as are the pediceled spikelets of the upper pairs.

Many genera of the Chloridoideae are dioecious, including *Scleropogon* and *Neeragrostis* of tribe Eragrosteae, *Buchloë* of tribe Chlorideae, and *Distichlis*, *Monanthochloë*, and *Allolepis* of tribe Aeluropodeae. In the small subfamily Oryzoideae, the genera *Zizania*, *Zizaniopsis*, and *Luziola* are monoecious, with the stamens and pistils often in the same inflorescence. Mainly on the basis of the unisexual spikelets, Hitchcock (1935, 1951) placed these genera in a separate tribe, the Zizanieae. In the present treatment, however, they are included in the tribe Oryzeae.

Number of florets per spikelet
The hypothetical primitive spikelet has several florets, with rudimentary florets, when present, above the fertile ones. Many grasses of the Pooideae, Chloridoideae, Bambusoideae, and Arundinoideae are of this type (Fig. 2-37A). In the Panicoideae, as has been noted, the spikelet is two-flowered, with a perfect upper floret and a staminate or sterile lower floret (Fig. 2-37B). A number of pooid and chloridoid grasses have spikelets with a single perfect floret and no reduced florets (Fig. 2-37C). Most of these, except those of the tribe Chlorideae, were included in the tribe Agrostideae (subfamily Festucoideae) by Hitchcock (1935, 1951). It now is assumed that reduction in the spikelet to a single floret has taken place independently in several unrelated groups. Following Ohwi (1942), Pilger (1954), Tateoka (1957), and Stebbins and Crampton (1961), grasses previously grouped in the Agrostideae are referred to the tribes Festuceae, Aveneae, Stipeae, Eragrosteae, and Aristideae in this text. The genus *Agrostis* is a characteristic member of the Aveneae.

Disarticulation
Disarticulation of the spikelet may be below the glumes, between the upper glume and lowermost floret, or at all nodes of the rachilla above the glumes. In the Panicoideae, disarticulation is below the glumes. The spikelets may fall separately, as in *Paspalum*, *Panicum*, and most other genera of the tribe Paniceae, or they may fall in pairs or groups. In *Cenchrus* and *Pennisetum* of the Paniceae, the spikelets are borne singly or in clusters, subtended by an

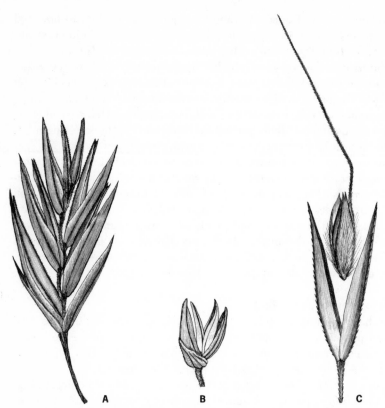

Fig. 2-37. Spikelet types. (A) *Festuca arundinacea* with several florets; (B) *Panicum hemitomon* with two florets, the upper perfect, the lower staminate or neuter; (C) *Agrostis avenacea* with one floret.

Fig. 2-38. Burr of *Cenchrus echinatus*, with fused spines and bristles enclosing a cluster of spikelets. (From Gould and Box, 1965.)

involucre of bristles and / or flattened spines, and disarticulation is below the involucre (Fig. 2-38). Spikelets of the tropical annual *Anthephora hermaphrodita* are borne in clusters of four, with the indurate first glumes united at the base and the cluster deciduous as a whole. In tribe Andropogoneae, the spikelets are typically borne in pairs of one sessile and one pediceled (two pediceled spikelets at the branch tips), and the pair of spikelets falls as a unit together with the pedicel and a section of the rachis (Fig. 2-39). Spikelets of the Pooideae and Chloridoideae for the most part disarticulate above the glumes, but there are many exceptions. In the pooid genera *Limnodea*, *Polypogon*, and *Alopecurus* with one-flowered spikelets, disarticulation is below the glumes. Similarly, in the pooid *Beckmannia syzigachne* and the

Fig. 2-39. Pair of sessile (and fertile) and pediceled (and sterile) spikelets of *Bothriochloa ischaemum* disarticulating as a unit. (From Gould and Box, 1965.)

chloridoid *Uniola paniculata*, with several-flowered spikelets, the spikelets fall entire. In a few genera, such as *Melica* and *Bouteloua*, disarticulation is below the spikelet in some species and above the glumes in others.

Inflorescences of the tribes Triticeae and Monermeae are spikes or spicate racemes, and in many species of the former, and all of the latter, disarticulation is at the nodes of the rachis. The spikelets thus fall entire and attached to sections of the rachis. In a number of chloridoid grasses, especially those of the tribe Chlorideae, the spikelets fall in clusters on short inflorescence branches. This is true of the small genera *Lycurus*, *Tragus*, *Aegopogon*, and *Cathestecum*; the female spikelets of *Buchloë dactyloides*; and about half of the species of *Bouteloua*. In *Schedonnardus paniculatus* and a number of species of *Eragrostis* and *Aristida*, the entire inflorescence breaks off at the base and functions as a tumbleweed in seed dispersal.

Glumes

Glumes, lemmas, and paleas are floral bracts and thus modified leaves. The glumes and lemmas were probably derived directly from leaf sheaths, but paleas appear to be modified prophyll structures. Most frequently, glumes

Fig. 2-40. Spikelet of *Alopecurus myosuroides* with glumes partially united.

are one-, three-, five-, or seven-nerved, but they may be nerveless or, at the other extreme, many-nerved. The size and thickness of the glumes relative to the development of the fertile lemma are important points of differentiation between the tribes Andropogoneae and Paniceae. In the Andropogoneae, the first glume is large and broad, and the glumes are thicker and firmer than the usually membranous lemma of the fertile floret. In contrast, the first glume of the Paniceae is characteristically short or absent, and the glumes are thinner and less firm than the lemma of the fertile floret. In *Digi-*

taria, Leptoloma, an species of *Panicum*, the first glume is minute or vestigial, and in *Anthaenantia, Eriochloa, Axonopus*, and most species of *Paspalum*, it is completely reduced. In *Reimarochloa*, both glumes are usually absent, with the second glume occasionally present on the terminal spikelet.

In *Sitanion, Taeniatherum*, and some species of *Hordeum*, the glumes are setaceous. In a few grasses with one-flowered spikelets, suppression of the internode between the glumes has brought them into an opposite rather than alternate position. In *Alopecurus* and some species of *Polypogon*, the glumes are not only opposite but also partially united at the margins (Fig. 2-40). Glumes typically are absent in members of the tribe Oryzoideae. Structures of the *Oryza* spikelet originally interpreted as being glumes are now believed to be vestigial florets.

Lemmas

The lemma is extremely important in classification, especially at the generic level. It is always present and has a high degree of stability within a genus. Lemma characters of taxonomic importance are shape, texture, size in respect to the glumes, nervation, awn development, and surface features such as pubescence, papillae, and markings. Some of the more important variations are as follows:

Lemma with nine or more nerves and awns: *Pappophorum, Enneapogon*, and related genera.

Lemma with three nerves: most genera of the Eragrosteae and Chlorideae, as *Eragrostis, Tridens, Chloris*, and *Bouteloua*, and a few genera unrelated to these grasses, including the reed grasses *Arundo* and *Phragmites*.

Lemma with a dorsal awn: in most genera of the tribe Aveneae, the midnerve of the lemma diverges to form an awn from the back or base, well below the usually notched apex. The dorsal awn is present in some or all species of *Avena, Helictotrichon, Trisetum, Deschampsia, Aira, Agrostis, Calamogrostis, Alopecurus, Corynephorus, Holcus*, and *Arrhenathrum*.

Lemma firm or hard, terete, completely enclosing the palea and flower, with a sharp-pointed callus at the base and one or three awns from an awn column at the apex: *Aristida, Stipa*, and related genera.

Lemma (of the fertile floret) firm or hard, rounded on the back and usually dorsally compressed, clasping the palea by its margins: genera of the tribe Paniceae (*Panicum, Paspalum, Digitaria, Setaria*, etc.).

Lemma (of the fertile floret) membranous, hyaline, greatly reduced, usually with a stout midnerve and geniculate awn: genera of the tribe Andropogoneae (*Andropogon, Schizachyrium, Bothriochloa, Sorghum, Sorghastrum*, etc.).

As is indicated above, the presence or absence of lemma awns provides a reliable character for differentiation of many grass groups. In some genera, however, such as *Agrostis, Bromus, Agropyron, Elymus, Muhlenbergia, Leptochloa, Chloris,* and *Bouteloua,* the awn is present in some species but not in others. Awns may be the mere hairlike projection of the nerves, or they may be long, stout, geniculate structures highly specialized for seed dispersal. In species of *Avena, Stipa, Heteropogon,* and other genera, the twisted, geniculate awns are hygroscopic and twist and untwist with moisture changes. This action helps bury the mature grain in the soil, with a more favorable position for germination, or in the hair or hide of an animal, for better seed dispersal.

Paleas

The palea is characteristically two-nerved and two-keeled. In *Guadua* and a few other bamboos, the palea has many fine nerves in addition to the strong nerves of the keels, a nervation pattern identical with that of the prophyll. The palea is oriented opposite the lemma and with its dorsal surface against the rachilla, when that structure is continued above the base of the floret. Together the lemma and palea enfold and protect the delicate grass flower. For the most part, paleas are less variable in size, texture, and other characteristics than glumes or lemmas. In a few genera, the palea nerves are awntipped, and others they are conspicuously ciliate or puberulent. Florets of the Andropogoneae, which are characteristically enclosed by large, broad, usually thick glumes, have membranous, reduced paleas, or the palea is absent. In *Agrostis,* with large but thin glumes and a relatively open spikelet, the palea is usually small or absent.

Most grasses of the Oryzoideae have paleas with three or more nerves. This condition is reported for *Leersia,* the pistillate spikelets of *Zizaniopsis* and *Luziola,* and the staminate spikelets of *Zizania.* The spikelets of these grasses are glumeless and one-flowered, bringing the palea into a prominent and exposed position and possibly stimulating the development of additional vasculation.

The probable homology of the palea and prophyll has been discussed by Arber (1925, 1934), Philipson (1935), and others. Arber considered the palea to be a typical two-nerved monocotyledonous bracteole at the base of the flower axis, comparable to the prophyll of the lateral shoot. Weatherwax (1920), Avery (1930), and other investigators have noted the probable foliar nature of the coleoptile of the embryo. If this structure can be considered the first leaf of the primary shoot, then the homology of coleoptile, prophyll, and palea is apparent. These structures are similar in (1) *position,* the first leaf on a vegetative or flower axis; (2) *form,* a modified leaf sheath, basically with many uniformly spaced nerves, the two nerves on the lateral folds or keels enlarged, and the other nerves usually lost from the palea and coleoptile; and (3) *function*, protection of the embryonic or immature axis and its appendages, the coleoptile with the additional function of enzyme secretion.

FLOWER AND FRUIT
The grass flower is terminal on a short axis developed in the axil of the lemma. Immediately below the flower on the same short axis is the palea, which, together with the lemma, encloses the flower.

Lodicules
The lowermost organs of the grass flower are pale green or white mounds of tissue termed *lodicules*. These are generally interpreted as perianth structures. The lodicules and stamens of the grass flower are cyclic, as contrasted with the alternate arrangement of the spikelet bracts (glumes, lemmas, paleas). The number of lodicules is variable, but three are characteristic of bamboos, and two are most frequent in other grasses. Significant differences have been noted in the shape and structure of lodicules of the subfamilies Panicoideae and Pooideae (Fig. 2-41). In the former they tend to be short, truncate, thick, and heavily vasculated. In the latter they are usually elongated, pointed, thick at the base and membranous above, and have little or no vasculation. Bamboo lodicules are similar to the pooid type, but are

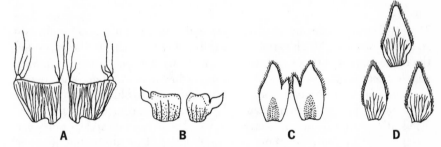

A   **B**   **C**   **D**

Fig. 2-41. Lodicules characteristic of (*A*) panicoid, (*B*) chloridoid, (*C*) pooid, and (*D*) bambusoid grasses. (Diagrams copied from Stebbins, 1956.)

heavily vasculated. Stebbins (1956) distinguished a fourth type of lodicule, that typical of the tribe Chlorideae. This is of the panicoid form, but has little vasculation. Jirásek and Jozífová (1968) surveyed the lodicules of approximately 150 species of grasses. While lodicular characteristics correlated well in numerous cases with other confirmed subfamily characteristics, they reported considerable variation in lodicular features within the subfamilies. This prompted Guédès and Dupuy (1976) to examine the comparative morphology of the lodicules. They concluded that all lodicule types are closely related, and the obvious variations in these structures are a result of developmental differences in the same basic portions of the lodicules.

Lodicules play a role in the opening of the flower at anthesis. At the proper stage of flower maturation, they rapidly become turgid and force the separation of lemma and palea for the exsertion of stigmas and anthers. After anthesis the lodicules return to a nonturgid condition.

Stamens

The flowers of most grasses have 3 stamens, these cyclically arranged above the lodicules. Bamboos and members of the Oryzoideae usually have 6 stamens, in two whorls of three each. Stamen number is high and variable in some bamboos. McClure (1966) noted records of 50 to 120 in individual flowers of *Ochlandra* species. In flowers of a few of the bamboos with 6 stamens, the filaments are united to form a tube of variable length (Arber, 1925). Reduction of stamen number to 2, as in *Anthoxanthum odoratum* and *Glyceria elata*, and 1, as in species of *Vulpia* and *Andropogon*, is occasional throughout the family. Grass anthers are mostly from 0.5 to 5.0 mm long. They are attached near the middle (versatile) or at the base (basifixed) to slender filaments that elongate rapidly at anthesis and then shrivel up when the pollen is released from the anthers. The development of microspores and pollen grains is discussed in Chap. 3.

Pistil

The pistil consists of a one-loculate ovary with a single ovule, and usually two styles and stigmas. Much variation is exhibited in the shape of the ovary and the development of styles and stigmas. The ovary may be globose, as in *Sporobolus heterolepis*; narrow and elongate, as in *Stipa, Aristida*, and species of *Bouteloua, Andropogon*, and *Schizachyrium*; or oblong or ovoid and more or less flattened, as in *Echinochloa crusgalli* and many other grasses. The feathery stigmas may be sessile, elevated on a common style, or on separated styles. In *Zea mays* there is a single long, filamentous style for each ovary or caryopsis of the thickened rachis or cob, and together these comprise the "silk" of the corn ear.

The number of carpels in the ovary has been the subject of considerable controversy. Lawrence (1951) noted that Hackel, Bews, Rendle, and Diels recognized the existence of a single carpel, while Lotsy, Weatherwax, Arber, and Belk maintained that there are three carpels joined edge to edge. Robbins (1931) supported the latter theory by pointing out that in all grasses the pistil contains three fibrovascular bundles, with two extending into style branches and the third continuing into either the dorsal lobe of the pistil or a style branch. As noted by Arber (1934), bamboos characteristically have three styles and stigmas. Since the grass ovule is supplied by only one of the vascular bundles, it does not seem improbable that each of the three bundles supported an ovule in ancestral types. The grass ovary thus appears to be a reduced form of the typical three-carpellate monocotyledonous ovary.

Fruit

As related by Arber, the ovule of the Gramineae has two integuments, each usually composed of two cell layers. The outer integument is gradually absorbed after fertilization, but the inner integument persists to form the seed coat. The inner portion of the *pericarp* (ovary wall) also breaks down and is absorbed, and the remaining pericarp tissue becomes united with the seed coat. The resulting dry, indehiscent, one-seeded fruit, in which the pericarp and seed coat are fused, is a *caryopsis*. The mature caryopsis is made up of the embryo, the endosperm that usually much exceeds the embryo in volume, and the fused seed coat and pericarp.

In a few genera, including *Sporobolus* and *Eleusine*, the seed coat remains separate from the pericarp. This type of fruit has been classified as an *achene*, but Lawrence (1951) referred to it as an *utricle*, defined as being a bladdery, one-seeded, indehiscent fruit. In *Sporobolus* the inner portion of the pericarp becomes mucilaginous on being moistened, and then, with slight drying, the seed is expelled from the pericarp.

In common terminology, grass fruits are generally referred to as *grains*. This is a general term that may apply to the ovary alone or to the ovary plus persistent spikelet bracts. Many relationships exist in different grasses between the mature ovary, the spikelet bracts and rachilla, and other inflorescence structures. In a few genera such as *Sporobolus* and *Eragrostis*, the lemma and palea spread widely or disarticulate separately to allow the ripened ovary to fall free. In many grasses, the lemma and palea loosely invest the mature ovary and disarticulate with it. The mechanical "cleaning" of such grain is relatively easy. In a number of genera, including *Bromus*, *Hordeum*, *Elymus*, and *Agropyron*, the mature caryopsis tends to adhere to the palea. Caryopses of *Aristida*, *Stipa*, *Phalaris*, *Oryza*, and all grasses of the Paniceae are tightly invested by the lemma and palea. In the Andropogoneae, it is the glumes that permanently enclose the caryopsis, and here a pedicel and section of the rachis often remain attached to the spikelet. In *Cenchrus* and *Buchloë*, the fruits are borne in burrs or burrlike clusters. As has been noted, the caryopses of *Coix* are enclosed in a hard, beadlike covering derived from a highly modified leaf sheath.

Embryo. The general structure of the *grass embryo* is shown in Fig. 2-42. At either end of the embryonic axis are the *plumule*, or embryonic shoot, and the *radicle*. Attached laterally at the scutellar node or transitory node is the *scutellum*. This functions in enzyme secretion and the absorption of food materials from the endosperm. The scutellum has been interpreted by Barnard (1964) and others as the cotyledon. Enclosing the plumule is the *coleoptile*, which is attached on the scutellar side of the embryonic axis. The probable homology of the coleoptile and the prophyll and palea has

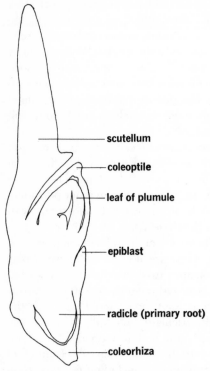

— scutellum

— coleoptile

— leaf of plumule

— epiblast

— radicle (primary root)

— coleorhiza

Fig. 2-42. The embryo of *Avena fatua* var. *sativa*. (Diagram based on figure in Avery, 1930.)

been discussed (p. 62). Protecting the root tip is a nonvasculated sheath, the *coleorhiza*. In the germinating embryo of some grasses, the coleorhiza develops hairs similar in structure and function to root hairs. Just above the coleorhiza and opposite the scutellar node, a flange of tissue termed the *epiblast* may be present. As interpreted by Brown (1959), this appears to be an outgrowth of the coleorhiza. Brown noted that the epiblast and coleorhiza have the same large-celled structure and form a continuous tissue. The few observations that have been made (Brown, 1959; LaRue, 1936) indicate that when hairs are present on the coleorhiza, they are also present on the epiblast. The epiblast is present on most pooid and chloridoid embryos, but is typically absent from the panicoid embryo.

In embryos of panicoid and choridoid grasses, there is a cleft between the coleorhiza and the scutellum and a distinct internode (*mesocotyl*) between the insertion of the scutellum and coleoptile. Pooid embryos do not have the cleft and also do not have a mesocotyl, the coleorhiza and scutellum both being attached at the scutellar node. The application of embryo characteristics to grass taxonomy has been dealt with by Yakovlev (1950), Reeder (1953, 1957, 1962), and Kinges (1961) and will be treated in more detail in Chap. 4.

The *endosperm*, the product of the fusion of two polar nuclei of the embryo sac and a sperm nucleus, provides the food supply for the embryo and young seedling. In most grasses, the endosperm is solid and starchy, but Brown (1955) reported a liquid endosperm in *Limnodea arkansana*, and Dore (1956) found this condition to be present in eighteen species of the genera *Koeleria*, *Trisetum*, *Sphenopholis*, and *Helictotrichon*, all of the tribe Aveneae.

EVOLUTIONARY TRENDS

Concepts of evolution within the grass family are based mainly on characteristics of living grasses. As has been noted, the fossil record is entirely too incomplete to be of much significance. The earliest grasses were probably herbaceous perennials, as are the majority of grasses today. The woody culms and tree habit of bamboos are believed to be secondary developments (Arber, 1934; Stebbins, 1956).

Vegetative structures

Many monocotyledonous plants have stem internodes with vascular bundles scattered in a continuous ground tissue or pith. This condition can be considered primitive in the grasses. Typical pooid culm internodes, with a central cavity and vascular bundles arranged in one to a few rings around its perimeter, would thus be an advancement.

With respect to leaf structure, the presence of bicellular microhairs in the epidermis appears relatively primitive, and the absence of these hairs in most pooid grasses would thus be more advanced. Mesophyll with elongated, closely packed, and regularly arranged chlorenchyma cells radiating out from the vascular bundles is more specialized than that with irregularly and loosely arranged cells. Further specialization in chlorenchyma tissue is seen in the development of large chloroplasts in the bundle sheath and the concentration of starch formation in these cells, which is most highly developed in grasses of the tribe Chlorideae of the subfamily Chloridoideae.

The inflorescence

The primitive grass inflorescence can be logically assumed to have been a relatively unspecialized panicle. From this branching system was derived the panicle with spicate or racemose branches, the raceme, and the spike. The inflorescence of a few grasses regularly or occasionally is reduced to a single spikelet. In two such taxa, *Danthonia unispicata* and *Brachypodium pringlei*, the single spikelet is pediceled, and the inflorescence is basically a raceme. Reduction of the inflorescence from a panicle to a spike has taken place independently in all three major subfamilies. In the Chloridoideae and Panicoideae, there has been a strong tendency for the spikelets to become sessile or short-pediceled on the primary inflorescence branches.

Fig. 2-43. Abnormal inflorescence of *Agropyron smithii*, the spikelets with glumes sessile on the inflorescence axis but the florets widely spaced on an elongated rachilla. See Fig. 2-36 for a normal inflorescence of the same species.

Table 2-2. Comparison of presumed primitive and advanced grass spikelet characters

Primitive	Advanced
Spikelet large, many-flowered	Spikelet small, few- or one-flowered
Glumes large, leaflike in texture, several-nerved, awnless or short-awned	Glumes variously modified, in some cases reduced or absent, in others highly developed for flower protection or seed dispersal
Lemma like the glumes	Lemma conspicuously different from the glumes
Palea present, two- to many-nerved	Palea nerveless, reduced or absent
Lodicules six or three	Lodicules two or one
Stamens six, in two whorls	Stamens three, two, or one, in one whorl
Stigmas three	Stigmas two or one

The primitive spikelet is believed to have been large and several-flowered and to have had perfect florets. The general pattern of modification of the spikelet from a bracteate flowering branch system is not too obscure. Stebbins (1950) discussed in detail the probable evolutionary derivation of the spikelet, using *Avena fatua* as an example. He concluded that the entire oat grain with its husks contains, in addition to the embryo and endosperm, all or part of eleven different appendages, which originally were modified from as many different branch systems. In teretological (abnormal) inflorescences, leafy branches may replace the florets of the spikelet. In so-called viviparous grasses (see Chap. 3), the spikelet becomes transformed into a miniature leafy shoot that can develop roots and become an independent plant. An interesting spikelet modification was observed in *Agropyron smithii* grown south of its natural range at College Station, Texas. This species has a spike inflorescence with several-flowered spikelets borne singly at the nodes of the main axis. In the College Station plants, the spikelets remained sessile, but the rachilla became greatly elongated above the glumes and between the florets. The florets thus were widely spaced, with the uppermost as much as 8 inches above the glumes (Fig. 2-43).

Table 2-2 presents a comparison of presumed primitive and advanced grass spikelet characters. The listing is partially from Hubbard (1948). No attempt has been made to include all the primitive or highly advanced characteristics of some members of the bamboo group.

Photosynthetic Pathways

Brown (1977) discussed in length the evolutionary significance of the C_4 photosynthetic pathway and the different subtypes of the Kranz syndrome. He proposed that the Kranz syndrome evolved at least seven times within the grasses: three separate times in the Paniceae (once each for the NAD-malic enzyme, NADP-malic enzyme, and PEP-carboxykinase subtypes of C_4 photosynthesis); twice in the Aristideae; at least once in the Danthonieae; and once in the Chloridoideae. It was assumed, because all the Chloridoideae have the P.S. subtype of Kranz anatomy, ancestral chloridoid grasses must have already been Kranz to allow the subsequent differentiation into specific tribes. He concluded that in most cases the evolution of the Kranz syndrome was an adaptation to regions with high temperatures, high light intensities, and low soil moisture. The P.S. subtype of Kranz leaf anatomy, with the NAD-malic enzyme subtype of C_4 photosynthesis, appears best adapted to these environments, since this is the most common form found in grasses of desert areas.

LITERATURE CITED

Arber, A. 1925. *Monocotyledons.* Cambridge University Press, New York.

————. 1934. *The Gramineae: A study of cereal, bamboo, and grass.* Cambridge University Press, New York.

Avery, G. S. 1930. Comparative anatomy and morphology of embryos and seedlings of maize, oats, and wheat. *Bot. Gaz.,* 89:1–39.

Barnard, C. 1955. Histogenesis of the inflorescence and flower of *Triticum aestivum* L. *Aust. J. Bot.,* 3:1–20.

————. 1964. Form and structure. In C. Barnard (ed.), *Grasses and Grasslands.* Macmillan & Co., Ltd., London, pp. 47–72.

Bjorkman, O. 1976. Adaptive and genetic aspects of C_4 photosynthesis. In R. H. Burris and C. C. Black (eds.), *CO_2 metabolism and plant productivity.* University Park Press, Baltimore, Md., pp. 287–309.

Booth, W. E. 1964. *Agrostology.* Edwards Brothers, Inc., Ann Arbor, Mich.

Branson, F. A. 1953. Two new factors affecting resistance of grasses to grazing. *J. Range Manag.,* 6:165–171.

Brown, W. V. 1955. A species of grass with liquid endosperm. *Bull. Torrey Bot. Club,* 82:284–285.

————. 1958. Leaf anatomy in grass systematics. *Bot. Gaz.,* 119:170–178.

————. 1959. The epiblast and coleoptile of the grass embryo. *Bull. Torrey Bot. Club,* 86:13–16.

————. 1974. Another cytological difference among the Kranz subfamilies of the Gramineae. *Bull. Torrey Bot. Club,* 101:120–124.

————. 1975. Variations in anatomy, associations, and origins of Kranz tissue. *Amer. J. Bot.,* 62:395–402.

————. 1977. The Kranz syndrome and its subtypes in grass systematics. *Mem. Torrey Bot. Club,* 23:1–97.

————; C. Heimsch; and W. H. P. Emery. 1957. The organization of the grass shoot apex and systematics. *Amer. J. Bot.,* 44:590–595.

Calderón, C. E., and T. R. Soderstrom. 1973. Morphological and anatomical considerations of the grass subfamily Bambusoideae based on the new genus *Maclurolyra. Smithsonian Contr. Bot.,* 11:1–55.

Calvin, M., and J. A. Bassham. 1962. *The photosynthesis of carbon compounds.* W. A. Benjamin, New York.

Cook, C.W. 1943. A study of the roots of *Bromus inermis* L. in relation to drought resistance. *Ecology*, 24:169–182.

Dore, W. G. 1956. Some grass genera with liquid endosperm. *Bull. Torrey Bot. Club*, 83:335–337.

Esau, K. 1953. *Plant anatomy.* John Wiley & Sons, Inc., New York.

Etter, A. G. 1951. How Kentucky bluegrass grows. *Ann. Mo. Bot. Gard.*, 38:293–375.

Evans, M. W. 1946. *The grasses: Their growth and development.* Ohio Agr. Exp. Sta. Agron. Mimeo. 105.

———, and F. O. Grover. 1940. Developmental morphology of the growing point of the shoot and the inflorescence in grasses. *J. Agr. Res.*, 61:481–520.

Gould, F. W., and T. W. Box. 1965. *Grasses of the Texas Coastal Bend (Calhoun, Refugio, Aransas, San Patricio, and northern Kleberg counties).* Texas A&M University, College Station.

Guédès, M., and P. Dupuy. 1976. Comparative morphology of lodicules in grasses. *Bot. J. Linnaean Soc.*, 73:317–331.

Gutierrez, M.; V. E. Gracen; and G. E. Edwards. 1974. Biochemical and cytological relationships in C_4 plants. *Planta*, 119:279–300.

Haberlandt, G. 1884. *Physiologischen Pflanzenanatomie.* Leipzig.

Harberd, D. J. 1961. Observations on population structure and longevity of *Festuca rubra. New Phytol.*, 60:184–206.

———. 1962. Some observations on natural clones in *Festuca ovina. New Phytol.*, 61:85–100.

Hatch, M. D. 1975. The C_4 pathway of photosynthesis: Mechanism and function. In R. H. Burris and C. C. Black, *CO_2 metabolism and plant productivity,* Proceedings of the fifth annual Harry Steenbock symposium, pp. 59–81. University Park Press, Baltimore.

———, and C. R. Slack. 1966. Photosynthesis by sugar cane leaves: A new carboxylation reaction and the pathway of sugar formation. *Biochem. J.*, 101:103–111.

Hattersley, P. W., and L. Watson. 1975. Anatomical parameters for predicting photosynthetic pathways of grass leaves: The "maximum lateral cell count" and the "maximum cell distant count." *Phytomorphology*, 25:325–333.

———, and ———. 1976. C_4 grasses: An anatomical criterion for distinguishing between NADP-malic enzyme species and PCK on NAD-malic enzyme species. *Aust. J. Bot.*, 24:297–308.

Hitchcock, A. S. 1935. *Manual of the grasses of the United States.* U.S. Dep. Agr. Misc. Publ. 200.

———. 1951. *Manual of the grasses of the United States.* 2d ed. (Revised by Agnes Chase.) U.S. Dep. Agr. Misc. Publ. 200.

Hoshikawa, K. 1969. Underground organs of the seedlings and the systematics of Gramineae. *Bot. Gaz.*, 130:192–203.

Hubbard, C. E. 1948. The genera of British grasses. In J. Hutchison, *British flowering plants,* pp. 284–348.

Jirásek, V., and M. Jozífová. 1968. Morphology of lodicules, their variability and importance in the taxonomy of the Poaceae family. *Boletin de la Sociedad Argentina de Botanica,* 12:324–349.

Kinges, H. 1961. Merkmale des Graminees embryos. *Bot. Jahrb.,* 81:50–93.

Knobloch, I. W. 1944. Development and structure of *Bromus inermis* Leyss. *Iowa State Coll. J. Sci.,* 19:67–98.

Kortschank, H. P.; C. E. Hartt; and G. O. Burr. 1965. Carbon dioxide fixation in sugar cane leaves. *Pl. Physiol.,* 40:209–213.

LaRue, C. D. 1936. The growth of plant embryos in culture. *Bull. Torrey Bot. Club,* 63:365–382.

Lawrence, G. H. M. 1951. *Taxonomy of vascular plants.* The Macmillan Company, New York.

McClure, F. A. 1925. Some observations on the bamboos of Kwangtung (China). *Lingnan Agr. Rev.,* 3:40–47.

———. 1963. A new feature in bamboo rhizome anatomy. *Rhodora,* 65:134–136.

———. 1966. *The bamboos: A fresh perspective.* Harvard University Press, Cambridge, Mass.

McNeal, F. H. 1961. Segregation for stem solidness in a *Triticum aestivum* x *T. Durum* wheat cross. *Crop Sci.,* 1:111–114.

Metcalfe, C. R. 1960. *Anatomy of the Monocotyledons.* I. *Gramineae.* Oxford University Press, London.

Moore, R. T. 1977. Gas exchange and photosynthetic pathways in range plants. In R. E. Sosebee, *Rangeland plant physiology,* Range Science Series no. 4, pp. 42–46. Peerless Printing Company, Denver.

Ohwi, J. 1942. Gramina Japonica. IV. *Acta Phylotax. Geobot.,* 261–274.

Philipson, W. R. 1935. The development of the spikelet in *Agrostis canina* L. *New Phytol.,* 34:421–436.

Pilger, R. 1954. Das System der Gramineae. *Bot. Jahrb.,* 76:281–284.

Rahn, J. E. 1980. *Biology: The Science of Life.* Macmillan Publ. Co., Inc., New York.

Rechenthin, C. A. 1956. Elementary morphology of grass growth and how it affects utilization. *J. Range Manag.,* 9:167–170.

Reeder, J. R. 1953. Affinities of the grass genus *Beckmannia* Host. *Bull. Torrey Bot. Club,* 80:187–196.

———. 1957. The embryo in grass systematics. *Amer. J. Bot.,* 44:756–768.

———. 1962. The bambusoid embryo: A reappraisal. *Amer. J. Bot.,* 49:639–641.

Robbins, W. W. 1931. *The botany of crop plants.* McGraw-Hill Book Company, New York.

Sharman, B. C. 1942. Shoot apex in grasses and cereals. *Nature (London),* 149:82–88.

———. 1945. Leaf and bud initiation in the Gramineae. *Bot. Gaz.,* 106:269–286.

———. 1947. The biology and developmental morphology of the shoot apex in the Gramineae. *New Phytol.,* 46:20–34.

Smith, B. N., and W. V. Brown. 1973. The Kranz syndrome in the Gramineae as indicated by carbon isotopic ratios. *Amer. J. Bot.,* 60:505–513.

Sprague, H. G. 1933. Root development of perennial grasses and its relation to soil conditions. *Soil Sci.,* 36:189–209.

Stebbins, G. L. 1950. *Evolution and variation in plants.* Columbia University Press, New York.

————. 1956. Cytogenetics and evolution of the grass family. *Amer. J. Bot.*, 43: 890–905.

————, and B. Crampton. 1961. A suggested revision of the grass genera of temperate North America. *Recent Advances in Bot.*, 1:133–145.

Stuckey, I. H. 1941. Seasonal growth of grass roots. *Amer. J. Bot.*, 28:486–491.

Tateoka, T. 1957. Miscellaneous papers of the phylogeny of Poaceae (10). Proposition of a new genetic system of Poaceae. *J. Jap. Bot.*, 48:565–573.

Waller, S. S., and J. K. Lewis. 1979. Occurrence of C_3 and C_4 photosynthetic pathways in North American grasses. *J. Range Manage.*, 32:12–28.

Weatherwax, P. 1920. Position of scutellum and homology of coleoptile in maize. *Bot. Gaz.*, 69:179–182.

————. 1923. *The story of the maize plant.* The University of Chicago Press, Chicago.

Weaver, J. E. 1919. *Ecological relations of roots.* Carnegie Inst. Wash. Publ. 286.

————. 1926. *Root development of field crops.* McGraw-Hill Book Company, New York.

————, and E. Zink. 1945. Extent and longevity of the seminal roots of certain grasses. *Plant Physiol.*, 20:359–379.

Yakovlev, M. S. 1950. Structure of endosperm and embryo in grasses as a systematic feature. *Morf. Anat. Trudy Bot. Akad. Nauk. S.S.S.R.*, ser. 7, 1:121–218.

3

Reproduction and the Cytogenetic Basis of Plant Differences

SEXUAL REPRODUCTION

As is typical of vascular plants, the life-cycle of grasses is characterized by two distinct phases, the *sporophyte generation* and the *gametophyte generation*. The sporophyte develops from the fertilized egg or zygote and, as in all seed-bearing plants, is the dominant phase. Thus, the "grass plant," with its roots, stems, leaves, and flowering parts, is sporophytic, with the $2n$ chromosome number in all its cells.

In the development of gametes, the chromosome number is reduced to n. The products of the *reduction division* (meiotic division) are male and female *spores*, which soon divide to form the *gametes*. The n spores and the resulting n gametes represent the gametophyte generation.

Microsporogenesis

Pollen grains, representing the male gametophyte generation, are developed in the anther of the stamen. The anther usually consists of four locules

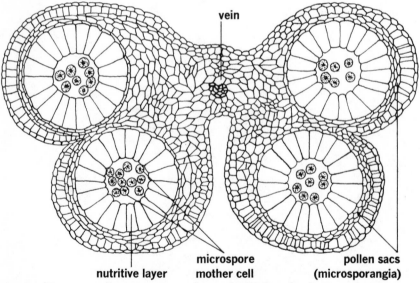

vein

microspore
mother cell

pollen sacs
(microsporangia)

nutritive layer

Fig. 3-1. Diagrammatic cross-section of an anther before the division of microspore mother cells. (Copied from Smith et al., *A Textbook of General Botany*, 5th ed., 1953, with permission of the Macmillan Company, New York.)

or pollen sacs (*microsporangia*) (Fig. 3-1). At an early stage of development, a number of *archesporial cells* are differentiated in the central region of the anther. Mitotic division of each archesporial cell gives rise to two cells, one of which continues to divide mitotically to form the wall tissue of the micro-

Contributed by A. H. Mohamed.

sporangium. The second cell is the pollen mother cell (pmc), which divides meiotically to form four *microspores*, immature pollen grains. Meiosis involves two divisions, one in which the chromosomes are reduced to half the somatic number, and a second involving no reduction of chromosome number (Fig. 3-2). The innermost layer of the microsporangium is the *tapetum*, whose function is apparently concerned with the nutrition of the developing pollen mother cells. In grasses, tapetal cells are generally binucleate, though in some cases they are mononucleate or multinucleate.

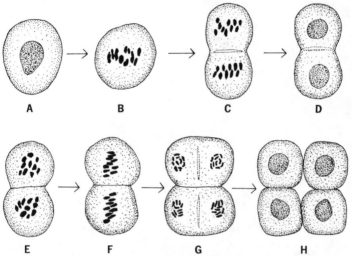

Fig. 3-2. Diagrammatic representation of stages of microspore formation. (A) Pollen mother cell (pmc); (B) metaphase of division I; (C) early telophase of division I; (D) dyad; (E) prophase of division II; (F) metaphase of division II; (G) telophase of division II; (H) tetrad of spores.

With the formation of microspores (Fig. 3-3), the sporophyte generation comes to an end. Each nuclear division of meiosis is followed immediately by cytokinesis, and at the completion of the second division, wall formation proceeds. The pollen grain wall consists of two layers, the *exine* (outer wall) and the *intine* (inner wall) and a pore through which the pollen tube emerges. Although the pollen of some angiosperms has multiple pores, in modern grasses there is always a single pore.

As the pollen grain matures, the nucleus moves to one pole and then divides mitotically to form two nuclei, a small *generative nucleus* and a somewhat larger *vegetative* or *pollen tube nucleus*. Condensed cytoplasm accumulates around each nucleus without the cell wall formation. At about the time of anthesis, the generative nucleus divides mitotically to give rise to two *sperm nuclei*. The mature pollen grain is thus three-nucleate, with all three nuclei containing the *n* chromosome number. In many species, the formation of the two sperm cells does not take place until after germination of the pollen grain. The pollen tube develops from a localized extension of the intine and, as has been noted, emerges through the pore. The vegetative nucleus is always located at the tip of the pollen tube.

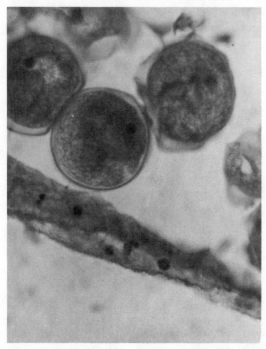

Fig. 3-3. Microspores or immature pollen grains in anther of *Echinochloa crusgalli*. At maturity the pollen grains will be three-nucleate.

Megasporogenesis

The ovule, or *megasporangium*, is developed in the ovary, the lower part of the pistil. The ovule primordium arises from the placenta as a conical protuberance with a rounded apex. By mitotic cell division and cell growth, the *nucellus*, *integuments*, and *funiculus*, or stalk of the ovule, are differentiated. Two integuments enclose the nucellar tissue, but not completely, and a canal-like opening, the *micropyle*, is left at the apex. The base of the ovule, where the integuments, nucellus, and funiculus merge, is termed the *chalaza*.

As the nucellus matures, one of the cells beneath the outer cell layer and at the apex becomes larger and exhibits cytoplasmic changes. This is the *archesporial cell*. The archesporial cell divides mitotically to give rise to two cells, the outer one of which continues to divide to form nutritive tissue around the embryo sac. The other (innermost) cell becomes the *megaspore mother cell* (mmc). This undergoes meiosis to form four cells (*n*) in a linear arrangement and separated by thin cell walls. These, the *megaspores*, are the final product of the sporophyte generation.

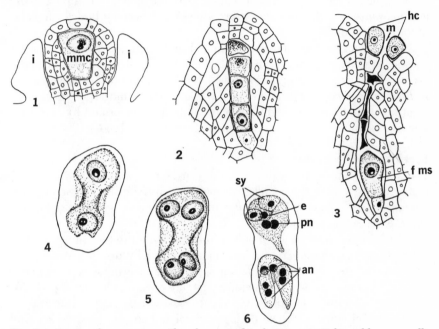

Fig. 3-4. Stages of megaspore and embryo sac development in *Echinochloa crusgalli* (1, 4, and 5) and in *E. muricata* (2, 3, and 6): 1. Megaspore mother cell (mmc) in a young ovule with incompletely formed integuments (i); 2. four megaspores, showing the deterioration of the micropylar megaspores; 3. embryo sac with functional megaspore (fms), remnants of the three degenerated megaspores, and hypertrophied cells (hc) around the micropyle (m); 4. two-nucleate embryo sac; 5. four-nucleate embryo sac; 6. mature embryo sac showing the egg cell (e), synergids (sy), polar nuclei (pn), and antipodal cells (an), each with two nuclei. (Camera lucida drawings by Arshad Ali.)

In the typical situation, the gametophyte generation (Fig. 3-4) starts by the enlargement of, and increase in, cytoplasmic material of the megaspore cell at the chalazal end of the ovule. This then becomes functional, the only functional megaspore of the ovule. The megaspore nucleus divides mitotically to form two nuclei, which migrate to the opposite ends of the enlarged megaspore cell or embryo sac. Each of the nuclei has the haploid (*n*) chromosome number. Subsequent mitotic divisions result in the formation of four nuclei at each end of the embryo sac, two of which then move to the center as *polar nuclei*. One of the three nuclei remaining at the micropylar end becomes the *egg*; the other two are the *synergids*. The three nuclei at the opposite end of the embryo sac are the *antipodals*. Thus, the mature embryo sac, or *megagametophyte*, typically consists of eight haploid nuclei, one of which is the egg, or female gamete. In some grasses, including *Echinochloa muricata*, the antipodals soon divide and become binucleate cells (Fig. 3-4, part 6).

The megagametophyte development briefly outlined above is characteristic of the majority of angiosperms. There are, however, numerous modifications and variations in grasses as well as other groups of flowering plants. In *Bouteloua curtipendula* and related species, all four megaspore nuclei remain functional, and the eight-nucleate embryo sac is formed by one division of each of the spore nuclei (Mohamed and Gould, 1966).

Fertilization

The sexual union of two gametes is known as *syngamy*. When, as in all higher plants, one of the gametes is smaller than the other and active, the syngamy is referred to as *fertilization*. The small gamete is usually the male, and the larger gamete the female.

Pollination, the transfer of pollen from the anther to the stigma, is necessary in sexual reproduction. Grass pollen is usually disseminated by air currents, but self- or cross-pollination of grass plants is influenced by many factors. Perennial grasses, for the most part, are cross-fertilized and partially or totally self-sterile. In contrast, annual grasses are mainly self-fertile. A number of self-fertile grasses, both perennial and annual, produce *cleistogamous* spikelets in which pollination takes place in closed florets. In *Sporobolus cryptandrus* and *Bothriochloa barbinodis*, the normal spikelets of the terminal inflorescences are cleistogamous and remain partially or totally enclosed in the uppermost leaf sheaths when developed under unfavorable climatic conditions. Many annuals such as *S. vaginiflorus* develop lateral inflorescences of cleistogamous spikelets late in the flowering season. The perennial *Leptochloa dubia* produces an abundance of fertile seed in short axillary inflorescences entirely enclosed in the leaf sheaths. *Stipa leucotricha* may develop cleistogamous spikelets among the normal *chasmogamous* spikelets of the terminal inflorescence and also at the base of the plant (Brown, 1952). The most extreme specialization is seen in grasses such as *Chloris chloridea*, *Amphicarpum purshii*, and *A. muhlenbergianum*, where highly modified cleistogamous spikelets are borne on underground branches.

Although most grasses are pollinated by the wind, several incidents of possible insect pollination of grasses have been reported for tropical species of *Olyra* and *Pariana* (Soderstrom and Calderón, 1971) and the temperate species *Paspalum dilatatum* (Adams et al., 1981).

As soon as a pollen grain comes in contact with the stigma, the pollen tube germinates and grows down through stylar tissue. Entrance to the ovule is through the micropyle, and the pollen tube continues growth into the embryo sac. On entering the embryo sac, the pollen tube "explodes" and the vegetative nucleus disintegrates, leaving the two sperm nuclei. One of these sperm nuclei unites with the egg to form the zygote ($2n$), and the other unites with the two n polar nuclei to form a $3n$ endosperm cell. This phenomenon of fertilization involving two sperm nuclei is known as *double fertilization*.

The fertilized egg or zygote divides and redivides repeatedly to form the embryo. Divisions of the original $3n$ endosperm cell result in the formation of extensive endosperm tissue, the principal source of food materials for the embryo when it germinates. Although the size of the embryo relative to that of the endosperm varies, the latter is normally many times the bulk of the former. Division of endosperm nuclei frequently is not accompanied by wall formation, and the resulting cells are multinucleate.

ASEXUAL REPRODUCTION

In many grasses, sexual reproduction is supplemented by asexual reproduction, and in a few taxa sexual reproduction appears to have been eliminated entirely. The term *apomixis* is applied to reproduction which involves structures commonly concerned in sexual reproduction but in which there is no actual fusion of male and female gametes. Apomixis falls into two main classes: *vivipary*, in which the progeny are produced by means other than by seed, and *agamospermy*, in which reproduction is by seed. Vegetative reproduction resulting from the development of stolons, rhizomes, and tillers is not considered to be apomictic by most workers, but vegetative reproduction involving spikelet structures is generally included as "vipipary." Vegetative reproduction often occurs as an accessory reproductive process along with seed production through sexual means. Stebbins (1950), however, pointed out that such asexual reproduction is not customarily considered to be a form of apomixis unless it is the only means of reproduction. The rather extensive literature on apomixis in flowering plants has been reviewed by several authors, including Stebbins (1941, 1950), Gustafsson (1946, 1947a, 1947b), Maheshwari (1950), and Burnham (1962).

Apomixis is relatively frequent in grasses, with *facultative apomixis* reported for many groups and *obligate apomixis* noted for others. Facultative apomicts are capable of producing offspring both sexually and asexually, while obligate apomicts have lost the capacity for sexual reproduction and produce offspring only by asexual means. Clausen (1954) noted that very few plant species are obligate apomicts. He emphasized the fact that balance between apomixis and sexuality is controlled by the interaction between the genotype and the environment. Extensive genetic studies with *Poa* indicate that, in general, sexuality is dominant to apomixis and is governed by a balance between a series of genetic factors.

Vivipary

Gustafsson (1946) has reviewed the types of vivipary reported in flowering plants. Apomictic vivipary—with flowers, lemmas, and paleas being transformed into, or replaced by, bulbils or bulblets—has been studied in species of *Poa*, *Festuca*, *Deschampsia*, and *Agrostis*. It has also been observed by the author in *Bouteloua hirsuta*. In American strains of *Poa bulbosa*, the lowermost floret of the spikelet usually has a normally developed lemma, but the palea is lacking. The lemmas of the second and third florets are long and leaflike and invest a bulbil which develops from the floral parts. In spikelets of *P. alpina*, one or more of the flowers are regular, while the others are rudimentary. Usually the second flower forms part of the bulbil. In this case, the androecium and gynoecium are absent, the nucellus aborts, and integuments, lodicules, and stamen primordia are transformed. Roots are usually formed while the bulbil is still in the spikelet and often are 1 to 2 cm long before it falls. The bulbils of *P. alpina* are released by the development of a special separation zone in the rachilla, just above the point of attachment of the upper glume.

Beetle (1980) has reported on three different forms of spikelet modifications for vegetative reproduction in grasses: (1) vivipary, (2) proliferation (the conversion of the spikelet above the glumes into a leaf shoot), and (3) phyllody (the metamorphosis of the glumes, lemmas, and/or paleas into leaves). Factors such as heredity, malformation resulting from physical or insect damage, adverse environmental conditions, and true vivipary were cited as the cause for these different modifications. These various conditions occurred in three subfamilies (Pooideae, Chloridoideae, and Panicoideae) but were most frequent in the pooid tribes Poeae and Aveneae.

Agamospermy

Types of agamospermy, involving seed production without fusion of gametes, may be considered in two groups.

Adventitious embryony. This is the simplest type of agamospermy, in which the embryo arises in the ovule but outside the embryo sac, either in the nucellus or in the integument. As the new sporophyte arises directly from sporophyte tissue, no gametophyte is developed, and alternation of generations is avoided. This has been reported in *Poa* by Tinney (1940).

Gametophytic apomixis. Apomixis involving morphological alternation of generations may occur in several ways. In this process there is, morphologically, a complete alternation of generations, but reduction division is avoided, and a diploid gametophyte is produced. With a diploid gametophyte, fertilization must be avoided if the apomictic plant is to be fertile and stable. Reduction division may be avoided either by *diplospory* or by *apospory*. In diplospory, the archesporial cell functions directly as a spore without going into meiosis. The nuclei of the embryo sac are thus all $2n$. Diplospory has been reported for species of *Poa* and *Calamagrostis* (Nygren, 1954). In apospory, the embryo sac is formed from a cell of the inner integument by a series of mitotic divisions. Aposporous four-nucleate embryo sacs have been reported for species of *Paspalum*, *Panicum*, and *Poa* (Nygren, 1954; Warmke, 1954). Mohamed and Gould (1966) found eight- and three-nucleated embryo sacs in *Bouteloua curtipendula* var. *caespitosa*. Similar observations on other grasses were previously reported by Warmke (1954) and Snyder (1957).

Formation of the embryo by mitotic division of the egg cell without fertilization is termed *parthenogenesis*. This may be coupled with either diplospory or apospory and may proceed without pollination (autonomous) or may require pollination (pseudogamous). *Pseudogamous parthenogenesis* (or *pseudogamy*, as referred to by many authors) is the common method of avoiding fertilization in agamospermic grasses. In this process, fertilization of the polar nuclei by a male gamete is necessary for development of the endosperm and the embryo.

Brown and Emery (1957, 1958) made a comprehensive survey of apomixis in the tribes Paniceae, Andropogoneae, Chlorideae, Eragrosteae, Pappophoreae, Zoysieae, and Aristideae. They reported that agamospermy is more frequent in the Paniceae and Andropogoneae than in other tribes of the grass family. They also noted that apospory is apparently characteristic of these two tribes and that the aposporous embryo sacs usually have only four nuclei at maturity.

CYTOGENETIC BASIS OF DIFFERENCES IN GRASSES

Chromosomes

Chromosomes are the principal vehicle of hereditary transmission and have an important role in determining the characteristics of individuals and of populations of individuals grouped into species. The discoveries of the earliest cytologists clearly indicated the importance of chromosomes in cell lineage and in species continuity. Chromosomes appear as distinct, recognizable units only during stages of cell division. Although they retain their individuality during other phases of the life cycle of the cell, they are much elongated, intermingled, and generally indistinct. In the early (prophase) stages of cell division, the chromosomes appear as fine, compact, cylindrical threads. At diakinesis they shorten and thicken, and during metaphase and anaphase they appear as short, cylindrical bodies. In these stages they readily take nuclear stains, such as iron hematoxylin, safranin, crystal violet, carmine, and orcein.

Each chromosome generally possesses a *centromere*, which is essential to its normal functioning. The centromere usually divides the chromosome into two arms. It has a definite position on a given chromosome and may be median, submedian, subterminal, or even terminal. Chromosomes are generally interpreted as being comprised of two long, spiral, threadlike *chromonemata*. Scattered along the length of the chromonema are minute beadlike granules. These are the *chromomeres*. The size and position of each chromomere are constant for a given chromosome.

Large chromomeres, called *knobs*, are present in *Zea mays* (corn) and the closely related genus *Tripsacum*. W. L. Brown and Anderson (1947) and Ibrahim (1960) found the high knob number in corn to be correlated with high kernel row number, row irregularity, many seminal roots, and absence of husk leaves. Mohamed and Shoeib (1965) determined that maize inbred lines with high knob frequency are significantly better combiners and produce higher yields than those with a low knob frequency. Other correlations with knob number have been reported (W. L. Brown, 1949; Mangelsdorf and Cameron, 1942; Reeves, 1944).

Chromosomes of different species or even those of the same species may differ in thickness, total length, position of the centromere, chromomere pattern, and number and position of secondary constrictions, satellites, and heterochromatin regions. Studies of chromosomes are made either during mitosis or during stages of meiotic division. Mitotic chromosomes may be observed in tissue sections prepared by the usual paraffin sectioning method. Squash or smear preparations, however, are usually more satisfactory and give better results. Meristematic tissue may be obtained from root tips, young leaves, or intercalary stem meristems. Meiotic chromosomes are usually studied from pmc (pollen mother cell) smear preparations. For the most part, chromosome counts are most satisfactorily made on metaphase or anaphase stages of pmc division I. In some grasses with numerous chromosomes, counts are possible only in diakinesis.

Chromosomes are composed of deoxyribonucleic acid (DNA) and a special class of proteins called *histones*, which form aggregates with DNA. The histone-DNA complexes are referred to as *nucleoproteins*. Chromosomes also contain other types of proteins and ribonucleic acid (RNA). It has been determined that DNA carries the primary genetic information.

Chromosomes are present in pairs in somatic cells of sexually reproducing organisms. The two members of each pair are essentially identical and are referred to as homologous chromosomes or *homologues*. One homologue is paternal, that is, derived from the male parent, and the other maternal. The total number of chromosomes in each individual cell is termed the *chromosome complement*. In general, the chromosome complement of the nucleus is constant for all cells of an individual plant and all plants of a given population. Occasionally, because of a breakdown of cell wall formation during cell division or the formation of restitution nuclei, chromosome numbers are doubled in the nuclei of some tissues. If these tissues are directly involved in reproduction, either vegetative or sexual, polyploid individuals may result.

The *basic chromosome* number (x) is an important criterion in the grouping of grass genera into tribes and subfamilies. This is actually or theoretically the lowest gametic chromosome number in the species or group of related species. Although polyploid chromosome series have been developed in most large grass genera, with few exceptions there is but a single basic chromosome number for the genus. As is discussed in Chap. 5, the great majority of grasses have chromosome numbers in multiples of 6, 7, 9, or 10. Notable exceptions are $x = 11$ in *Stipa* of the Pooideae, $x = 8$ in *Erioneuron*, and $x = 11$ in *Aristida* of the Chloridoideae. As presently interpreted, there are two basic numbers in *Panicum*, $x = 9$ and $x = 10$.

In many grass species, a variable number of short *supernumerary* or *accessory chromosomes* are present in addition to the normal longer ones (A-chromosomes). Among the species in which these have been reported are *Zea mays, Poa pratensis, Secale cereale*, and species of *Sorghum*. The extra chromosomes are generally designated as B-chromosomes. Such chromosomes have terminal or subterminal centromeres (Darlington and Upcott, 1941; McClintock, 1953) and do not pair with any of the A-chromosomes or with one another in meiosis.

Individual variation

It is probably a safe assumption that no two individual plants or animals are exactly alike. The most nearly identical plants are those propagated from a single clone or from a normally self-pollinated plant and grown under uniform environmental conditions. Variations among offspring of two parents may be attributed to genetic recombinations, mutations, and/or environmental factors.

Mutations. Mutations are sudden changes in the hereditary constitution of an individual. In the broadest interpretation, mutations include all changes in the hereditary material that affect the nature of the individual phenotype. This would include not only changes in the structure of chromosomes but also changes in the number of chromosomes. Usually, however, the term *mutation* is used to specify only submicroscopic changes termed *gene mutations* or *point mutations*.

Mutations occur with either low or high frequency in all plants. Most have little or no effect on the nature of the plant population in which they occur. On the other hand, mutations are the only known way by which the changes responsible for organic evolution are initiated. Mutations in somatic tissue are usually eliminated by the death of the individual plant in which they occur. Somatic mutations, however, may be maintained indefinitely in plants that propagate vegetatively by rhizomes, stolons, or cuttings. Recessive mutations, termed *concealed heterozygous mutants*, may occur without affecting the phenotype of the individual. Such mutants probably exist in all populations of sexually reproducing organisms. According to Dobzhansky and Spassky (1953), concealed mutations may be an important source of genetic raw material from which new adaptive genotypes are built in the progress of evolution.

Chromosomal changes. While chromosomes have definite organization, they may undergo spontaneous alteration in their structure. Also, as has been noted, the number of chromosomes in the cells of a plant may be changed. Such modifications may affect the structure, genetic behavior, or adaptability of the plant. The frequency of natural chromosomal changes is low, but the incidence of change may be greatly speeded up by the use of mutagenics, including ionized and nonionized radiation, chemicals, and high and low temperatures.

STRUCTURAL CHANGES

Structural changes of chromosomes may be intrachromosomal or interchromosomal. *Intrachromosomal changes* are deficiencies (deletions), duplications, and inversions. A *deficiency* is a chromosomal aberration in which a segment, either terminal or interstitial, is missing. This deleted segment, with the genes located on it, will be lost in the cytoplasm if it lacks a centromere. Deficiencies can be either homozygous or heterozygous. Homozygous deficiencies are usually lethal or semilethal, and thus the majority are of little consequence with respect to changes in species characteristics.

Duplication is the replication of a chromosomal segment with the genes it carries. The resulting nucleus, cell, tissue, or individual is said to be hyperploid for that segment. As in the case of deficiencies, duplications can be either homozygous or heterozygous. In the fruit fly, *Drosophila*, duplication has been found to induce phenotype differences.

Inversion is the condition in which a segment of the chromosome is inverted with respect to the rest of the chromosome. Inversions are believed to occur in a large number of plant species. They can be easily recognized in the heterozygous condition in the pachynema stage of cell division by the formation of a loop configuration. Heterozygous inversions tend to lead to partial sterility, as a result of the formation of dicentric and acentric chromosomes. Dicentric chromosomes form bridges that break at anaphase I or II to form acentric fragments.

Interchromosomal changes are interchanges or translocations in which end segments of nonhomologous chromosomes have exchanged positions. Translocations have been noted in many grasses including corn, barley, and sorghum. In heterozygous translocations, ring or chain configurations are found at diakinesis and at the first metaphase of meiosis. Homozygous translocations, however, have normal pairing during meiosis and may have no observable cytological peculiarities. Relatively little is known about the effects of translocations on the plant, but heterozygous translocations are often associated with partial sterility. As stated by Stebbins (1950), inversions and translocations do not form the raw material for selection. Because of the reduced fertility of heterozygous individuals, however, they can form isolating mechanisms that separate plant species.

Reciprocal translocations can lead to the reduction of the basic chromosome number. For a discussion of this point, the reader is referred to Stebbins (1950).

NUMERICAL CHANGES

Changes in chromosome number may involve *euploidy*, the gain or loss of complete sets of chromosomes, or *aneuploidy*, the gain or loss of one or more chromosomes of a set. Euploids thus have a balanced set or complement of chromosomes, while aneuploids have an unbalanced complement.

Euploidy

In *monoploid (haploid)* organisms, each chromosome of the basic set is represented only once in the nucleus. Monoploid plants are usually smaller and less vigorous than their diploid prototypes and are sterile because of irregularity of synapsis (pairing) and disjunction of the chromosomes during meiosis. Occasionally, all chromosomes of the genome pass to one pole at the first meiotic division, forming a normal gamete. The union of two gametes of this type may give rise to a diploid plant.

The *diploid* chromosome complement, with twice the basic or haploid number of chromosomes, supposedly is characteristic of sexually breeding organisms. In many groups of angiosperms, however, polyploids, with more than two of the basic sets of chromosomes, are frequent. Polyploidy is especially prevalent in Gramineae, and less than one-fourth of the grass species are diploid. Diploids tend to have a different distributional pattern and ecological adaptation from those of the related polyploid taxa. In general, the diploids seem to be less adaptable to environmental changes, such as those which result from man's use or abuse of land areas. In their regions of adaptation, however, diploids are usually not replaced by tetraploids or plants of higher ploidy, except under conditions of environmental disturbance or change. Few of the larger genera of grasses are represented by diploid species only. Exceptional in this respect is *Melica*, in which all North American species are diploid, with $2n = 18$. Also of note is the genus *Dichanthelium*. Of over twenty-six southeastern United States species of this group for which chromosome numbers have been reported, only a few tetraploids ($2n = 36$) are known, and the remainder are diploids ($2n = 18$).

Plants that have three or more of the basic sets of chromosomes are termed *polyploids*. If these sets, or *genomes*, are essentially the same, the plant is termed an *autopolyploid*. Autopolyploids generally arise by doubling of the chromosome number of diploid plants with only one type of genome. This increase in chromosome numbers may result from fusion of unreduced gametes or from doubling of chromosome numbers in somatic tissue from which flowering and fruiting shoots are eventually produced. Autopolyploids are frequent in plants that reproduce asexually. Meiotic divisions in autopolyploids are frequently irregular, and sterility in such plants tends to be relatively high.

Allopolyploids result from the crossing of plants with relatively distinct chromosomes. The term is usually applied to polyploids that have arisen from interspecific crosses. Hybridization followed by doubling of the chromosome number is known to be an extremely important mechanism in the evolution and speciation of many grass groups.

When crosses are made or occur naturally between distinct taxonomic species, the F_1 hybrid is usually highly sterile. This sterility is generally due to the inability of the nonhomologous chromosomes to synapse and to the breakdown of meiosis. In such hybrids, meiotic chromosomes are usually distributed unevenly to the two poles, or they lag and are lost in the cytoplasm. Sterile hybrids may propagate asexually or, by a doubling of chromosome number in the gametes or in somatic tissue, as in a seedling, may produce fertile offspring. As Stebbins has noted (1941), in natural populations sterile hybrids frequently exist in considerable numbers beside their parents. This particularly is true of grasses which tend to grow in large stands, often with related species intermingled. Though sterile, these hybrids are often more vigorous than the parent plants.

The polyploid series of species, varieties, and apparent hybrid derivatives in the North American *Bouteloua curtipendula* complex provides a fine example of evolution through hybridization and allopolyploidy. As reported by Gould and Kapadia (1964), ten species of the group are known to be or are suspected of being diploid, with $2n = 20$. One species and one variety of another species are basically tetraploids $(2n = 40)$ but also include aneuploids derived from hybridizations. Both tetraploid types are rhizomatous, and one, *B. curtipendula* var. *curtipendula*, is the "sideoats grama" of the western and central North American prairies. Frequent on well-drained, usually loose soils in the southwestern United States and northern and central Mexico (also in Argentina) are cespitose, usually tall and coarse aneuploids with chromosome numbers from $2n = 58$ to 103, which reproduce by obligate or facultative apomixis. These cespitose aneuploids, taxonomically grouped in *B. curtipendula* var. *caespitosa*, appear to have developed through hybridization of combinations of diploids and possibly tetraploids. Most differences exhibited by populations of this variety appear traceable to specific diploid taxa.

Old and New World species of the genus *Bothriochloa* are related in a complicated pattern of evolutionary development that involves many levels of polyploidy (Gould, 1959). Old World species, for the most part, have chromosome numbers of $2n = 40$, 50, and 60, whereas New World species have chromosome numbers of $2n = 60$, 80, 120, and 180. No diploids $(2n = 20)$ have been reported in this genus, but they do occur in the closely related Old World genera *Capillipedium* and *Dichanthium*.

Aneuploidy

A *monosomic* is a plant in which the somatic cells possess $2n - 1$ chromosomes, that is, one less than the normal diploid complement. Monosomics do not occur at the diploid level but only at the polyploid level. At meiosis all but one of the chromosomes form *bivalents*, the single chromosome remaining as a *univalent*. Thus, two types of gametes are expected, n and $n - 1$. The univalent tends to pass at random to either pole at the first anaphase of meiosis. Frequently, however, it fails to move to either pole and gets lost in the cytoplasm. For this reason, $n - 1$ gametes will be more frequent than n gametes. The $n - 1$ nuclei usually do not form viable gametes, but occasionally, as in *Triticum*, they are viable, and their fusion results in $2n - 2$ zygotes. Plants developing from this type of zygote are called *nullisomics*. These plants are much weaker than monosomics and are highly sterile.

Trisomics are complementary to monosomics. The somatic cells of the trisomic individual have $2n + 1$ chromosomes. In meiosis, the pairs of chromosomes form *bivalents*, and the extra chromosome, which is usually homologous to the members of one pair, commonly forms a *trivalent* with this pair. Occasionally the extra chromosome remains as a univalent. Although the n and $n + 1$ gametes might be expected in equal numbers, the extra chromosome frequently fails to move to either pole and, as in the case of the monosomics, is often lost in the cytoplasm.

In the *tetrasomic*, all chromosomes of the somatic cells are in pairs, but one pair is duplicated, forming four homologues. The chromosome number is thus $2n + 2$. Phenotypically, tetrasomics may be more robust than their corresponding diploids, but often there is no observable difference. The four homologous chromosomes of the tetrasomic tend to form a *quadrivalent* configuration at meiosis. When disjunction by two is regular, a fairly stable genetic system may be maintained. Crosses between tetrasomic and diploid plants may result in trisomics.

Environmental variation

Environment plays an important role in the development of the organism. Modifications or alterations of the *phenotype* of an individual induced by environmental agents, however, cannot be passed on from one generation to the next unless the effects have been on the germ plasm. To determine whether or not specific phenotypic characters are genetically controlled or are merely the result of environmental factors, the plant must be grown under a controlled environment. Differences due to environment tend to be quantitative rather than qualitative. Absolute size of the plant or of its organs is greatly influenced by climate and soil factors. On the other hand, the presence or absence of such structures as awns on a grass spikelet is inherited. The expression of genetically controlled characters, however, may be greatly affected by factors of the environment such as exposure to sunlight. The "suntan" character in *Zea mays* is governed by a series of genes which interact with one another to develop the pigments. The color does not appear except on plant parts that have been exposed to sunlight. Accordingly, the outer husk leaves become red, while the inner ones lack this coloration. The tassel normally becomes red, but when shielded from the light it remains colorless.

LITERATURE CITED

Adams, E. E.; W. E. Perkins; and J. R. Estes. 1981. Pollination systems in *Paspalum dilatatum* Poir. (Poaceae): An example of insect pollination in a temperate grass. *Amer. J. Bot.*, 68:389–394.

Beetle, A. A. 1980. Vivipary, proliferation, and phyllody in grasses. *J. Range Manage.*, 33:256–261.

Brown, W. L. 1949. Number and distribution of chromosome knobs in U.S. maize varieties. *Genetics*, 45:811–817.

————, and E. Anderson. 1947. The northern flint corns. *Ann. Mo. Bot. Gard.*, 34:1–28.

Brown, W. V. 1952. The relation of soil moisture to cleiostogamy in *Stipa leucotricha*. *Bot. Gaz.*, 113:438–444.

————, and W. H. P. Emery. 1957. Apomixis in the Gramineae, tribe Andropogoneae: *Themeda triandra* and *Bothriochloa ischaemum*. *Bot. Gaz.*, 118:246–253.

————, and ————. 1958. Apomixis in the Gramineae: Panicoideae. *Amer. J. Bot.*, 45:253–263.

Burnham, C. R. 1962. *Discussions in cytogenetics*. Burgess Publishing Company, Minneapolis.

Clausen, J. 1954. Partial apomixis as an equilibrium system in evolution. *Caryologia*, 1:469–479.

Darlington, C. D., and M. B. Upcott. 1941. The activity of inert chromosomes in *Zea mays. J. Genet.*, 41:275–296.

Dobzhansky, T., and B. Spassky. 1953. Genetics of natural populations. XXI. Concealed variability in two sympatric species of *Drosophila*. *Genetics*, 38:471–484.

Gould, F. W. 1959. The glume pit of *Andropogon barbinodis*. *Brittonia*, 11:182–187.

————, and Z. J. Kapadia. 1964. Biosystematic studies in the *Bouteloua curtipendula* complex. II. Taxonomy. *Brittonia*, 16:182–207.

Gustafsson, A. 1946. Apomixis in higher plants. I. The mechanism of apomixis. *Lunds Univ. Arsskr. N.F. Avd.* 2, 43:1–66.

————. 1947a. Apomixis in higher plants. II. The casual aspect of apomixis. *Lunds Univ. Arsskr. N.F. Avd.* 2, 43:71–178.

————. 1947b. Apomixis in higher plants. III. Biotype and species formation. *Lunds Univ. Arsskr. N.F. Avd.* 2, 43:183–370.

Ibrahim, M. A. 1960. A survey of chromosome knobs in maize varieties. *Genetics*, 45:811–817.

McClintock, G. 1953. The association of non-homologous parts of chromosomes in the mid-prophase of meiosis in *Zea mays. Z. Zellforsch. u. Mikr. Anat.*, 19:191–237.

Maheshwari, P. 1950. *An introduction to the embryology of the angiosperms*. McGraw-Hill Book Company, New York.

Mangelsdorf, P. C., and J. W. Cameron. 1942. Western Guatemala: A secondary center of origin of cultivated maize varieties. *Bot. Mus. Leafl. (Harvard Univ.)*, 10:217–252.

Mohamed, A. H., and F. W. Gould. 1966. Biosystematic studies in the *Bouteloua curtipendula* complex. V. Megasporogenesis and embryo sac development. *Amer. J. Bot.*, 53:166–169.

————, and E. A. Shoeib. 1965. Inheritance of quantitative characters in *Zea mays*. IV. Relationship between knob frequency and combining ability. *Can. J. Genet. & Cytol.*, 7:388–393.

Nygren, A. 1954. Apomixis in angiosperms. II. *Bot. Rev.*, 20:577–649.

Reeves, R. G. 1944. Chromosome knobs in relation to the origin of maize. *Genetics*, 29:141–147.

Smith, G. M., et al. 1953. *A textbook of general botany*, 5th ed. The Macmillan Company, New York.

Snyder, L. A. 1957. Apomixis in *Paspalum secans. Amer. J. Bot.*, 44:318–324.

Soderstrom, T. R., and C. E. Calderón. 1971. Insect pollination in tropical rain forest grasses. *Biotropica*, 3:1–16.

Stebbins, G. L. 1941. Apomixis in the angiosperms. *Bot. Rev.*, 7:507–542.

———. 1950. *Evolution and variation in plants.* Columbia University Press, New York.

Tinney, F. W. 1940. Cytology of parthenogenesis in *Poa pratensis. J. Agr. Res.*, 60:351–360.

Warmke, H. E. 1954. Apomixis in *Panicum maximum. Amer. J. Bot.*, 41:5–11.

4
Grass Classification

The primary objective of plant classification is the grouping of plants and populations into recognizable units with reasonably well-defined boundaries and stable names. Modern taxonomists strive to establish a phylogenetic arrangement of the taxa, based on known or presumed genetic relationships.

MORPHOLOGICAL CRITERIA

External morphology

For practical purposes, characters of external (gross) morphology provide the prime bases for recognition of genera, species, and subspecies or varieties. In many families of flowering plants, these are the only characters that have been employed in the differentiation of taxa. It has long been recognized that inflorescence and flower structures vary less with temporary environmental changes than most vegetative structures and thus are more reliable in taxonomy. In the system of Bentham (1881) as modified by Hitchcock (1951), characters of spikelet structure and arrangement were used almost exclusively in the delimitation of grass subfamilies, tribes, and genera. Vegetative shoot characteristics such as culm height; leaf length, width, and pubescence; and plant longevity (annual or perennial) have been used in species differentiation, but here again the spikelet is considered the most important single organ in classification. It is now known that external morphology frequently is not a reliable indicator of phylogenetic relationships in the higher categories of classification (subfamilies, tribes, genera). Visible, external plant structure, however, remains the necessary basis for practical plant differentiation and identification. By necessity, morphological characteristics are and will continue to be used as the basis of species recognition.

Anatomy and Histology

Investigation of the epidermis and internal structure of grass roots, stems, and leaves and of the anatomical structure of the embryo has disclosed a number of characteristics of taxonomic value.

   Leaf Epidermis. Prat (1932, 1936) discussed in detail the differences in shape of silica-cells of the pooid, panicoid, chloridoid, and oryzoid grasses. He observed the absence of *bicellular microhairs* in pooid grasses and the differences between typical panicoid and chloridoid microhairs. Recently, Johnston and Watson (1976) have discovered bicellular microhairs on several species of *Stipa* that are considered pooid grasses. Tateoka et al. (1959) examined the taxonomic significance of bicellular microhairs based on the characteristics of the (1) ratio of total length to maximal width, (2) ratio of the length of upper cell to that of the lower cell, (3) hair angle, and (4) difference in the thickness of the cell wall between the upper and lower cells. They found quantitative differences between the microhairs of the Panicoideae and the Chloridoideae and described the microhairs of the Bambusoideae, with nearly equal basal and maximal width throughout, as a unique form. Metcalfe (1960) noted diagnostic subfamily differences in the shape of the subsidiary cells of the stomata and the arrangement of long-cells and short-cells over the veins.

Underground organs of the seedling. The importance of the seedling underground organs in grass taxonomy has been reported by Hoshikawa (1969). Six types of underground seminal organs have been described based on the presence or absence of transitory node roots (TNR), elongation or nonelongation of the mesocotyl, and presence or absence of mesocotyl roots (MR):

Type A. TNR present, mesocotyl does not elongate, MR absent
Type B. TNR present, mesocotyl elongates, MR absent
Type C. TNR present, mesocotyl elongates, MR present
Type D. TNR absent, mesocotyl elongates, MR absent
Type E. TNR absent, mesocotyl elongates, MR present
Type F. TNR absent, mesocotyl does not elongate, MR absent

Also, seven types of seedling establishment have been identified based on the length and number of the TNR, MR, and crown node roots (CNR):

Type C. CNR longest and develop early
Type C'. CNR longest and develop later
Type M. MR longest and eight to ten in number
Type M'. MR longest but fewer and smaller than in type M
Type P. Only a primary root is present
Type T. TNR are longest
Type T'. TNR are longest but fewer and smaller than in type T

These types have been combined into formulas and correlated with the grass subfamilies. Hoshikawa (1969) reported the following formulas for these tribes of U.S. grasses:

Arundinoideae	Arundineae	E-M'
	Centosteceae	D-C'
	Danthonieae	E-M'
Bambusoideae	Bambuseae	F-P
Chloridoideae	Chlorideae	E-C'
	Eragrosteae	E-C, E-C'
	Zoysieae	D-C
Oryzoideae	Oryzeae	D-C, D-C'
Panicoideae	Andropogoneae	E-M, E-M'

	Paniceae	E-M'
Pooideae	Aveneae	B-C, B-T, D-C, D-C'
	Meliceae	B-C, D-C', D-C
	Poeae	B-C, B-C', D-C, D-C'
	Stipeae	D-C, D-C', C-C
	Triticeae	A-T, B-T', B-C

Root epidermis. Investigations by Sinnott (1939), Sinnott and Bloch (1939), Bloch (1943), Reeder and von Maltzalen (1953), and Row and Reeder (1957) have disclosed differences in typical pooid and panicoid root epidermis cell divisions, the position of root hairs in the epidermal cell, and the angle at which the root hairs emerge (Fig. 4-1 and Table 4-3). Although considerable variation was observed between genera and species, and even root epidermal areas of the same plant, the majority of the grasses studied were typical in these root characters. Row and Reeder found that the alternation of long- and short-cells (pooid) versus equal-sized cells (panicoid) is a much more reliable character than either the position or angle of the root hair.

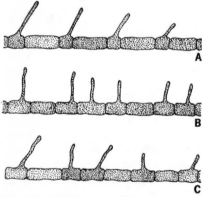

Fig. 4-1. Epidermal cells in region of root hairs of grass root tips. (A) *Dactylis glomerata*, a typical pooid species; (B) *Setaria glauca*, a typical panicoid species; (C) *Eleusine indica*, a chloridoid species. (Redrawn from Row and Reeder, 1957.)

Stem in transverse section. Brown, Harris, and Graham (1959) examined the stem internodes of 133 species of grasses from eighty genera and twenty-one tribes. Correlating the results of this study with data obtained by previous workers, they reported that while about 93 percent of the pooid grasses examined had hollow internodes, the number of grasses with solid or semisolid internodes in tribes of the Chloridoideae and Panicoideae ranged from 49 to 100 percent. Previously, Canfield (1934) had reported that 74 percent of the grasses examined from the Jornada Experimental Range Station near Las Cruces, New Mexico, had solid internodes. Some 90 percent of the grasses examined by Canfield were of the chloridoid and panicoid subfamilies. Brown et al. (1959) concurred with Canfield in the conclusion that grasses of hot, arid regions tend to have solid internodes.

Culm node and leaf sheath pulvini. Brown, Pratt, and Mobley (1959) studied the relationship of hollow and solid stem internodes to the development of meristematic swellings at the base of the culm internode and at the base of the sheath. These meristematic regions, referred to by Brown et al. as *pulvini*, make it possible for a culm that has been forced down or "lodged" to resume an erect position. Typical pooid stems with hollow internodes were found to lack culm pulvini but to have well-developed sheath pulvini. In contrast, panicoid and chloridoid grasses with solid or semisolid stems tend to have well-developed culm pulvini and no sheath pulvini.

Leaf in transverse section. Both Avdulov (1931) and Prat (1936) recognized two types of leaf blade anatomy, one associated with the pooid leaf and the other with the panicoid leaf (see Table 4-2). Brown (1958) suggested the recognition of four additional types: bambusoid, arundinoid, aristidoid, and chloridoid. Subfamily and tribal differences involve mainly the number and type of vascular bundle sheath cells, the type and arrangement of the cells of the mesophyll, the type and location of plastids in cells of the mesophyll and of the parenchyma bundle sheath, and the number of chlorophyll-containing cells between adjacent parenchyma bundle sheaths. As stated in Chap. 2, various combinations of leaf blade anatomical characteristics are indicative of the C_3 or C_4 photosynthetic pathways. Differences also exist in the number and position of sclerenchyma cells above and below the vascular bundles.

Hattersley and Watson (1975, 1976), correlated the number of chlorophyll-containing mesophyll cells between adjacent parenchyma sheaths, the lateral cell count, with the C_3 and C_4 photosynthetic pathways. C_4 plants typically have four or fewer cells between parenchyma sheaths, while C_3 plants have more than four (usually about seven) intervening cells. This lateral cell count agrees with observations of intervascular interval or interveinal distance by Lommasson (1961), Prat (1936), and Kanai and Kashiwagi (1975). Brown (1977) views this increase in intercalary veins as an evolutionary adaptation to increase the volume of Kranz tissue to maximize efficiency in C_4 photosynthesis.

Embryo anatomy and embryo length relative to that of the endosperm. The significance of the embryo in grass taxonomy has been investigated by Bruns (1892), Van Tieghem (1897), Yakovlev (1950), Reeder (1953, 1957, 1962), and Kinges (1961). Characters of taxonomic importance have, for the most part, been indicated as pooid or panicoid (Fig. 4-2). Reeder (1957) proposed the formula F + FF for the pooid embryo and P − PP for the panicoid embryo, based on the following four features:

F Vascular trace to scutellum and coleoptile diverging at approximately the same point
+ Epiblast present
  F Coleorhiza and scutellum not separated by a cleft
    F Embryonic leaf margins not overlapping
P Vascular traces to scutellum and coleoptile separated by an internode
− Epiblast absent
  P Coleorhiza and scutellum separated by a cleft
    P Embryonic leaf margins overlapping

Bambusoid, oryzoid, arundinoid, and chloridoid embryos are considered "pooid" or "panicoid" in respect to the four basic characters listed above, and the formula F + FF or P − PP, or other combinations of these symbols, will be used to indicate embryo types in all groups.

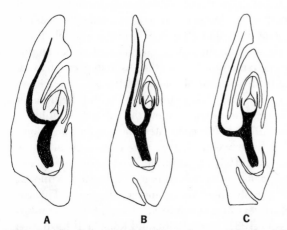

A   **B**   **C**

Fig. 4-2. Median sagittal sections of grass embryos. (A) *Dactylis glomerata*, representing the pooid type; (B) *Brachiaria platyphylla*, representing the panicoid type; (C) *Buchloë dactyloides*, representing the chloridoid type. (Redrawn from Reeder, 1953.)

A fifth embryo character of importance is that of embryo length relative to caryopsis length. For the most part, embryos of panicoid grasses characteristically are one-third to nine-tenths the length of the caryopsis (Fig. 4-3*A*), whereas embryos of pooid grasses are less than one-third the length of the caryopsis (Fig. 4-3*B*).

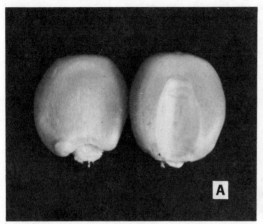 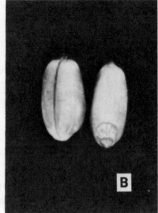

Fig. 4-3. (*A*) Caryopsis of *Zea mays*, with an embryo more than one-third the length of the endosperm (panicoid type); (*B*) caryopsis of *Triticum aestivum*, with embryo less than one-third the length of the endosperm (pooid type).

Starch grains of the endosperm. Following the early observation of Harz (1880) that the type of starch grains of the grass endosperm had systematic application, numerous systematists, including Stebbins and Crampton (1961), have attempted to correlate simple and compound types of starch grains with tribal and even subfamily groupings. Tateoka (1962), however, after examining starch grains of 766 grasses, concluded that the type of starch grains varied even within some species and could not be used for differentiation of the major taxonomic categories. Tateoka distinguished four types of starch grains (Fig. 4-4):

Type A. Simple grains whose shape is broadly elliptic, elliptic-round, or rarely reniform. This type is consistent in the tribe Triticeae and in *Bromus* and *Brachypodium* of the Poeae.
Type B. Simple grains with hexagonal, pentagonal, round, or rarely rectangular shape. This type is frequent in the Paniceae, Andropogoneae, and a few species of the Eragrosteae, Chlorideae, Pappophoreae, and other tribes.
Type C. Simple and compound grains in the same endosperm, the simple grains usually round, the compound grains with few, usually two to four,

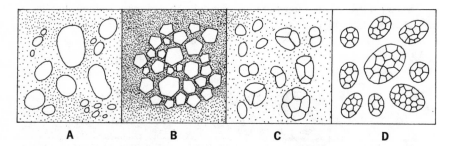

Fig. 4-4. Four types of starch grains as delimited by Tateoka: (A) *Agropyron* sp.; (B) *Panicum* sp.; (C) *Perotis* sp.; (D) *Festuca* sp. (From Tateoka, 1962.)

granules. This type is frequent in the Panicoideae and occasional in the Chloridoideae and Arundinoideae. It has not been observed in the Pooideae, Oryzoideae, or Bambusoideae.

Type D. Compound grains only. This type is characteristic of the Pooideae (excluding *Bromus* and *Brachypodium* and genera of the Triticeae). It is also frequent in the Chloridoideae and occasional in the other subfamilies.

Cytology

Among the first cytological studies of grasses were those of Kuwada (1919) on *Zea mays*, Bremer (1923, 1924, 1925) on *Saccharum*, Longley (1924) on *Zea mays* and hybrids involving that species, and Evans (1926) on *Festuca*. It was not until the publication by Avdulov (1931), however, that the significance of chromosome numbers in grass taxonomy became apparent. Avdulov made chromosome counts for 232 species. He recognized three chromosome series, one with a basic number of $x = 9$ or 10 and small chromosomes, the second with a basic number of $x = 12$ and small chromosomes, and the third with a basic number of $x = 7$ and large chromosomes. Correlated with other characters, these cytological features provided the basis for Avdulov's arrangement of grass tribes in three major groups: subfamily Sacchariferae ($x = 9$ or 10) and the Phragmitiformes ($x = 12$) and Festuciformes ($x = 7$) of subfamily Poatae.

Tateoka (1960) reported that the number of grasses whose chromosome numbers were determined by Avdulov and preceding investigators totaled 370 species. Exactly thirty years after Avdulov's publication appeared, Carnahan and Hill (1961) stated that chromosome numbers were known for some twenty-three hundred species. Grass chromosome numbers vary from $2n = 4$, reported for species of *Zingeria*, to $2n = 194$ in *Saccharum* and $2n = 220$ in *Bothriochloa*, of the Andropogoneae. The typical basic (x) chromosome numbers for the six subfamilies of grasses are:

Pooideae	$x = 7$
Panicoideae	$x = 5, 9, 10$
Chloridoideae	$x = 9, 10$
Bambusoideae	$x = 12$
Oryzoideae	$x = 12$
Arundinoideae	$x = 6, 12$

Exceptions to the general rule are numerous, and basic numbers of $x = 4, 6, 8, 11, 13, 17, 19,$ and 23 were also listed by Carnahan and Hill (1961). Grass chromosome numbers have been reported by many investigators, and the records continue to accumulate. Summaries of these records have been presented by Myers (1947), Darlington and Wylie (1955), and Carnahan and Hill (1961).

Polyploidy. The presence of multiple sets of chromosomes, polyploidy, is frequent in grasses. Whereas the incidence of polyploidy in flowering plants as a whole has been estimated to be 30 to 35 percent (Stebbins, 1950), Carnahan and Hill (1961) stated that about 80 percent of the grasses are polyploid. Polyploidy may or may not be associated with hybridization, but hybridization followed by doubling of chromosome number and return to fertility is recognized as an important evolutionary process in the Gramineae. Two or more chromosome numbers are known for many species and groups of closely related species, and the knowledge of polyploidy is a valuable aid in classification. Frequently associated with the differences in numbers of chromosomes are differences in pollen grain and epidermal cell size. Pollen size correlated with chromosome numbers has been used as a taxonomic character in distinguishing species and varieties in *Bothriochloa* (Gould, 1957a, 1957b), *Digitaria* (Gould, 1963), and *Bouteloua* (Kapadia and Gould, 1964).

Chromosome size. Chromosome size is dependent on many factors. The type of cell, the stage of mitosis or meiosis, and the type of fixative used in the microscope slide preparation all have an effect on chromosome measurements. It is a well-established fact, however, that grasses of the Pooideae have larger chromosomes than those of the other subfamilies. Extremely small chromosomes are characteristic of several genera of the Panicoideae and Chloridoideae.

Significant differences in chromosome size may exist between genera of a single tribe and species of the same genus. Essad (1954) used size and shape of chromosomes in *Lolium* for a suggested division of the genus, and

Naylor and Rees (1958) reported that chromosomes of *L. temulentum* average about one-third longer than those of *L. perenne*.

Persistent nucleoli in root-tip cells. In the division of somatic cells, the nucleolus usually disappears before the metaphase stage, and nucleoli are re-formed in the daughter cell nuclei. Zirkle (1928), Frew and Bowen (1929), Ramanujam (1938), and more recent workers have noted the persistence of the nucleolus at mitotic metaphase and later mitotic stages. As described by Zirkle, Ramanujam, and others, the nucleolus is included on the metaphase plate, constricted in the middle to become dumbbell-shaped, and then divided into two parts. The two parts move toward the opposite spindle poles, but do not become incorporated in the telophase nuclei and eventually disintegrate in the cytoplasm (Ramanujam, 1938). Brown and Emery (1957) investigated root-tip cell divisions of forty-five grass species with reference to the persistence or nonpersistence of nucleoli. In all twelve species of pooid grasses studied, not a single persistent nucleolus was noted. On the other hand, all twenty-eight species of grasses of the subfamilies Panicoideae and Chloridoideae examined had persistent nucleoli in some or all of the cells. *Danthonia spicata* and *Arundo donax*, of the Arundinoideae, were of the pooid type, with no persistent nucleoli, but *Chasmanthium latifolium*, of the same subfamily, had persistent nucleoli. *Zizaniopsis miliacea* and an *Arundinaria* species, of the subfamilies Oryzoideae and Bambusoideae, respectively, had persistent nucleoli in the majority of the cells examined. The persistence or nonpersistence of nucleoli further demonstrates the difference between panicoid and pooid grasses and the close relationship between the subfamilies Panicoideae and Chloridoideae.

NONMORPHOLOGICAL CRITERIA

Genetics

The taxonomic grouping and differentiation of taxa is based on characteristics known or assumed to be genetically controlled. Unfortunately, little genetic information is available concerning the majority of the world's grasses. Most data regarding the inheritance and transmission of characteristics, hybridization, and reproduction relate to cultivated crop plants such as corn, wheat, sugar cane, and rice. Experimental taxonomists dealing with live plants and living populations have contributed significantly to our knowledge of the genetics of a few genera of native North American grasses. Perhaps outstanding among these are the contributions of Stebbins and his students and of Dewey on the genera *Agropyron*, *Elymus*, and *Sitanion*, of the tribe Triticeae (see Chap. 5, "Literature Cited," for contributions of these investigators).

Biochemistry

Serology. Kraus (1897) first made the observation that the combination of antigen and antibody results in a visible precipitate, and this phenomenon was employed by the English zoologist Nuttall (1901) for indicating possible relationships between species. Although the precipitin reaction has been used as a tool for indicting serological correspondence in proteins in many plant groups, the application of serology to problems of grass taxonomy has been extremely limited. Fairbrothers and Johnson (1961) reported definite serological differences between grasses of the Pooideae and Chloridoideae. They found good serological correspondence between species of the pooid genera *Festuca*, *Bromus*, *Lolium*, *Dactylis*, *Melica*, and *Briza* and also between chloridoid grasses of the genera *Eragrostis*, *Tridens*, *Triplasis*, *Distichlis*, and *Spartina*. Further, the tests indicated a difference of ranking in respect to serological correspondence among the groups of related genera.

Germination response to IPC. Al-Aish and Brown (1958) investigated the seed germination response of a number of grasses to isopropyl-*n*-phenyl carbamate. The treated seeds of pooid grasses were completely inhibited from germination, whereas germination of none of the treated seeds of panicoid grasses was seriously inhibited.

Germination response to low-oxygen atmospheres. Al-Aish and Brown also germinated seeds of a number of grasses in low-oxygen atmospheres. The few panicoid grasses tested were found to have a much greater capacity for good germination in low-oxygen atmospheres than the pooid grasses. Seeds of *Oryza sativa* (subfamily Oryzoideae) far surpassed those of all other grasses tested in ability to germinate under low-oxygen tensions.

Electrophoretic analysis. Hall (1959) and Hall and Johnson (1963) have used electrophoretic patterns from seed extracts to trace the relationships of protein fractions of hybrids to the parent types. Hall investigated a "ryewheat" hybrid and its parent species (*Secale cereale* and *Triticum aestivum*), and Hall and Johnson worked with relationships of the natural amphiploid *Stiporyzopsis caduca* and its *Stipa viridula* and *Oryzopsis hymenoides* parents. In both ryewheat and *Stiporyzopsis*, the protein specificity of the parents as measured by electrophoresis was retained. The potential significance of electrophoretic analysis in determining taxonomic and phylogenetic relationships of grass taxa at all levels is apparent.

Photosynthetic pathways. Smith and Brown (1973), Brown (1977), and Waller and Lewis (1979) have correlated differences in photosynthetic path-

Table 4-1. Distribution of C_3 and C_4 photosynthetic pathways and Kranz anatomical subtypes according to grass tribes

Subfamily	Tribe	C_3	C_4	Anatomical Subtype
Arundinoideae	Arundineae	*		
	Centosteceae	*		
	Danthonieae	*	*	P.S.
Bambusoideae	Bambuseae	*		
	Phareae	*		
Chloridoideae	Aeluropodeae		*	P.S.
	Aristideae	*	*	D.S., P.S.
	Chlorideae		*	P.S.
	Eragrosteae		*	P.S.
	Orcuttieae		*	P.S.
	Pappophoreae		*	P.S.
	Unioleae		*	P.S.
	Zoysieae		*	P.S.
Oryzoideae	Oryzeae	*		
Panicoideae	Andropogoneae		*	M.S.
	Paniceae	*	*	M.S., P.S.
Pooideae	Aveneae	*		
	Brachyelytreae	*		
	Diarrheneae	*		
	Meliceae	*		
	Monermeae	*		
	Nardeae	*		
	Poeae	*		
	Stipeae	*		
	Triticeae	*		

Source: Adapted from Brown, 1977; Smith and Brown, 1973; and Waller and Lewis, 1979.

ways with the grass subfamilies and tribes (Table 4-1). The Bambusoideae, Pooideae, and Oryzoideae are exclusively C_3. Also, all members of the Arundinoideae are C_3 with the exception of several Danthonieae genera (*Alloeochaete*, *Asthenatherum*, and *Pheidochloa*), which are C_4 (Brown, 1977). Conversely, the C_4 pathway occurs in all genera of the Cloridoideae with the exception of two small African genera (*Stipagrostis* and *Sartidia*) of the Aristideae. Within the Panicoideae, all Andropogoneae genera use the C_4 pathway, while there are numerous C_3 and C_4 genera in the Paniceae.

All three C_4 subtypes (NAD-malic enzyme, NADP-malic enzyme, and PEP-carboxykinase) are found within the Panicoideae. All Andropogoneae genera have the NADP-malic enzyme, while all three subtypes occur in the

Paniceae. Typically, Paniceae genera contain only one of the decarboxylating subtypes; however, Gutierrez et al. (1974, 1976) and Brown (1977) reported all three subtypes in the genus *Panicum*. The only other genera in the Paniceae which have the PEP-carboxykinase enzyme are *Brachiaria, Eriochloa, Urochloa, Pseudobrachiaria*, and one species of *Panicum* (*P. maximum* Jacq.). Brown (1977) suggested transferring all *Panicum* species which contain PEP-carboxykinase to this "Brachiaria" group. Both the NAD-malic enzyme and PEP-carboxykinase subtypes occur throughout the genera and tribes of the Chloridoideae (Gutierrez et al., 1974, 1976; Brown, 1977).

Ecology and plant geography
Distribution patterns and adaptation to specific habitats have long been recognized as significant characteristics of plant taxa. Grasses of the Pooideae are consistently present in the cool or cold regions of the earth and grow and reproduce in the subtropics only during the cool season of the year. Conversely, members of the Bambusoideae are, for the most part, restricted to warm, frost-free areas, although a few have become cold-tolerant and survive in regions subjected to annual freezing. Most panicoid grasses are distributed in tropical-subtropical habitats, although many have become adapted to temperate climates. None is present in the vegetation of polar and high mountain regions, where long periods of frigid temperatures are experienced.

Grasses of the small subfamily Oryzoideae characteristically are adapted to moist or marshy sites. This group, apparently closely related to the bamboos, is widely distributed in tropical and subtropical regions of the world. Most successful of the grasses in warm, semiarid regions are those of the Chloridoideae. Some of the characteristics of this subfamily, such as the solid stem internodes and narrow, frequently involute or acicular leaf blades, appear to be habitat modifications.

SUBFAMILY DIFFERENTIATION
In the classification system of Bentham (1881), thirteen grass tribes were grouped in two subfamilies, the Festucoideae (Pooideae) and Panicoideae. Bews (1929) used the Bentham system as a basis for his treatment of the world's grasses. Hitchcock (1920, 1935, 1951) followed it with minor modifications in the classification of U.S. grasses. Hitchcock recognized fourteen tribes, ten in the Festucoideae (Bambuseae, Festuceae, Hordeae, Aveneae, Agrostideae, Phalarideae, Chlorideae, Zoysieae, Oryzeae, and Zizanieae) and four in the Panicoideae (Paniceae, Melinideae, Andropogoneae, and Tripsaceae). In the Bentham system, differentiation of subfamilies, tribes, and genera was based almost exclusively on morphological characters of the

Table 4-2. A comparative chart of the grass subfamily groupings of Avdulov, Prat, Parodi, and Stebbins and Crampton

Avdulov (1931)	Prat (1936)	Prat (1960)	Parodi (1961)	Stebbins and Crampton (1961)
POATAE				
Festuciformes	FESTUCOIDEAE	FESTUCOIDEAE	FESTUCOIDEAE	FESTUCOIDEAE
	BAMBUSOIDEAE	BAMBUSOIDEAE	BAMBUSOIDEAE	BAMBUSOIDEAE
		ORYZOIDEAE	ORYZOIDEAE	ORYZOIDEAE
Phragmitiformes		PHRAGMITIFORMES	PHRAGMITOIDEAE	ARUNDINOIDEAE
SACCHARIFERAE	PANICOIDEAE			
	Eupanicoideae	PANICOIDEAE	PANICOIDEAE	PANICOIDEAE
	Chloridoideae	CHLORIDOIDEAE	ERAGROSTOIDEAE	ERAGROSTOIDEAE

inflorescence. The woody culms and arborescent habit of the bamboos, however, were recognized as significant and distinctive features of this group.

The course of grass taxonomy following the publications of Avdulov (1931) and Prat (1936) has been outlined in Chap. 1. Prat (1960) attempted a phylogenetic arrangement of tribes and subfamilies on a world basis. Although authorities are not in entire agreement as to the most satisfactory grouping of the higher taxa, six subfamilies are recognized by Prat (1960), Parodi (1961), and Stebbins and Crampton (1961). As is shown in Table 4-2, the subfamily arrangements of these men correspond closely with the arrangement of Avdulov and the earlier one of Prat.

Because of the numerous striking differences that have been noted between grasses of the Pooideae and those of the Panicoideae, these groups have been used as a general "standard of comparison" for all grasses. Table 4-3 presents in summary form a number of the contrasting features of typical grasses of the two subfamilies.

Table 4-3. A comparison of some contrasting features of the Pooideae and Panicoideae

Subfamily Pooideae	Subfamily Panicoideae
Roots	
1. Epidermal cells in region of root hairs alternately long and short, only the short-cells giving rise to root hairs (Fig. 4-1A)	1. Epidermal cells in region of root hairs all alike and capable of giving rise to root hairs (Fig. 4-1B)
2. In some species, root hairs are produced toward the root-tip end of the cell and at an angle toward the root tip	2. Root hairs typically developed near the middle of the cell and perpendicular to the cell axis
Stems	
1. Culm internodes typically hollow, with vascular bundles in one to few rings at the inner margin of the cortex (Fig. 2-6)	1. Culm internodes predominantly solid or semisolid, the vascular bundles scattered in the ground tissue or in two to few rings at the inner margin of the cortex (Fig. 2-5)
2. Shoot apex usually with two layers of tunica-cells (Fig. 2-9)	2. Shoot apex usually with one distinct layer of tunica-cells
3. No meristematic swellings, or "pulvini," at base of internode, but well-developed pulvini at base of the associated leaf sheath	3. Well-developed culm pulvini, poorly developed sheath pulvini
Leaf epidermis	
1. Silica-cells round, elliptic, long and narrow, or crescent-shaped (Fig. 2-16)	1. Silica-cells mostly dumbbell-, cross-, or saddle-shaped (Fig. 2-16)

2. Microhairs typically absent, except in *Stipa*

3. Short-cells over the veins usually solitary or paired, seldom if ever in long rows

4. Stomata commonly with low, dome-shaped or parallel-sided subsidiary cells

Leaf anatomy

1. Vascular bundle sheath typically double, with an inner sheath of small, thick-walled cells and an outer sheath of large parenchyma cells (Fig. 2-18)

2. Chloroplasts of the parenchyma sheath cells similar to those of the chlorenchyma cells of the mesophyll

3. Chlorenchyma cells of the mesophyll loosely and irregularly arranged, with relatively large intercellular air spaces

4. Lateral cell count more than four

5. Non-Kranz

Photosynthetic pathway

1. C₃

Spikelet and Flower

1. Spikelet with one to several fertile florets, the reduced florets usually above the fertile ones when present (Fig. 2-37A, C)

2. Spikelet disarticulation usually above the glumes

3. Lodicules usually elongated, pointed, thick at the base and membranous above, with little or no vasculation (Fig. 2-41C)

Embryo

1. Vascular traces to scutellum and coleoptile diverging at approximately the same point (Fig. 4-2A)

---

2. Bicellular microhairs nearly always present (Fig. 2-17)

3. Short-cells over the veins usually in rows that are more than five cells long

4. Stomata commonly with triangular or tall, dome-shaped subsidiary cells.

1. Vascular bundles typically with a single sheath of large parenchyma cells (Fig. 2-21)

2. In many genera, the chloroplasts of the parenchyma sheath cells specialized for starch storage, this accompanied by a loss of ability to store starch in the plastids of the mesophyll chlorenchyma cells

3. Chlorenchyma cells of the mesophyll with reduced intercellular air spaces, in some cases with the cells more or less radially arranged around the vascular bundles (but not so much as in chloridoid grasses)

4. Lateral cell count two to eight or more

5. Kranz or non-Kranz

1. C₄, C₃

1. Spikelet with one fertile floret and with one reduced (neuter or staminate) floret below (Fig. 2-37B)

2. Spikelet disarticulation below the glumes

3. Lodicules usually short, truncate, thick, and heavily vasculated (Fig. 2-41A)

1. Vascular traces to scutellum and coleoptile separated by a more or less elongated internode (Fig. 4-2B)

2. Epiblast usually present
3. Lower part of scutellum fused to the coleorhiza

4. Embryonic leaf margins not overlapping. Vascular bundles of embryonic leaves few
5. Embryo relatively small, mostly one-sixteenth to one-fourth the size of the caryopsis (Fig. 4-3*B*)

Cytology
1. Basic chromosome number, $x = 7$ in most genera
2. Chromosomes typically large

3. Nucleoili not persistent in root-tip cell divisions

Reserve food storage
1. Levulosides present in several organs

Seed germination
1. No germination after immersion in a solution of 0.5 percent IPC (isopropyl-*n*-phenyl carbamate) for twenty-four hours

2. Low germination in low-oxygen atmospheres in the few grasses observed

Distribution
1. Grasses of cold, temperate, and subtropical regions of both hemispheres, cool-season in the subtropics

2. Epiblast absent
3. Lower part of scutellum separated from the coleorhiza by a notch or groove

4. Embryonic leaf margins overlapping. Vascular bundles of embryonic leaves numerous
5. Embryo relatively large, mostly one-third to four-fifths as long as the caryopsis (Fig. 4-3*A*)

1. Basic chromosome number, $x = 9$ or 10 in most genera
2. Chromosomes small to medium-sized

3. Persistent nucleoili in some but usually not all root-tip cell divisions

1. Levulosides generally lacking

1. Good germination after immersion in 0.5 percent IPC for twenty-four hours in the few grasses tested. Growth generally better than for controls in H_2O.
2. Good germination in low-oxygen atmospheres in the few grasses observed

1. Grasses mainly of tropical-subtropical regions, warm-season in the subtropics

SUBFAMILIES AND TRIBES OF UNITED STATES GRASSES

In the present treatment, the arrangement of U.S. genera and tribes for the most part follows that of Stebbins and Crampton (1961). Changes include the return of the monotypic tribe Aristideae to the subfamily Chloridoideae, where it was originally placed by Pilger (1954), and the shifting of the tribe Phareae to the subfamily Bambusoideae, as suggested by Parodi (1961). *Vaseyochloa* is included in the tribe Eragrosteae rather than in the tribe Aeluropodeae. On the basis of characters noted by Yates (1966a, 1966b, 1966c), the tribe Unioleae is moved to the Chloridoideae, and the reconstructed genus

Chasmanthium, made up of species previously referred to *Uniola*, is tentatively referred to the tribe Centosteceae, of the subfamily Arundinoideae.

The six subfamilies of the Gramineae are represented in the United States by twenty-two tribes, 124 genera, and 1,053 species of native grasses. An additional 300 adventive or introduced species are temporarily or permanently established outside of cultivation. Some of these, such as *Agropyron repens*, *Cynodon dactylon*, and *Sorghum halepense*, have become the dominant grasses of disturbed sites over wide areas.

Over 97 percent of all native U.S. grass species belong to the three subfamilies Pooideae, Panicoideae, and Chloridoideae. Representation by genus and species is as shown in Table 4-4.

Table 4-4. Approximate number of genera and species of U.S. grasses

Subfamily	Number of U.S. Genera	Number of U.S. Species	Percentage of U.S. Species
Pooideae	41	391	37.1
Panicoideae	32	325	30.9
Chloridoideae	42	310	29.4
Bambusoideae	2	3	0.3
Oryzoideae	4	11	1.1
Arundinoideae	3	13	1.2
Total	124	1,053	100.0

Although each of the three major subfamilies contributes about one-third of the species, their representation throughout the country is strikingly different (Figs. 4-5 to 4-7). The region of maximum diversity and species representation of the Pooideae is the Northwest, where the group has developed in a cool to cold climate. Moisture has been a secondary limiting factor, but most species are present in regions with a good moisture supply during the growing season.

Grasses of the Panicoideae have been particularly successful in moist, humid, tropical, or subtropical habitats. Representation in the United States is best in the Southeast and poorest in the Northwest. Although temperature is the primary limiting factor, few species have become adapted to conditions of aridity, and moisture is an important secondary limiting factor.

The Chloridoideae, with slightly over three hundred species in the United States, is well distributed over the North American continent. The best representation, however, is in the southwestern United States and northern Mexico. There, in a warm, dry climate, over 50 percent of the grass species of some areas are chloridoid.

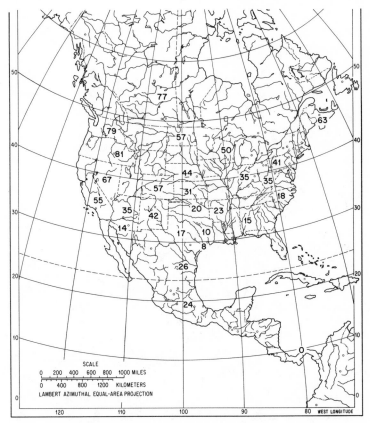

Fig. 4-5. Regional representation of native grasses of subfamily Po- oideae in the United States and areas of Canada, Mexico and Panama. The figures are percentages of the total number of native species in the area. (Goode Base Map 102, copyright by the Univ. of Chicago, with permission of Dept. of Geography, Univ. of Chicago.)

Subfamily I. Pooideae

ROOTS. Epidermal cells in region of root hairs alternately long and short, only the short-cells giving rise to root hairs. In some species, root hairs are produced toward the root-tip end of the cell and at an angle toward the root tip.

STEMS. Culm internodes typically hollow, with vascular bundles in one to few rings at the inner margin of the cortex. Shoot apex usually with two layers of tunica-cells. No meristematic swellings or "pulvini" at base of inter- node, but well-developed pulvini at base of associated sheath.

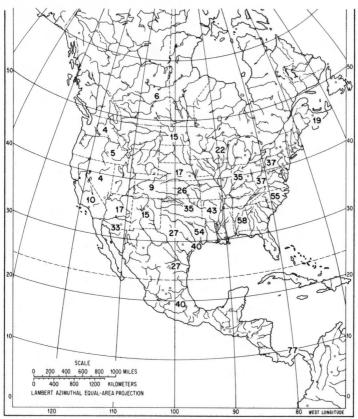

Fig. 4-6. Regional representation of native grasses of subfamily Pan-icoideae in the United States and areas of Canada, Mexico and Panama. The figures are percentages of the total number of native species in the area. (Goode Base Map 102, copyright by the Univ. of Chicago, with permission of Dept. of Geography, Univ. of Chicago.)

LEAF EPIDERMIS. Microhairs absent except in some species of *Stipa*. Silica-cells usually round, elliptical, long and narrow, or crescent-shaped. Short-cells over the veins usually solitary or paired, seldom if ever in long rows. Stomata commonly with low, dome-shaped or parallel-sided subsidiary cells. LEAF ANATOMY. Vascular bundle sheath typically double, with an inner sheath of small, thick-walled cells and an outer sheath of large parenchyma cells. Bundle sheath frequently interrupted on one or both sides by scle-renchyma tissue. Chloroplasts of the parenchyma sheath similar to those of the chlorenchyma cells of the mesophyll. Chlorenchyma cells of the mesophyll loosely and irregularly arranged, with relatively large intercellular air spaces. Non-Kranz.

PHOTOSYNTHETIC PATHWAY. C_3

SPIKELET AND FLOWER. Spikelet with 1 to several fertile florets, the reduced ones usually above the fertile ones when present. Spikelet disarticulation usually above the glumes. Lodicules usually elongate, pointed, thick at the base and membranous above, with little or no vasculation.

EMBRYO. Vascular traces to scutellum and coleoptile diverging at approximately the same point. Epiblast usually present. Lower part of scutellum fused to the coleorhiza. Embryonic leaf margins not overlapping. Vascular

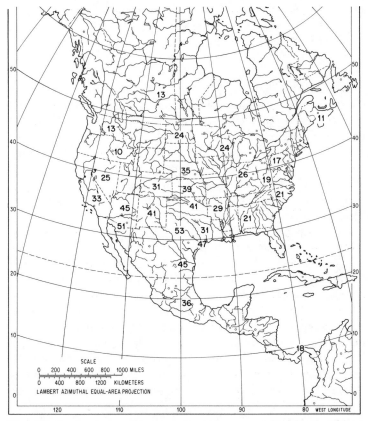

Fig. 4-7. Regional representation of native grasses of subfamily Chloridoideae in the United States and areas of Canada, Mexico and Panama. The figures are percentages of the total number of native species in the area. (Goode Base Map 102, copyright by the Univ. of Chicago, with permission of Dept. of Geography, Univ. of Chicago.)

bundles of embryonic leaves few. Embryo relatively small, mostly one-sixth to one-fourth the length of the caryopsis.

CYTOLOGY. Basic chromosome number, $x = 7$ in most tribes. Chromosomes typically large. Nucleoli not persisting in root-tip cell divisions.

RESERVE FOOD STORAGE. Levulosides present in several organs.

SEED GERMINATION. No germination after immersion in a solution of 0.5 percent IPC (isopropyl-n-phenyl carbamate) for twenty-four hours, in the several grasses tested. Low germination in low-oxygen atmospheres in the few grasses observed.

A large proportion of the grasses adapted to cool and cold climates of both hemispheres belong to the Pooideae. Seventy to 85 percent of the species of Canada and the northwestern United States are pooid (Fig. 4-5). The group contributes 40 to 50 percent of the grass species in the middle latitudes of the United States and about one-fourth of the species of the Southeast. In the semiarid Southwest the percentage drops to less than 15. Pooid grasses are absent at low elevations in both humid and dry tropical areas. Of some 144 grasses listed by Standley (1928) for the Panama Canal Zone, none is pooid in the modern interpretation of the group. Species of *Bromus, Poa, Festuca, Agropyron,* and other typical pooid grasses do grow at high altitudes in mountainous regions of tropical latitudes. In subtropical regions, such as the southern United States, pooid grasses grow and flower during the winter and spring months and are referred to as *cool-climate* or *cool-season* grasses. In terms of day length, these are "short-day" grasses, as opposed to the warm-season grasses of the Panicoideae and Chloridoideae, which are "long-day" plants.

Pooid grasses native to the United States may be grouped into seven tribes: the Poeae, Aveneae, Triticeae, Meliceae, Stipeae, Brachyelytreae, and Diarrheneae (Stebbins and Crampton, 1961). Two additional tribes, the Nardeae and Monermeae, are represented by adventive species. Grasses of the tribe Agrostideae of Hitchcock (1935, 1951) are dispersed into six tribes, four (Poeae, Aveneae, Brachyelytreae, and Stipeae) in the subfamily Pooideae and two (Eragrosteae and Aristideae) in the subfamily Chloridoideae. About half of the genera are included in the Aveneae. Following Tateoka (1957) and Stebbins and Crampton (1961), the tribe Phalarideae is abandoned, and the genera *Phalaris, Hierochloë,* and *Anthoxanthum* are placed in the Aveneae. Decker (1964) has characterized a number of the tribes and genera retained in, or removed from, the "classical" subfamily Pooideae.

Tribe 1. Poeae. Ligule membranous. Inflorescence a panicle or infrequently a raceme or spike. Spikelets 2- to several-flowered, rarely 1-flowered. Glumes unequal, one or both shorter than the lowermost lemma. Lemmas

5- to several-nerved, awnless or awned from a notched or entire apex. Basic chromosome number, $x = 7$.

U.S. genera*: *Bromus*, (*Brachypodium*), *Vulpia*, *Festuca*, (*Lolium*), *Leucopoa*, (*Scolochloa*), (*Sclerochloa*), (*Catapodium*), *Puccinellia*, *Poa*, (*Briza*), *Phippsia*, (*Coleanthus*), (*Dactylis*), (*Cynosurus*), (*Lamarckia*).

Tribe 2. Aveneae. Ligule membranous. Silica-cells of leaf epidermis elongate, with sinuous walls. Stomata rectangular or, in some cases, with slightly dome-shaped subsidiary cells. Inflorescence a panicle, infrequently a raceme. Spikelets 1- to several-flowered; the reduced florets, when present, above or infrequently below the fertile one or ones. Glumes, at least the second, longer than the lowermost lemma. Lemmas 3- to several-nerved, awnless or more frequently awned from the base, back or bifid apex. Basic chromosome number, $x = 7$.

U.S. genera: *Koeleria*, *Sphenopholis*, *Trisetum*, (*Corynephorus*), (*Aira*), *Deschampsia*, *Scribneria*, (*Avena*), (*Ventanata*), *Helictotrichon*, (*Arrhenatherum*), (*Holcus*), *Dissanthelium*, *Calamagrostis*, *Ammophila*, (*Apera*), *Agrostis*, *Polypogon*, (*Mibora*), *Cinna*, *Limnodea*, (*Anthoxanthum*), *Hierochloë*, *Phalaris*, *Alopecurus*, *Phleum*, (*Gastridium*), (*Lagurus*), *Milium*, *Beckmannia*.

Tribe 3. Triticeae. Ligule membranous. Inflorescence a bilateral spike or spicate raceme with both sessile and pediceled spikelets at the rachis nodes. Spikelets 2- to several-flowered except in *Hordeum*, which typically has only 1 floret per spikelet. Lemmas 5- to several-nerved, awnless or awned from the tip. Basic chromosome number, $x = 7$. A small but economically important group, generally referred to as the *cereal tribe*.

U.S. genera: *Elymus*, *Sitanion*, (*Taeniatherum*), *Hystrix*, *Hordeum*, *Agropyron*, (*Triticum*), (*Secale*).

Tribe 4. Meliceae. Ligule membranous. Silica-cells of leaf epidermis absent or scarce. Cells of inner bundle sheaths of leaf blade with uniformly thickened walls. Inflorescence a panicle or raceme of several-flowered spikelets. Disarticulation above or below the glumes. Lemmas several-nerved. Lodicules thick, fleshy, truncate, without vasculation. Chromosomes of mixed large and intermediate size (*Melica*), all intermediate (species of *Pleuropogon*), or all small (*Pleuropogon*, species of *Glyceria*) (Stebbins and Crampton, 1961). Basic chromosome number, $x = 8$, 9, or 10.

U.S. genera: *Melica*, *Glyceria*, *Catabrosa*, *Pleuropogon*, *Schizachne*.

*Genera represented by introduced or adventive species are in parentheses.

Tribe 5. Stipeae. Ligule membranous. Silica-cells of leaf epidermis cuboidal or oblong to oval, crescent-shaped, cross-shaped, or dumbbell-shaped (Metcalfe, 1960). Vascular bundles of the leaf blade with two bundle sheaths, but the outer often obscure or of small cells. Inflorescence a panicle of 1-flowered spikelets. Disarticulation above the glumes. Lemma indistinctly several-nerved. Embryo pooid (F + FF). Basic chromosome numbers, $x = 7$ and 11.

U.S. genera: *Stipa, Oryzopsis, Piptochaetium.*

Tribe 6. Brachyelytreae. Ligule membranous. Leaf epidermis with mostly dumbbell-shaped silica-cells, these in rows. Inflorescence a panicle. Spikelets 1-flowered, disarticulating above the glumes. Glumes minute, the first often absent. Basic chromosome number, $x = 11$.

U.S. genera: *Brachyelytrum.*

Tribe 7. Diarrheneae. Ligule membranous. Inflorescence a panicle. Spikelets mostly 3- to 5-flowered. Lemmas 3-nerved. Stamens 3, 2, or 1. Embryo anatomy pooid (F + FF). Chromosome numbers reported, $2n = 38$ and 60.

U.S. genera: *Diarrhena.*

Tribe 8. Nardeae. Ligule membranous. Leaf blade epidermis with linear bicellular microhairs and dumbbell-shaped silica-cells. Inflorescence a unilateral spike of 1-flowered spikelets. Disarticulation above the single glume. Lodicules absent. Chromosome numbers reported, $2n = 26$ and 30.

U.S. genera: (*Nardus*).

Tribe 9. Monermeae. Ligule membranous. Leaf epidermis with dumbbell-shaped silica-cells. Inflorescence a bilateral spike of 1-flowered spikelets. Disarticulation in the rachis, each spikelet falling attached to a section of the rachis. Basic chromosome number, $x = 7$.

U.S. genera: (*Monerma*), (*Parapholis*).

Subfamily II. Panicoideae

ROOTS. Epidermal cells in region of root hairs all alike, capable of giving rise to root hairs. Root hairs typically developed near the middle of the cell and perpendicular to the cell axis.

STEMS. Culm internodes predominantly solid or semisolid, with vascular bundles scattered in the ground tissue or in two to few rings at the inner margin of the cortex. Shoot apex usually with one distinct layer of tunica-cells. Well-developed "pulvini" or meristematic swellings at the base of the internodes, but poorly developed pulvini at the base of the associated leaf sheath.

LEAF EPIDERMIS. Bicellular microhairs nearly always present, these usually linear or slightly fusiform. Silica-cells mostly dumbbell-shaped, cross-shaped, or saddle-shaped. Short-cells over the veins usually in rows of more than five cells. Stomata commonly with triangular or tall, dome-shaped subsidiary cells.

LEAF ANATOMY. Vascular bundles typically with a single sheath of large parenchyma cells. Bundle sheath usually continuous or interrupted on only one side by sclerenchyma tissue. In many genera, the chloroplasts of the parenchyma sheath specialized for starch storage, this accompanied by a loss of the ability to store starch in the plastids of the mesophyll chlorenchyma cells. Chlorenchyma cells of the mesophyll with reduced intercellular air spaces, in some cases with the cells more or less radially arranged around the vascular bundles (but not so much so as in chloridoid grasses). Non-Kranz and Kranz.

PHOTOSYNTHETIC PATHWAY. Both C_3 and C_4. All C_4 subtypes (NADP-me, NAD-me and PEP-ck) are present.

SPIKELET AND FLOWER. Spikelet with 1 fertile floret and with 1 reduced (neuter or staminate) floret below. Spikelet disarticulation below the glumes. Lodicules usually short, truncate, thick, and heavily vasculated.

EMBRYO. Vascular traces to the scutellum and coleoptile separated by a more or less elongated internode. Epiblast absent. Lower part of scutellum separated from the coleorhiza by a notch or groove. Embryonic leaf margins overlapping. Vascular bundles of embryonic leaves numerous. Embryo relatively large, mostly one-third to four-fifths as long as the caryopsis.

CYTOLOGY. Basic chromosome number, $x = 9$ or 10 in most genera. Chromosomes small to medium-sized. Persistent nucleoli in some but usually not all root-tip cell divisions.

RESERVE FOOD STORAGE. Levulosides generally lacking in the plant organs.

SEED GERMINATION. Good germination after immersion in 0.5 percent IPC (isopropyl-n-phenyl carbamate) for twenty-four hours in the few grasses tested. Growth generally better than for the controls in H_2O. Good germination in low-oxygen atmospheres in the few grasses observed.

The subfamily Panicoideae provides the majority of the grasses of tropical-subtropical regions throughout the world. The group remains little changed in the "newer" systems of classification from the way it appeared in the systems of Bentham and Hitchcock. Following Stebbins and Crampton in their treatment of North American grasses (1961), the tribe Melinideae is not recognized, and the genus *Melinus* is included in the tribe Paniceae. Also, *Zea, Tripsacum*, and *Coix* are grouped in the Andropogoneae rather than being segregated into a separate tribe (Tripsaceae or Maydeae). The subfamily thus is reduced to two large tribes, the Paniceae and the Andropogoneae.

The Panicoideae are represented in the United States by thirty-two genera and 325 species of native grasses. Over half of the grasses in the southern states are panicoid, as are about a third of those in the Northeast (Fig. 4-6). In the far Northwest, only about 4 to 7 percent of the species belong to this subfamily. The Panicoideae are by far the dominant grass group of the American tropics. As listed by Standley (1928), 77 percent of the grasses of the Panama Canal Zone are panicoid, these mainly of the genera *Panicum* and *Paspalum*.

Tribe 10. Paniceae. Ligule membranous, a fringe of hairs, or absent. Inflorescence a panicle. Glumes and lemma of sterile floret similar in texture, commonly soft. First glume usually small or absent. Lemma of fertile floret typically firm or hard, smooth and shiny, indistinctly nerved, awnless (short-awned in *Eriochloa*), and with the margins overlapping or inrolled over the palea. Disarticulation below the glumes in *Cenchrus* and *Pennisetum* the spikelets falling enclosed in bristly or spiny involucres. Basic chromosome number, $x = 9$ or 10 for U.S. genera.

U.S. genera: *Digitaria, Leptoloma, Anthaenantia, Stenotaphrum, Brachiaria, Axonopus, Reimarochloa, Eriochloa, Paspalum, Paspalidium, Panicum, Lasiacis, (Oplismenus), Echinochloa, Sacciolepis, (Rhynchelytrum), Setaria, (Pennisetum), Cenchrus, Amphicarpum, (Melinus), (Anthephora), Steinchisma, Phanopyrum, Dichanthelium.*

Tribe 11. Andropogoneae. Ligule membranous, a fringe of hairs, or absent. Spikelets in pairs of 1 sessile and 1 pediceled, the pediceled spikelet reduced in most genera. Glumes, at least the first, large, firm or hard, permanently enclosing the florets. Lemmas of sterile and fertile florets, and palea of fertile floret membranous, usually greatly reduced or absent, the lemma of the fertile floret often consisting of little more than a stout, geniculate awn. Spikelets of a pair usually disarticulating as a unit together with the pedicel and a section of the rachis. Basic chromosome number, $x = 5$, 9, or 10.

U.S. genera: *Imperata, (Miscanthus), (Saccharum), Erianthus, (Sorghum), Sorghastrum, Andropogon, (Arthraxon), (Microstegium), (Dichanthium), Bothriochloa, Chrysopogon, (Hyparrhenia), Schizachyrium, (Eremochloa), Trachypogon, Elyonurus, Heteropogon, Coelorachis, (Hackelochloa), Tripsacum, (Zea), (Coix).*

Subfamily III. Chloridoideae

ROOTS. Epidermal cells in region of root hairs similar in size and all capable of giving rise to root hairs.

STEMS. Culm internodes predominantly solid or semisolid, with vascular bundles scattered in the ground tissue or in two to few rings at the inner margin of the cortex. Shoot apex usually with one distinct layer of tunica-cells. Well-developed "pulvini" or meristematic swellings at the base of the internodes, but poorly developed pulvini at the base of the associated leaf sheath.

LEAF EPIDERMIS. Bicellular microhairs nearly always present, these variable but most frequently broadly bullet-shaped or clavate. Silica-cells mostly cross- or saddle-shaped. Short-cells over the veins usually in rows of more than five cells. Stomata with triangular or low, dome-shaped subsidiary cells.

LEAF ANATOMY. Vascular bundles with an outer sheath of large, well-developed parenchyma cells and a usually indistinct inner sheath of small, thick-walled cells or the inner sheath absent. Sheath of the vascular bundle usually continuous around the bundle or interrupted on one side by sclerenchyma tissue. Chlorenchyma cells of the mesophyll large or small, typically elongate, radiating out on their longitudinal axis from the bundle sheath, this radial arrangement especially pronounced in grasses of the tribe Chlorideae. Each vascular bundle with its associated ring of chlorenchyma cells frequently separated from the adjacent bundles by bands of colorless cells, these continuous with bulliform cells of the adaxial leaf epidermis.

PHOTOSYNTHETIC PATHWAY. All C_4.

SPIKELET AND FLOWER. Spikelets with 1 to several florets. Reduced florets when present, usually above the perfect ones. In *Uniola*, reduced florets are present both above and below the perfect ones. Lemmas typically 3-nerved, but 1-nerved in *Sporobolus* and *Calamovilfa* and several-nerved in *Vaseyochloa*, of the Eragrosteae, and all genera of the small tribes Aeluropodeae, Pappophoreae, and Orcuttieae. Lodicules usually small and cuneate or obcuneate, infrequently oblong.

EMBRYO. (P + PF) Vascular traces to scutellum and coleoptile separated by an internode. Epiblast present. Lower part of scutellum free from the coleorhiza. Embryonic leaf margins not overlapping. Embryo relatively large, typically one-third to five-sixths the length of the caryopsis.

CYTOLOGY. Basic chromosome number for most genera, $x = 9$ or 10. In *Erioneuron*, $x = 8$. Chromosomes small. Nucleoli in root-tip cells persistent during cell divisions in about half of the cells, never totally absent or always present.

SEED GERMINATION. Good germination for two months after treatment with IPC (isopropyl-*n*-phenyl carbamate) for twenty-four hours in the few grasses tested. Growth generally better than for controls in H_2O.

Genera grouped in this subfamily were included in the pooid tribes Festuceae and Chlorideae by Hitchcock (1920, 1935, 1951). The Chloridoideae are represented in the United States by forty-two genera and 310 species of native grasses. The group has reached its best development in the warm, semiarid Southwest (Fig. 4-7), where in some areas it comprises more than 50 percent of the grass vegetation. Poorest representation is in the Northwest, where the number of species drops to less than 10 percent of the total for the area. In the more mesic situations and in the cooler climates, the subfamily is represented mainly by species of *Eragrostis*, *Muhlenbergia*, *Sporobolus*, and *Aristida*.

Although pooid in most spikelet characters, grasses of the Chloridoideae are panicoid in respect to chromosome numbers and size and in most features of the embryo, stem, leaf, and other of the "new" taxonomic criteria. Members of the two large and closely related tribes Eragrosteae and Chlorideae quite consistently have 3-nerved lemmas (1-nerved in *Sporobolus* and *Calamovilfa*), whereas both pooid and panicoid grasses, with few exceptions, have 5- to several-nerved lemmas.

Tribe 12. Eragrosteae. Ligule membranous, a fringe of hairs or absent. Inflorescence an open or contracted panicle, at least some of the primary branches rebranched. Spikelets 1- to several-flowered. Disarticulation mostly above the glumes but below the spikelet in *Lycurus* and species of *Muhlenbergia*. Lemmas 3-nerved, except in *Sporobolus* and *Calamovilfa*, which have 1-nerved lemmas, and *Vaseyochloa*, with several-nerved lemmas. Embryo P + PF. Basic chromosome numbers, $x = 8$, 9, and 10.

U.S. genera: *Eragrostis*, *Neeragrostis*, *Tridens*, *Triplasis*, *Erioneuron*, *Munroa*, *Vaseyochloa*, *Redfieldia*, *Scleropogon*, *Blepharidachne*, *Calamovilfa*, *Lycurus*, *Muhlenbergia*, *Sporobolus*, *Blepharoneuron*, (*Crypsis*), (*Eleusine*), (*Dactyloctenium*), *Leptochloa*, *Trichoneura*, *Gymnopogon*, *Tripogon*.

Tribe 13. Chlorideae. Ligule usually a ring of hairs. Inflorescence a unilateral spike or of few to several unilateral spicate primary branches. Spikelets with a single floret, with a single fertile floret and one or more reduced florets, or with several perfect florets. Spikelets unisexual in several genera. Disarticulation above or below the glumes. Lemmas 3-nerved. Embryo P + PF. Basic chromosome number, $x = 9$ or 10 in most genera.

U.S. genera: *Willkommia*, *Schedonnardus*, (*Cynodon*), *Microchloa*, *Chloris*, *Bouteloua*, *Buchloë*, *Cathestecum*, *Aegopogon*, *Spartina*, *Ctenium*, *Hilaria*.

Tribe 14. Zoysieae. Ligule a ring of hairs. Inflorescence a contracted raceme of 1-flowered spikelets, these short-pediceled and irregularly spaced on a continuous rachis. Disarticulation at the base of the spikelet. Lemma and palea thin, membranous, the palea usually reduced or absent. Basic chromosome number, $x = 10$.

U.S. genera: (*Zoysia*), (*Tragus*).

Tribe 15. Aeluropodeae. Stoloniferous or rhizomatous grasses of saline or alkaline soils. Ligule usually a short, fringed membrane. Leaves with short, pungent blades. Leaf epidermis with many papillae and small, rounded, often sunken bicellular microhairs. Inflorescence a contracted panicle or raceme, the spikelets few- to several-flowered, mostly unisexual. Disarticulation above the glumes. Lemmas several-nerved. Embryo P + PF. Basic chromosome number $x = 10$.

U.S. genera: *Distichlis, Allolepis, Monanthochloë, Swallenia.* The genus *Swallenia* does not fit well into this group, and its placement here is only tentative.

Tribe 16. Unioleae. Ligule a dense ring of soft hairs. Leaf epidermis with club-shaped bicellular microhairs and square or saddle-shaped silica-cells. Inflorescence a large panicle. Spikelets several- to many-flowered, both the lowermost and the uppermost florets sterile. Disarticulation below the glumes. Embryo P − PF. Basic chromosome number, $x = 10$.

U.S. genera: *Uniola*.

Tribe 17. Pappophoreae. Ligule a ring of hairs. Leaf epidermis with bicellular microhairs, these long and bulbous-tipped in *Cottea* and *Enneapogon* but short in *Pappophorum*. Inflorescence a panicle of 3- to several-flowered spikelets. Disarticulation about the glumes, the florets falling together. Lemmas with 9 or more nerves and awns. Embryo P + PF. Basic chromosome number, $x = 10$.

U.S. genera: *Pappophorum, Enneapogon, Cottea.*

Tribe 18. Orcuttieae. Leaves lacking ligules and not differentiated into sheath and blade. Epidermal bicellullar microhairs small, sunken, the terminal cells swollen and glandular. Inflorescence a panicle of several-flowered spikelets. Spikelets persistent, with no disarticulation in the rachilla or below the glumes. Lemmas 5- to many-nerved. Embryo P + PF. Basic chromosome number, $x = 10$.

U.S. genera: *Orcuttia, Neostapfia.*

Tribe 19. **Aristideae.** Ligule a ring of hairs or a short, ciliate membrane. Leaf epidermis with linear bicellular microhairs and dumbbell-shaped silica-cells. Vascular bundles of the leaf blade surrounded by two parenchyma sheaths, the cells of both containing specialized chloroplasts. No inner sheath of small cells developed. Chlorenchyma cells of mesophyll elongated and radially arranged around the vascular bundles. Inflorescence a panicle of 1-flowered spikelets. Disarticulation above the glumes. Glumes large, thin. Lemma thick, usually firm, narrowing at the apex to an awn column bearing 3 awns, the lateral awns sometimes reduced or lacking. Lodicules 2, vascu-lated and with membranous appendages like those of the Arundinoideae (Stebbins and Crampton, 1961). Embryo P − PF, differing from the typical chloridoid embryo in the lack of an epiblast. Basic chromosome number, $x = 11$.

U.S. genera: *Aristida.*

Subfamily IV. Bambusoideae

ROOTS. Epidermal cells in region of root hairs all alike and capable of giving rise to root hairs. Root hairs typically developed near the middle of the cell.

STEMS. Culm internodes hollow, semisolid, or solid, the vascular bundles scattered in the ground tissue or arranged in rings at the inner margin of the cortex.

LEAF EPIDERMIS. Leaf epidermis with bicellular microhairs. Short-cells over the veins paired or in rows. Silica-cells mostly cross-, saddle-, or dumb-bell-shaped. Stomata usually with low or tall, dome-shaped subsidiary cells.

LEAF ANATOMY. Vascular bundles of blade with a double sheath, the outer of large, thin-walled cells, and the inner of thick-walled cells. Mesophyll chlo-renchyma of arm-cells, these not radiating out from the vascular bundles. Large fusoid-cells present in the mesophyll,, oriented perpendicular to the vascular bundles and appearing as cavities. Non-Kranz.

PHOTOSYNTHETIC PATHWAY. All C_3.

SPIKELET AND FLOWER. Spikelets 1- to many-flowered, awned or awnless. Disarticulation usually above the glumes and between the florets. Palea 2-nerved or with 2 strong nerves and numerous fine nerves. Flowers bisexual or unisexual, ordinarily with 3 lodicules. Stamens usually 6 or 3, rarely many. Pistil with a single style terminating in 1, 2, or 3 (rarely more) stigma-tic branches.

EMBRYO. Mostly F + PP. Vascular traces to the scutellum and coleoptile di-verging at the same point. Epiblast present (absent in the Streptochaeteae). Lower part of scutellum separated from the coleorhiza by a notch or groove. Embryonic leaf margins overlapping. Embryo relatively small, usually less than one-third the length of the caryopsis.

CYTOLOGY. Chromosomes small, the basic number $x = 12$ in the tribe Bambuseae, $x = 10$ and 11 in the Olyreae, and $x = 11$ in the Streptochaeteae. No chromosome records have been obtained for the Phareae and Parianeae.

In the recent interpretation by Parodi (1961), the bamboo subfamily includes not only the classical bamboo group, tribe Bambuseae, but also the tribes Streptochaeteae, Olyreae, Phareae, and Parianeae. Only the tribes Bambuseae and Phareae are represented by native United States grasses, the former by two species of *Arundinaria*, and the latter by one species of *Pharus*.

Tribe 20. Bambuseae. Culms woody, lignified, usually freely branching above and with a highly developed rhizome system below. Leaves of the leafy branches different in form from those of the main shoot axis, the latter usually bladeless or with short, early deciduous blades. Leaves of the branches usually with a petiole-like constriction between the sheath and blade. Ligule membranous or a thickened rim or collar. Inflorescence usually a large, many-flowered panicle but reduced to a single spikelet in those genera with pseudospikelets. Disarticulation usually above the glumes and between the florets. Spikelet generally large and several-flowered but with a single floret in a few species and with reduced florets both above and below the fertile ones in others. Spikelets rarely unisexual. The most frequent spikelet types may be characterized as follows: Glumes lanceolate, thin, shorter than the lemmas, the first often greatly reduced. Lemmas 5- to many-nerved, usually awnless and similar to the glumes. Palea large, strongly 2-keeled, and with 2 strong nerves on the keels. Numerous, fine, evenly spaced nerves may be present between and lateral to the keels (as in the prophyll of the shoot). Lodicules mostly 3, usually distinctly veined, variable in size and texture, one usually symmetrical and lanceolate, and the other two broader, asymmetrically obtuse or truncate at the apex, and paired. Stamens usually 6, occasionally reduced to 3 or very numerous. McClure (1966) noted records of as high as 50 to 120 stamens in individual flowers of *Ochlandra* species. Filaments united to form a tube of variable length in some genera with 6 stamens per flower. Ovary with a single locule and a single ovule, the fruit a caryopsis, berry, nut, or utricle. The largest fruits are as much as 12 cm long and 8 cm wide. Three vascular strands, interpreted as representing the midribs of three carpels, generally pass from the ovary into the styles, which are also typically three in number. When the fruit is a berry, the ovary is packed with food, and the endosperm is of minor importance. The fruit of *Melocanna bambusoides* becomes as large as a walnut and may germinate and grow as much as 6 inches before dropping from the shoot. Basic chromosome number, $x = 12$.

U.S. genera: *Arundinaria*.

Bamboos are distributed on the continents of Asia, Africa, and North and South America and throughout the islands of the Pacific and Indian Oceans. The more primitive types have their center of distribution in southeastern Asia and the islands of the Pacific and Indian Oceans, a region of warm, moist, or monsoon climates. Some of the more advanced climbing and herbaceous types have become adapted to relatively dry and cold climates. Bamboos of this nature are characteristic of the Himalaya and Andes mountain ranges. *Arundinaria aristata* occurs at 11,000 feet in the eastern Himalayas. In the United States, hardy, exotic bamboos used as ornamentals remain evergreen at temperatures down to 5° F. The only bamboos native to the United States are two species of *Arundinaria*, *A. tecta* and *A. gigantea*.

Bamboo culms generally grow in clumps or clusters from a stout, rhizomatous base. In tropical regions, they form dense undergrowths or jungles of interlaced culms, branches, leaves, and thorny shoots or adventitious roots. Almost all other plants are excluded from these areas. The length of the culm internode varies greatly, reaching 5 feet in *Ochlandra travancorica*. The diameter of the culm varies from the size of a lead pencil to more than a foot in thickness. Some of the bamboos, such as *Dendrocalamus strictus*, have solid or nearly solid culms. These are referred to as *male bamboo* and are used for spear and lance staves and walking sticks. Culm heights of 120 to 130 feet are commonly attained by *D. giganteus* and *D. brandigii*, the tallest of the bamboos. On the other hand, Arber (1934) noted that pygmy bamboos of Japan and the Kuriles may be only 10 to 15 cm in height and "no thicker than a crowquill."

Bamboo culms are generally cylindrical, but may be grooved on the internodes. This groove is due to pressure initiated by the axillary bud, the imprint being retained and extended with the elongation of the internode. An exception to the cylindrical-shaped culm is that of *Bambusa angulata*, which is 4-angled.

In the common tropical bamboos, twelve to fifty culms are produced annually, depending upon the species. The culms may be dark green, bluish, yellow black, or variegated. Buds that solidify into recurved spines are developed in some species, while in others a whorl of rootlets which harden into spines is produced below the sheath.

New culms normally develop from the rhizomatous plant base during the rainy season. These rapidly attain full diameter and appear like great scaly cones clad in large embracing sheaths. The culms generally complete their growth in a single season, but are not considered mature until about the third year. The rate of culm growth is phenomenal. At Kew Gardens, *Bambusa arundinacea* grew 91 cm every twenty-hour hours for a period of one month (Arber, 1934).

The branch system of many of the bamboos is complicated. *Arundinaria simonii* may have as many as twenty-five branchlets arising from one side of a node, with the opposite side of the node being completely naked. *Arthrostylidium multispicatum* has long, whiplike shoots 2 to 4 meters in length which become attached to some support and then develop leaves and branches from clusters of sharp, scale-covered branch buds. *Arthrostylidium sarmentosum* is an herbaceous climber which dies back each year, an extreme case of departure from the arboreal habit. Culms of *Arthrostylidium* species climb high, branch frequently, and in their greatest development swing down in great curtains from the trees.

LEAVES

The leaves of the bamboo shoot are alternate in two rows, as in other grasses. The foliage leaves, however, differ from those of most other grasses in the well-defined petiole-like constriction at the base of the blade and in being borne on branches rather than directly on the main culm. Leaf size in bamboos varies greatly. Arber credits *Neurolepis nobilis (Planotia nobilis)* with having the largest leaves of all grasses, those being as much as 30 cm wide and 4.5 meters long. The other extreme is found in *Arthrostylidium capillitifolium*, whose leaves are almost hairlike. The foliage of some species of the genus *Bambusa* has a fernlike character resulting from an increase in number and reduction in size of the leaves and a decrease in the length of the culms.

Scale leaves may attain a large size. Because of the difficulty of obtaining flowers or even the mature foliage of bamboos, the scale leaves are often used in distinguishing species. Leaf sheaths in a number of bamboos bear sharp-pointed hairs that are highly irritating to the human skin.

INFLORESCENCE—FLOWERING

Records of periodicity of flowering and of reproduction by seed in bamboos are extremely scanty. It is a well-established fact, however, that most bamboos enter a reproductive state only after many years of vegetative growth. McClure (1966) noted that in some bamboos the vegetative state shows signs of persisting indefinitely and with undiminished vigor. The commonly cultivated *Bambusa vulgaris* has been observed for over 150 years, during which time there has been occasional flowering but no records of seed production. In contrast with the long-continued vegetative state in most bamboos, some, such as *B. lineata*, are reported as "constant-flowering." For about 100 years, *B. lineata* has been observed in cultivation at Bogor, Calcutta, and Peradeniya, but despite the constant flowering of this species, no record of the production of fruit or the death of a flowerng plant has been published (McClure, 1966).

For the most part, the reproductive cycle of bamboos follows one of two patterns. Bamboos with long reproductive cycles usually flower gregariously, set an enormous amount of seed, and then die within one or two years. Others, including the temperate North American species of *Arundinaria*, do not die after flowering, but are retarded in vegatative growth during this period.

The inflorescence of numerous bamboos, including *A. tecta* and *A. gigantea*, is a panicle, which is large, much-branched, and many-flowered or small and few-flowered. In other bamboos, however, the individual inflorescences are reduced to structures which have been termed *pseudospikelets* by McClure (1934). The pseudospikelet consists of a shortened axis bearing a single spikelet at the apex and few to several bracts below (Fig. 4-8). The lowermost bract characteristically is a 2-keeled, modified prophyll. Unlike

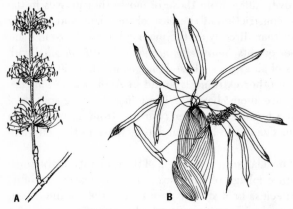

A   **B**

Fig. 4-8. *Dendrocalamus sikkimensis.* (A) Specialized flowering shoot branch bearing cluster of pseudospikelets at each node; (B) pseudospikelet with bracteate leaves at the base and a terminal spikelet with 3 florets. (Redrawn from Arber, 1926.)

glumes, the lower bracts of pseudospikelets often enclose buds in their axils, and some bear blades at their apex. The buds differ from those borne in the axils of lemmas in that they develop or have the potential of developing into secondary, tertiary, or higher-order pseudospikelets. In some bamboos, the pseudospikelets are many times compounded, and their increase in numbers continues over a period of several months or even a few years. If the basic inflorescence concept presented in Chap. 2 is followed, each pseudospikelet may be considered a highly modified leafy branch, complete with prophyll at its base and terminating in an inflorescence of one spikelet.

McClure (1966), in agreement with the concepts of Holttum (1958), considered the entire flowering branch with pseudospikelets to be an "indeterminate inflorescence" and thus different from the typical grass inflorescence, which is determinate. He noted, however, that in species of *Schizostachyum, Bonia*, and *Arundinaria* the buds subtended by bracts on the rachis of the primary pseudospikelet may remain dormant; he stated: "In such cases the individual inflorescence consists of but a single pseudospikelet."

Tribe 21. Phareae Herbaceous perennials with broad, flat blades and a petiole-like constriction between the sheath and blade. Inflorescence a panicle, with unisexual, 1-flowered spikelets borne in pairs of 1 large, pistillate, and sessile and 1 small, staminate, and long-pediceled. Lemma of pistillate spikelet bearing uncinate hairs at least near the tip. No chromosome records reported.

U.S. genera: *Pharus*.

Subfamily V. Oryzoideae

ROOTS. No studies of root epidermis or root hairs reported.

STEMS. Culm internodes usually hollow, with a large central cavity and with large or small intercellular air spaces in the ground tissue.

LEAF EPIDERMIS. Bicellular microhairs present, mostly narrow and cylindrical or fusiform. Silica-cells oblong, cross-, saddle-, or dumbbell-shaped, often broader than long in respect to the longitudinal rows of cells. Short-cells over the veins mostly in rows of more than five, but pairs also present.

LEAF ANATOMY. Midrib of blade with 2 or more vascular bundles distributed near both the adaxial and abaxial surfaces and sometimes also in the interior (Tateoka, 1963). Vascular bundles with two bundle sheaths, the inner of small, inconspicuous cells, and the outer of large parenchyma cells. Mesophyll as in bambusoid leaves, with arm-cell chlorenchyma and large, elongated fusiform cells. Sclerenchyma both adaxial and abaxial, occasionally grading into sclerenchyma or colorless cells extending out from the vascular bundles. Non-Kranz.

PHOTOSYNTHETIC PATHWAY. All C_3.

SPIKELET AND FLOWER. Spikelets with a single floret, this often subtended by 2 scales or bristles that previously have been interpreted as glumes but now are believed to represent reduced florets. Lemma 5- to several-nerved. Palea 2-nerved in some genera but 3- to several-nerved in others. Spikelets bisexual or unisexual; when unisexual, the plants are usually monoecious. Stamens 6, 3, or 1. Ovary with 1 or 2 styles.

EMBRYO. Embryo F + FP or F + PP (Tateoka, 1964), the embryo relatively small with respect to the caryopsis.

CYTOLOGY. Chromosomes small, the basic number $x = 12$ for most genera but $x = 15$ for *Zizania*.

In the present interpretation, the subfamily Oryzoideae is made up of a few small genera of herbaceous annuals and perennials adapted to moist, marshy, or aquatic habitats. The general distribution is in tropical and sub-tropical regions of the world. In temperate North America, however, the group is represented as far north as Canada by *Zizania aquatica* and a few species of *Leersia*.

Following the concepts of Parodi (1961) and Tateoka (1963), the North American genera are all included in the tribe Oryzeae. As treated by Stebbins and Crampton (1961), the subfamily is much larger and more heterogeneous and includes the tribes Phareae, Olyreae, and the Asiatic Ehrharteae. Furthermore, following Hitchcock (1920, 1935, 1951), Stebbins and Crampton grouped the North American genera with unisexual flowers (*Zizania, Zizaniopsis, Luziola*) into a separate tribe, the Zizanieae.

Tribe 22. Oryzeae. Since a single tribe is recognized in the Oryzoideae, the tribal characteristics are as stated for the subfamily. In all genera, the ligule is membranous, and disarticulation is at the base of the spikelet.

U.S. genera: (*Oryza*), *Leersia, Zizania, Zizaniopsis, Luziola*.

Subfamily VI. Arundinoideae

ROOTS. Epidermal cells in region of root hairs are all alike and capable of giving rise to root hairs. Position of root tip in cell and angle of root tip in respect to main root axis variable, often intermediate between the panicoid and chloridoid condition.

STEMS. Internodes with a large or small central cavity, or solid.

LEAF EPIDERMIS. Epidermis with or without bicellular microhairs. Short-cells over the veins solitary, paired, or in rows of three to many. Silica-cells various, tall and narrowly rectangular to crescent, cross-, saddle-, or dumb-bell-shaped.

LEAF ANATOMY. Vascular bundles of blade with two sheaths, the inner of small cells with slightly thickened walls, and the outer of large, thin-walled parenchyma cells that do not contain chloroplasts. Chlorenchyma cells of the mesophyll relatively small and tightly packed. Arm-cells occasionally present, but fusoid-cells apparently completely lacking. Non-Kranz.

PHOTOSYNTHETIC PATHWAY. C_3 or C_4 in three Danthonieae genera.

SPIKELET AND FLOWER. Spikelets 1- to many-flowered, awned or awnless. Disarticulation usually above the glumes. Flower bisexual or unisexual. Lodicules 2, stamens 3, pistil with 2 styles.

EMBRYO. Embryo P − PF in arundinoid-danthonoid grasses, P + PP in the tribe Centosteceae. Embryo small to moderately large but never more than half as long as the caryopsis (at least in the Danthonieae).

CYTOLOGY. Chromosomes small, the basic chromosome number $x = 6$ or 12.

The Arundinoideae is an ancient and somewhat heterogeneous assemblage. Grasses of this group are widespread in the world, but the majority, especially those of the tribe Danthonieae, are distributed in the Southern Hemisphere. In North America they are present in cool, temperate, and subtropical regions. Like the pooid grasses, they are adapted to cool-season growth in the warmer climates. This group was first recognized by Dumortier as early as 1823. Avdulov (1931) placed the genera now referred to Arundinoideae, Bambusoideae, and Oryzoideae in the subdivision Phragmitiformes of his subfamily Poatae. Despite the regular occurrence of chromosome numbers in multiples of six or twelve, the Arundinoideae appear to have as much in common with the Pooideae and Chloridoideae as with the bambusoid-oryzoid grasses.

Stebbins and Crampton (1961) included in the Arundinoideae the tribes Arundineae, Danthonieae, Unioleae, and Aristideae. In the present treatment, the Aristideae and the Unioleae (in the restricted sense of Yates, 1966a, 1966c) are referred to the subfamily Chloridoideae. The tribe Centosteceae, represented by the genus *Chasmanthium* (Yates, 1966c), is provisionally grouped in this subfamily.

Tribe 23. Arundineae. Tall, robust perennials, most with stout, rhizomatous, or densely clumped bases. Ligule a fringed membrane (*Arundo*) or a dense ring of hairs (*Cortaderia*). Spikelets few- to several-flowered, bisexual or unisexual. Disarticulation above the glumes. Embryo P − PF. Basic chromosome number, $x = 12$. A small group of small genera, worldwide in distribution but with only one species, *Phragmites australis*, native to the United States.

U.S. genera: (*Arundo*), *Phragmites*, (*Cortaderia*), (*Molinia*).

Tribe 24. Danthonieae. Annuals or perennials. Ligule a dense ring of hairs (in *Danthonia* and *Schismus*). Inflorescence a panicle, raceme, or in a few species of *Danthonia* occasionally or frequently reduced to a single pediceled spikelet. Disarticulation above the glumes. Lemmas bifid or toothed at the apex, awned or mucronate. Embryo P − PF. Basic chromosome number, $x = 6$ or 12 ($x = 7$ in species of *Pentaschistis*).

Based on evidence from morphology, anatomy, and cytology, Hubbard in 1948 proposed the tribe Danthonieae for *Danthonia* and related genera previously included in the pooid tribe Aveneae. These grasses are distributed predominantly in the Southern Hemisphere. In spikelet characteristics they are similar to the Aveneae. Leaf anatomy characteristics may be either pooid or panicoid (de Wet, 1954, 1956), and bicellular epidermal microhairs may be present or absent. Prat (1936), de Wet, and Metcalfe (1960) all have noted, however, that some species of this group have totally pooid leaf anatomy and epidermis. Metcalfe observed that relationships between *Danthonia* and the Aveneae do not seem impossible for those species in which the leaf structure is wholly pooid but is clearly incorrect for those species in which the leaf epidermis is panicoid. Thus, there is considerable uncertainty as to the proper placement of some taxa now referred to the Danthonieae.

U.S. genera: *Danthonia*, (*Sieglingia*), (*Schismus*).

Tribe 25. Centosteceae. Ligule a fringed membrane (in *Chasmanthium*). Leaf epidermis with bicellular microhairs, these from rod- to acorn-shaped. Silica-cells cross- to dumbbell-shaped, these over the veins, usually in rows but also solitary or paired in some species. Vascular bundles of the blade mostly small, with a double sheath, the inner of small, thick-walled cells, and the outer of large parenchyma cells. Chloroplasts present in the outer sheath cells, at least in *Chasmanthium*. Chlorenchyma cells of mesophyll tightly packed, more or less radially arranged around the vascular bundles in some types and differentiated into well-marked palisade and spongy parenchyma in others (Metcalfe, 1960). Inflorescence an open or contracted panicle. Spikelets bisexual, several-flowered, disarticulating above the glumes and between the florets. Embryo P + PP. Basic chromosome number, $x = 12$.

The tribe Centosteceae is represented in the United States by five species of *Chasmanthium*, these included in *Uniola* by Hitchcock. Yates (1966a, 1966b, 1966c) has shown that these grasses differ from *Uniola* in several important characters, and the genera are now referred not only to different tribes but also to different subfamilies.

U.S. genera: *Chasmanthium*.

LITERATURE CITED

Al-Aish, and W. V. Brown. 1958. Grass germination responses to isopropyl-*n*-phenyl carbamate and classification. *Amer. J. Bot.*, 45:16–23.

Arber, A. 1926. Studies in the Gramineae. I. The Flowers of certain Bambuseae. *Ann. Bot.*, 40:447–469.

———. 1934. *The Gramineae: A study of cereal, bamboo, and grass.* Cambridge University Press, New York.

Avdulov, N. P. 1931. Karyo-systematische Untersuchungen der Familie Gramineen. (Russian with German summary.) *Bull. Appl. Bot. Suppl.*, 44.

Bentham, G. 1881. *Notes on Gramineae. J. Linn. Soc. Bot.*, 19:14–134.

Bews, J. G. 1929. *The world's grasses: Their differentiation, distribution, economics, and ecology.* Longmans, Green & Co., Ltd., London.

Bloch, R. 1943. Differentiation in red root-tips of *Phalaris arundinacea. Bull. Torrey Bot. Club*, 70:182–183.

Bremer, G. 1923. A cytological investigation of some species and species hybrids within the genus *Saccharum. Genetica*, 5:97–148, 273–326.

———. 1924. The cytology of the sugarcane. II. A cytological investigation of some cultivated kinds and their parents. *Genetica*, 6:497–525.

————. 1925. The cytology of the sugarcane. III. The chromosomes of primitive forms of the genus *Saccharum. Genetica*, 7:293–322.

Brown, W. V. 1958. Leaf anatomy in grass systematics. *Bot. Gaz.*, 119:170–178.

————. 1977. The Kranz syndrome and its subtypes in grass systematics. *Mem. Torrey Bot. Club*, 23:1–97.

————, and W. H. P. Emery. 1957. Persistent nucleoli and grass systematics. *Amer. J. Bot.*, 44:585–590.

————; W. F. Harris; and J. D. Graham. 1959. Grass morphology and systematics. I. The internode. *Southw. Naturalist*, 4:126–130.

————; G. A. Pratt; and H. M. Mobley. 1959. Grass morphology and systematics. II. The nodal pulvinus. *Southw. Naturalist*, 4:115–125.

Bruns, E. 1892. Der Grasembryo. *Flora*, 76:1–33.

Canfield, R. H. 1934. Stem structure of grasses on the Jornada Experimental Range. *Bot. Gaz.*, 94:636–648.

Carnahan, H. L., and H. D. Hill. 1961. Cytology and genetics of forage grasses. *Bot. Rev.*, 27:1–162.

Darlington, C. D., and A. P. Wylie. 1955. *Chromosome atlas of flowering plants.* George Allen & Unwin, Ltd., London.

Decker, H. F. 1964. An anatomic-systematic study of the classical tribe Festuceae (Gramineae). *Amer. J. Bot.*, 51:453–463.

de Wet, J. M. J. 1954. The genus *Danthonia* in grass phylogeny. *Amer. J. Bot.*, 41:204–211.

————. 1956. Leaf anatomy and phylogeny in the tribe Danthonieae. *Amer. J. Bot.*, 43:175–182.

Essad, S. 1954. Contribution a la systématique du genu *Lolium. Ann. de l'Amélior. des Plantes*, 4:325–351.

Evans, G. 1926. Chromosome complements in grasses. *Nature*, 118:841.

Fairbrothers, D. E., and M. A. Johnson. 1961. The precipitin reaction as an indicator of relationship in some grasses. *Recent Advances in Bot.*, 1:116–120.

Frew, P., and R. H. Bowen. 1929. Nucleolar behavior in the mitosis of plant cells. *Quart. J. Microscop. Sci.*, 73:197–214.

Gould, F. W. 1957a. New andropogons with a key to the native and naturalized species of section Amphilophis in the U.S. *Madroño*, 14:18–29.

————. 1957b. Pollen size as related to polyploidy and speciation in the *Andropogon saccharoides-A. barbinodis* complex. *Brittonia*, 9:71–75.

————. 1960. Chromosome numbers in southwestern grasses. II. *Amer. J. Bot.*, 47:873–877.

————. 1963. Cytotaxonomy of *Digitaria sanguinalis* and *D. adscendens. Brittonia*, 15:241–244.

Gutierrez, M.; V. E. Gracen; and G. E. Edwards. 1974. Biochemical and cytological relationships in C_4 plants. *Planta* (Berl.) 119:279–300.

————; G. E. Edwards; and W. V. Brown. 1976. PEP carboxykinase containing species in the Brachiaria group of the subfamily Panicoideae. *Biochem. Syst. and Ecol.* 4:47–49.

Hall, O. 1959. Immuno-electrophoretic analyses of allopolyploid ryewheat and its parental species. *Hereditas*, 45:495–504.

————, and B. L. Johnson. 1963. Electrophoretic analysis of the amphiploid of *Stipa viridula* × *Oryzopsis hymenoides* and its parental species. *Hereditas*, 48: 530–535.

Harz, C. O. 1880. Beitrage zur Systematik der Gramineen. *Linnaea*, 43:1–30.

Hattersley, P. W., and L. Watson. 1975. Anatomical parameters for predicting photosynthetic pathways of grassleaves: The "maximum lateral cell count" and the "maximum cell distant count." *Phytomorphology* 25:325–333.

————, and ————. 1976. C_4 grasses: an anatomical criterion for distinguishing between NADP-malic enzyme species and PCK and NAD-malic enzyme species. *Aust. J. Bot.* 24:297–308.

Hitchcock, A. S. 1920. *The genera of grasses of the United States, with special reference to the economic species*. U.S. Dep. Agr. Bull. 772.

————. 1935. *Manual of the grasses of the United States*. U.S. Dep. Agr. Misc. Publ. 200.

————. 1951. *Manual of the grasses of the United States*, 2nd ed. (Revised by Agnes Chase.) U.S. Dept. Agr. Misc. Publ. 200.

Holttum, R. E. 1958. The bamboos of the Malay Peninsula. *Gard. Bull.*, 16:1–35.

Hoshikawa, K. 1969. Underground organs of the seedling and the systematics of the Gramineae. *Bot. Gaz.* 130:192–203.

Hubbard, C. E. 1948. The genera of British grasses. In J. Hutchison, *British flowering plants*, pp. 284–348.

Johnston, C. R., and L. Watson. 1976. Microhairs: A universal characteristic of nonfestucoid grass genera? *Phytomorphology* 26:297–301.

Kanai, R., and M. Kashiwagi. 1975. *Panicum miliodes*, a Gramineae plant having Kranz leaf anatomy without C_4 photosynthesis. *Pl. Cell Physiol.* 16:669–679.

Kapadia, Z. J., and F. W. Gould. 1964. Biosystematic studies in the *Bouteoua curtipendula* complex. III. Pollen size as related to chromosome numbers. *Amer. J. Bot.*, 51:166–172.

Kinges, H. 1961. Merkmale des Graminees embryos. *Bot. Jahrb.*, 81:50–93.

Kraus, R. 1897. Uber specifische Reactionen in keimfreien Filtraten aus Cholera, Typus, and Pestbouillon Culturen, erzeugt durch homologes Serum. *Wien. klini. Wochensch.*, 10:736–738.

Kuwada, Y. 1919. Die Chromosomenzahl von *Zea mays* L. *J. Coll. Sci. Imp. Univ. Tokyo*, 39:1–148.

Lommasson, R. C. 1961. Grass leaf venation. *Recent Advances in Botany*, 1: 108–111.

Longley, A. E. 1924. Chromosomes in maize and maize hybrids. *J. Agr. Res.*, 28:673–682.

McClure, F. A. 1934. The inflorescence in *Schizostachyum* Nees. *J. Wash. Acad. Sci.*, 24:541–548.

————. 1966. *The bamboos: A fresh perspective*. Harvard University Press, Cambridge, Mass.

Metcalfe, C. R. 1960. *Anatomy of the monocotyledons*. I. *Gramineae*. Oxford University Press, London.

Myers, W. M. 1947. Cytology and genetics of forage grasses. *Bot. Rev.*, 13:318–421.

Naylor, B., and H. Rees. 1958. Chromosome size in *Lolium temulentum* and *L. perenne. Nature*, 181:854–855.

Nuttall, G. H. F. 1901. The new biological test for blood in relation to zoological classification. *Proc. Royal Soc. London*, 69:150–153.

Parodi, L. R. 1961. La taxonomia de las gramineae Argentinas a la luz de las investigaciones más recientes. *Recent Advances in Bot.*, 1:125–130.

Pilger, R. 1954. Das system der Gramineae. *Bot. Jahrb.*, 76:281–384.

Prat, H. 1932. L'épiderme des Graminées; étude anatomique et systématique. *Ann. Sci. Nat. Bot.*, 14:117–324.

———. 1936. La systématique des Graminées. *Ann. Sci. Nat. Bot.*, 18:165–258.

———. 1960. Vers une classification naturelle des Graminées. *Bull. Soc. Bot. Fr.*, 107:32–79.

Ramanujam, S. 1938. Cytogenetical studies in the Oryzeae. I. Chromosome studies in the Oryzeae. *Ann. Bot. N.S.* 2:107–125.

Reeder, J. 1953. Affinities of the grass genus *Beckmannia* Host. *Bull. Torrey Bot. Club*, 80:187–196.

———. 1957. The embryo in grass systematics. *Amer. J. Bot.*, 44:756–768.

———. 1962. The bambusoid embryo: A reappraisal. *Amer. J. Bot.*, 49:639–641.

———, and K. von Maltzalen. 1953. Taxonomic significance of root-hair development in the Gramineae. *Proc. Nat. Acad. Sci. (U.S.)*, 39:593–598.

Row, H. C., and J. R. Reeder. 1957. Root hair development as evidence of relationships among genera of Gramineae. *Amer. J. Bot.*, 44:596–601.

Sinnott, E. W. 1939. Growth and differentiation in living plant meristems. *Proc. Nat. Acad. Sci. (U.S.)*, 25:55–58.

———, and R. Bloch: 1939. Cell polarity and the differentiation of root hairs. *Proc. Nat. Acad. Sci. (U.S.)*, 25:248–252.

Smith, B. N., and W. V. Brown. 1973. The Kranz syndrome in the Gramineae as indicated by carbon isotope ratios. *Amer. J. Bot.* 60:505–513.

Standley, P. C. 1928. Flora of the Panama Canal Zone. *Contrib. U.S. Nat. Herb.*, 27:1–416.

Stebbins, G. L. 1950. *Variation and evolution in plants.* Columbia University Press, New York.

———. 1956. Cytogenetics and evolution of the grass family. *Amer. J. Bot.*, 43:890–905.

———, and B. Crampton. 1961. A suggested revision of the grass genera of temperate North America. *Recent Advances in Bot.*, 1:133–145.

Tateoka, T. 1957. Miscellaneous papers on the phylogeny of Poaceae (10). Proposition of a new phylogenetic system of Poaceae. *J. Jap. Bot.*, 32:275–287.

———. 1960. Cytology in grass systematics: A critical review. *Nucleus*, 3:81–110.

———. 1962. Starch grains of endosperm in grass systematics. *Bot. Mag. Tokyo*, 75:377–383.

———. 1963. Notes on some grasses. XIII. Relationship between Oryzeae and Ehrharteae, with special reference to leaf anatomy and histology. *Bot. Gaz.*, 124:264–270.

———. 1964. Notes on some grasses. XVI. Embryo structure of the genus *Oryza* in relation to the systematics. *Amer. J. Bot.*, 51:539–543.

————; S. Inoue; and S. Kawano. 1959. Notes on some grasses, IX: Systematic significance of bicellular microhairs of leaf epidermis. *Bot. Gaz.*, 121:80–91.

Van Tieghem, P. 1897. Morphologie de l'embryon et de la plantule chez les Graminées et les Cyperacées. VIII. *Ann. Sci. Nat. Bot.*, 3:259–309.

Waller, S. S., and J. K. Lewis. 1979. Occurrence of C_3 and C_4 photosynthetic pathways in North American grasses. *J. Range Manage.*, 32:12–28.

Yakovlev, M. S. 1950. Structure of embryo and endosperm in grasses as a systematic feature. *Morf. Anat. Trudy Bot. Akad. Nauk. S.S.S.R.*, ser. 7, 1:121–218.

Yates, H. O. 1966a. Morphology and cytology of *Uniola* (Gramineae). *Southw. Naturalist*, 11:145–189.

————. 1966b. Revision of grasses traditionally referred to *Uniola*. I. *Uniola* and *Leptochloöpsis. Southw. Naturalist*, 11:372–394.

————. 1966c. Revision of grasses traditionally referred to *Uniola*. II. *Chasmanthium. Southw. Naturalist*, 11:415–455.

Zirkle, C. 1928. Nucleolus in root tip mitosis in *Zea mays. Bot. Gaz.*, 86:402–418.

5
United States
Grasses

KEY TO THE
GENERA

The following key is based on readily observable inflorescence characters and general lines of evolutionary modification. It is constructed strictly for identification purposes and is not intended to reflect natural groupings. Because of the relatively unique and consistent spikelet structure of the Panicoideae, however, it has been possible to treat the two tribes of this subfamily as natural groups (Groups I and II). Fig. 5-1, a synopsis of the basic characters of the six major groups of the key, is presented to facilitate the rapid placement of most grasses into their proper group.

1. Leaf blades about 1 cm long; stoloniferous, mat-forming perennial with fascicled leaves and inconspicuous, unisexual spikelets
                                                    MONANTHOCHLOË (p. 311)
1. Leaf blades more than 1 cm long; plants not as above
  2. Spikelets unisexual, the staminate and pistillate conspicuously different
    3. Plants monoecious (staminate and pistillate spikelets on the same plant)
      4. Staminate and pistillate spikelets in separate inflorescences
        Leaf blades 3 cm or more broad; the cultivated corn
                                                    ZEA (p. 264)
        Leaf blades much less than 1 cm broad   **LUZIOLA (p. 325)**
      4. Staminate and pistillate spikelets in the same inflorescence
        5. Pistillate spikelets becoming indurate or enclosed in a hard, bony involucre
          6. Pistillate spikelets 1 per spike or branch, associated with 2 reduced (sterile) spikelets in a bony, beadlike involucre; coarse annual   **COIX (p. 264)**
          6. Pistillate spikelets usually 2 to 6 per spike or branch, becoming indurate but not enclosed in a bony, beadlike involucre; tall, stout perennials   **TRIPSACUM (p. 262)**
        5. Pistillate spikelets neither indurate nor in bony, beadlike involucres
          7. Inflorescence a spicate raceme
                                                    HETEROPOGON (p. 261)
          7. Inflorescence a panicle   **GROUP III (p. 144)**
    3. Plants usually dioecious; low, stoloniferous, mat-forming perennials
      8. Pistillate and staminate spikelets awnless, pistillate spikelets in burrlike clusters hidden in the leafy portion of the plant, staminate spikelets sessile on 1 to 4 short spicate branches of a well-exserted inflorescence   **BUCHLOË (p. 301)**
      8. Pistillate spikelets long-awned (3-awned from the nerves of each

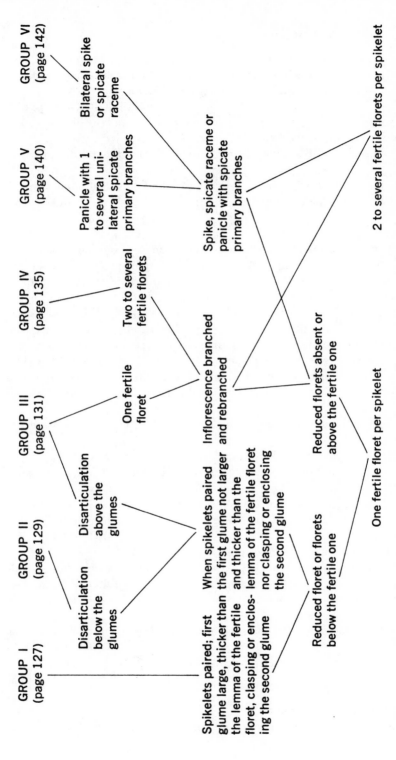

Fig. 5-1. Diagrammatic guide to the six major key groups of grass. All groups except I (tribe Andropogoneae) and II (tribe Paniceae) are "artificial" and contain distantly related as well as closely related genera.

lemma) staminate spikelets awnless; both pistillate and staminate spikelets in contracted, usually spikelike racemes

SCLEROPOGON (p. 275)

2. Spikelets perfect, or if unisexual, then staminate and pistillate spikelets not conspicuously different

9. Spikelets with a single perfect floret, with or without reduced florets

A (p. 138)

9. Spikelets with 2 or more perfect florets, or if unisexual, then with 2 or more staminate or pistillate florets   **AA (p. 138)**

A (1 Perfect floret)

10. Spikelets in pairs of 1 sessile or subsessile and 1 pediceled (2 pediceled at branch tips), infrequently both spikelets short-pediceled; pediceled spikelet like the sessile one or more often reduced or rudimentary, occasionally represented by pedicel alone; first glume large and firm, tightly clasping or enclosing second glume; florets basically 2, but the lowermost usually absent or represented by a thin membrane; lemma of fertile floret thin, membranous, the midnerve often extended into a stout, geniculate awn (tribe Andropogoneae)   **GROUP I (p. 139)**

10. Spikelets in pairs or not; when paired, then first glume not larger and firmer than lemma of fertile floret and not clasping or enclosing second glume

11. Reduced floret or florets (staminate or sterile) present below the fertile one

12. Reduced floret 1; lemma of reduced floret similar to second glume in size and texture (in *Reimarochloa*, both glumes usually absent); disarticulation below glumes (tribe Paniceae)   **GROUP II (p. 141)**

12. Reduced florets 1 or 2; lemma of reduced florets not similar to second glume in size and texture; disarticulation above glumes

GROUP III (p. 144)

11. Reduced florets absent or present above the fertile one

13. Inflorescence a panicle or open raceme, primary branches spreading or contracted but not spicate   **GROUP III (p. 144)**

13. Inflorescence a spike, spicate raceme, or with 2 to several spicate primary branches

14. Inflorescence of 1 to several unilateral spicate primary branches

GROUP V (p. 153)

14. Inflorescence a terminal, bilateral spike or spicate raceme

GROUP VI (p. 155)

AA (2 or more perfect florets)

15. Inflorescence an open or contracted panicle, or raceme with spikelets on

well-developed pedicels   **GROUP IV (p. 148)**
15. Inflorescence a spike or spicate raceme, or with spicate primary branches
  16. Inflorescence with 2 (infrequently 1) to several unilateral primary branches   **GROUP V (p. 153)**
  16. Inflorescence a terminal, bilateral spike or spicate raceme
                                          GROUP VI (p. 155)

GROUP I (Tribe Andropogoneae)

1. Spikelets all alike and fertile   **B (p. 139)**
1. Spikelets not all alike, the pediceled ones, or less frequently the sessile or subsessile ones, staminate or sterile   **BB (p. 139)**

B (Spikelets all alike, fertile)

2. Pediceled spikelets present, like sessile spikelets
  3. Inflorescence branches 2 to 6; annual with leaf blades 3 to 10 cm long
                              MICROSTEGIUM (p. 253)
  3. Inflorescence branches numerous; stout perennials with long leaf blades
    4. Spikelets falling in pairs together with sections of a disarticulating rachis
      5. Spikelet awned   **ERIANTHUS (p. 248)**
      5. Spikelet awnless   **SACCHARUM (p. 248)**
    4. Spikelets falling separately from a persistent rachis
      6. Spikelets (lemma of fertile floret) awned; panicle relatively long and slender, with appressed branches   **IMPERATA (p. 246)**
      6. Spikelet awnless; panicle short and broad, with erect-spreading branches   **MISCANTHUS (p. 248)**
2. Pediceled spikelets completely reduced, represented by pedicel only
  7. Blades ovate, not over 6 cm long; low annual   **ARTHRAXON (p. 253)**
  7. Blades lanceolate or linear, at least some more than 6 cm long; perennials
    8. Spikelets mostly 7 to 9 mm long, borne on numerous scattered branches of a large terminal panicle   **SORGHASTRUM (p. 249)**
    8. Spikelets 5 mm or less long, borne in small clusters on branchlets of a much-divided, broomlike flowering culm **ANDROPOGON (p. 250)**

BB (Spikelets not all alike, sessile
or pediceled one staminate or sterile)

9. Spikelet awnless
  10. Flowering culm terminating in a panicle with numerous branches
                                SORGHUM (p. 249)
  10. Flowering culm or leafy branch terminating in a spicate raceme

11. Raceme less than 3 cm long; pedicels fused to rachis; sessile spikelet globose, conspicuously alveolate; annual

HACKELOCHLOA (p. 262)

11. Raceme mostly 4 to 10 cm long; pedicels not fused to rachis; sessile spikelet not globose; perennials

    12. Rachis and pedicels glabrous; pediceled spikelet rudimentary, greatly reduced

        13. Sessile spikelets sunken in thickened rachis

COLEORACHIS (p. 261)

        13. Sessile spikelets not sunken in rachis; rachis slender

EREMOCHLOA (p. 258)

    12. Rachis and pedicels puberulent or ciliate; pediceled spikelet well developed   **ELYONURUS (p. 259)**

9. Spikelet (lemma of fertile floret) awned

    14. Awns 3 to 8 cm long

        15. Spikelets on long, slender branches of a few-flowered terminal panicle   **CHRYSOPOGON (p. 255)**

        15. Spikelets in unilateral spicate racemes, these single at culm and branch tips

            16. Perfect (awned) spikelets sessile; glumes and awn of perfect spikelet dark brown at maturity   **HETEROPOGON (p. 261)**

            16. Perfect spikelets pediceled; glumes and awn of perfect spikelet light-colored   **TRACHYPOGON (p. 258)**

    14. Awns less than 3 cm long

        17. Flowering culms much-branched above, terminating in numerous short, leafy branchlets, each bearing 1 to 6 pedunculate, spicate flower clusters above uppermost leaf or bract

            18. Branchlets terminating in single spicate raceme

SCHIZACHYRIUM (p. 258)

            18. Branchlets terminating in small panicle with 2 to 6 spicate branches

                19. Inflorescence branches 2, the lowermost pair of spikelets of each branch awnless and sterile   **HYPARRHENIA (p. 256)**

                19. Inflorescence branches 2 to 6, sessile spikelet of the lowermost pair awned and usually fertile   **ANDROPOGON (p. 250)**

        17. Flowering culms not branched into numerous short, leafy branchlets; spikelets in large or small panicles

            20. Pedicels, at least those above, and usually the upper rachis internodes, with a central groove or membranous area

BOTHRIOCHLOA (p. 255)

            20. Pedicels and rachis internodes flat or rounded, without a central groove or membranous area

                21. Panicle axis above the lowermost branch usually 15 to 30 cm

long; panicle branches numerous, freely rebranched, and not conspicuously spicate in appearance   **SORGHUM (p. 249)**
21. Panicle axis above the lowermost branch less than 15 cm long; panicle branches not or sparingly rebranched, conspicuously spicate in appearance
22. Pediceled spikelet shorter or narrower than the sessile one, tapering to a narrow apex; native species
ANDROPOGON (p. 250)
22. Pediceled spikelet about as large as the sessile one, broadly rounded at apex; introduced species
DICHANTHIUM (p. 254)

Group II (Tribe Paniceae)

1. Spikelets in involucres of bristles or flattened spines, these disarticulating with spikelets
2. Bristles or spines fused together, at least at the base
CENCHRUS (p. 243)
2. Bristles and spines not fused together   **PENNISETUM (p. 241)**
1. Spikelets not in bristly or spiny involucres, bristles persistent when present
3. Spikelets all or in part (at least those terminating the branchlets) subtended by 1 to several bristles   **SETARIA (p. 238)**
3. Spikelets not subtended by bristles
4. Inflorescence a spike, spikelets partially embedded in a thick, flattened rachis; first glume present   **STENOTAPHRUM (p. 215)**
4. Inflorescence not a spike, or if so, then spikelets lacking first glume
5. Glumes both absent (the second occasionally present on terminal spikelet); inflorescence of 1 to 5 unilateral spicate branches
REIMAROCHLOA (p. 220)
5. Glumes, at least the second, well developed
6. Second glume awned; first glume awned or awnless
7. First glume minute; second glume and lemma of sterile floret about equal, silky-villous   **RHYNCHELYTRUM (p. 238)**
7. First glume well developed; spikelets not silky-villous
8. First glume much shorter than the second, awnless or with an awn shorter than the body   **ECHINOCHLOA (p. 235)**
8. First glume about as long as the second, with an awn to 3 times the length of the body   **OPLISMENUS (p. 233)**
6. Second and first glumes both awnless
9. Lemma of sterile floret bearing a delicate awn 1 to 10 mm long between lobes of a notched apex   **MELINIS (p. 245)**
9. Lemma of sterile floret awnless or with a stout awn from an entire apex

10. Plants with cleistogamous subterranean spikelets as well as aerial spikelets, the former developing seed, the latter perfect but rarely fertile; New Jersey to Florida
AMPHICARPUM (p. 244)

10. Plants with aerial spikelets only

    11. Lemma of fertile floret thin and flexible, margins membranous and not inrolled over palea; first glume minute or absent **C (p. 142)**

    11. Lemma of fertile floret relatively thick and rigid, margins typically inrolled over palea; first glume present or absent **CC (p. 142)**

C (Lemma margins membranous and not inrolled over palea)

12. Spikelets on long or short pedicels in an open or loosely contracted panicle; panicle branches not spicate

    13. Spikelets in a contracted panicle much longer than broad; pedicels mostly shorter than spikelets **ANTHAENANTIA (p. 215)**

    13. Spikelets in an open panicle about as broad as long at maturity; pedicels mostly 2 to several times as long as spikelets
LEPTOLOMA (p. 214)

12. Spikelets subsessile or on short, appressed pedicels in a panicle of few to several spicate primary or secondary inflorescence branches
DIGITARIA (p. 212)

CC (Lemma margins thick and inrolled over palea)

14. First glume absent on some or all spikelets

    15. Lemma of fertile floret mucronate or short-awned; cuplike or disklike ring present at base of spikelet **ERIOCHLOA (p. 221)**

    15. Lemma of fertile floret not mucronate or awned; cuplike or disklike ring not present at base of spikelet

        16. Lemma of fertile floret with rounded back turned away from rachis; spikelets narrowly oblong, borne singly and widely spaced in 2 rows
AXONOPUS (p. 220)

        16. Lemma of fertile floret with rounded back turned toward the rachis; spikelets broadly ovate to oblong, closely placed and often paired in 2 or 4 rows **PASPALUM (p. 223)**

14. First glume present on all spikelets

    17. Second glume densely long-hairy; first glume glabrous, more than half

the length of spikelet   **BRACHIARIA** (*B. ciliatissima*) (**p. 219**)
17. Second glume not densely long-hairy; first glume less than half the length of spikelet when second glume pubescent
  18. Inflorescence of 2 to several spicate, unbranched primary branches; spikelets in regular rows; first glume much shorter than second glume and lemma of sterile floret
    19. Second glume and lemma of sterile floret scabrous-pubescent with short, stiff hairs   **ECHINOCHLOA** (**p. 235**)
    19. Second glume and lemma of sterile floret glabrous
      20. Plants annual; rachis of inflorescence branches winged, 1 to 2 mm broad   **BRACHIARIA** (**p. 219**)
      20. Plants perennial; rachis less than 1 mm broad
            PASPALIDIUM (**p. 225**)
  18. Inflorescence with some or all of the primary branches rebranched, or if not, then first glume about as long as second glume and lemma of sterile floret
    21. Culms woody, freely branched above, often viny; lemma and palea of fertile floret with minute tufts of hair at the tips; Florida
            LASIACIS (**p. 233**)
    21. Culms not woody and freely branched above, except in *Panicum antidotale*; lemma and palea of fertile floret without tufts of hair at the tips
      22. Second glume gibbous at the base, thin, strongly 7- to 11-nerved, 3 to 4 times as long as first glume; fertile floret on a short stipe   **SACCIOLEPIS** (**p. 237**)
      22. Second glume not gibbous; fertile floret not stipitate
        23. Tip of lemma and palea of fertile floret usually abruptly pointed, tip of palea free from lemma
            ECHINOCHLOA (**p. 235**)
        23. Tip of lemma and palea of fertile floret usually rounded, tip of palea enclosed by lemma
          24. Palea of sterile floret enlarged and inflated, firm, obovate
            STEINCHISMA (**p. 230**)
          24. Palea of sterile floret not enlarged or inflated
            25. Lemma and palea of fertile floret strongly transversely rugose   **BRACHIARIA** (**p. 219**)
            25. Lemma and palea of fertile floret rough or smooth, but not transversely rugose
              26. Plant annual or perennial, not developing a rosette of short, broad basal leaves during the cool season; plants flowering in warm season only

27. First glume nearly as long as the spikelet; fertile floret less than 1/3 the length of the spikelet
       PHANOPYRUM (p. 233)
27. First glume 1/2 or less the length of the spikelet; if second glume as long as the spikelet then fertile floret more than 1/2 the length of the spikelet
       PANICUM (p. 226)
26. Plant perennial, most species developing a rosette of short, broad basal leaves during the cool season; plants flowering first during the cool season; small axillary inflorescences commonly produced on much-branched and reduced lateral shoots during the warm season
       DICHANTHELIUM (p. 228)

Group III (Panicle, with rebranched primary branches; fertile floret 1)

1. Glumes absent or rudimentary
 2. Lemma with an awn 1 or 2 cm long **BRACHYELYTRUM (p. 209)**
 2. Lemma awnless or with awn much less than 1 cm long
  3. Spikelets perfect
   4. Lemma about 1 mm long, short-awned, not laterally compressed; tufted annual with culms less than 10 cm tall; West Coast
       COLEANTHUS (p. 171)
   4. Lemma more than 1 mm long, awnless, strongly compressed laterally
    5. Spikelets 7 to 10 mm long; annual, the cultivated rice
       ORYZA (p. 321)
    5. Spikelets less than 6 mm long; native perennials
       LEERSIA (p. 322)
  3. Spikelets unisexual
   6. Leaf blades 1 to 4 cm long; staminate and pistillate spikelets in separate inflorescences, these inconspicuous; plant with trailing or floating culms and usually floating leaves **LUZIOLA (p. 325)**
   6. Leaf blades much more than 4 cm long; staminate and pistillate spikelets in the same inflorescence, this large and conspicuous
    7. Staminate spikelets pendulous on the spreading lower branches of panicle, pistillate spikelets erect on the stiffly erect upper branches **ZIZANIA (p. 323)**
    7. Staminate and pistillate spikelets on the same branches, pistillate at the tip, staminate below **ZIZANIOPSIS (p. 325)**
1. Glumes, at least the second, well developed
 8. Leaf blades 1.5 to 4 cm broad, twisted so that the abaxial surface is up-

permost; spikelets in pairs, one large, sessile, and pistillate, the other small, pediceled, and staminate; Florida, infrequent
PHARUS (p. 321)

8. Leaf blades usually less than 1.5 cm broad; spikelets not as above
9. Spikelets with 1 perfect floret and 1 or more staminate or rudimentary florets; glumes both large, equaling or exceeding the perfect floret
10. Lemma of perfect floret with nine subequal, plumose awns
ENNEAPOGON (p. 315)
10. Lemma of perfect floret awnless or with a single awn
11. Spikelets with 3 well-developed florets, the lower two staminate
12. Glumes and florets about equal in length
HIEROCHLOË (p. 188)
12. Glumes unequal in length, the second much larger than the first and the florets   **ANTHOXANTHUM (p. 188)**
11. Spikelets with 1 or 2 well-developed florets
13. Well-developed floret 1, lemma awnless
14. One or two scalelike rudiments present below fertile floret
PHALARIS (p. 189)
14. One rudiment present above fertile floret
LAMARCKIA (p. 172)
13. Well-developed florets 2, lemma of either the upper or lower awned
15. Lemma of lower floret awnless; lemma of upper floret awned
16. Lemma of upper floret awned from the back; annual
VENTENATA (p. 179)
16. Lemma of upper floret awned from between the lobes or teeth of a cleft apex; annual or perennial
HOLCUS (p. 180)
15. Lemma of lower floret with a stout geniculate awn; lemma of upper floret awnless   **ARRHENATHERUM (p. 180)**
9. Spikelets with 1 perfect floret and no reduced or rudimentary florets
17. Spikelets suborbicular, laterally compressed, subsessile, and crowded on short branches of a narrow, elongate panicle
BECKMANNIA (p. 193)
17. Spikelets not as above
18. Glumes and lemmas awnless   **D (p. 145)**
18. Glumes or lemmas awned   **DD (p. 146)**
D (Glumes and lemmas awnless)
19. Disarticulation below glumes; low, mat-forming annual; California   **CRYPSIS (p. 285)**

19. Disarticulation above glumes
    20. Lemma with a tuft of hair at the base; spikelets 5 mm or
        more long
        21. Spikelets mostly 10 to 15 mm long; panicles dense and
            tightly contracted   **AMMOPHILA (p. 182)**
        21. Spikelets 5 to 8 mm long; panicles open or loosely con-
            tracted   **CALAMOVILFA (p. 277)**
    20. Lemma without a tuft of hair at the base when spikelets
    are 5 mm or more in length
        22. Glumes both as long as, or longer than, lemma
            AGROSTIS (p. 183)
        22. Glumes, at least the first, shorter than lemma
            23. Floret dorsoventrally compressed; lemma becoming
                hard and shiny, margins tightly clasping palea, as in
                Panicum   **MILIUM (p. 192)**
            23. Floret terete; lemma relatively thin, not as above
                24. Lemma 3-nerved
                    25. Nerves of lemma densely pubescent
                        BLEPHARONEURON (p. 283)
                    25. Nerves of lemma glabrous or scabrous
                      26. First glume minute or wanting; lemma about
                        1.5 mm long, broadly acute at apex; low, tufted
                        alpine perennial   **PHIPPSIA (p. 170)**
                    26. First glume usually well developed; if minute,
                      the lemma longer than 1.5 mm or plant annual
                        MUHLENBERGIA (p. 279)
                24. Lemma 1-nerved
                  27. Panicle spicate, densely flowered, 1 to 4 cm long
                    and 4 to 10 mm thick; annuals
                      CRYPSIS (p. 285)
                  27. Panicle open or spicate; when densely flowered
                    and 4 mm or more thick, then much more than 4
                    cm long   **SPOROBOLUS (p. 280)**

DD (Glumes or lemmas awned)
28. First glume usually 2- or 3-awned, second glume usually 1-awned; spikelets in pairs, the lower of the pair sterile, the two falling together
        LYCURUS (p. 278)
28. First and second glumes not as above, or if so, then spikelets not falling
in pairs
    29. Disarticulation below glumes; glumes equal or nearly so, as long as, or
    longer than, lemma

30. Glumes awned   **POLYPOGON (p. 184)**
30. Glumes awnless
  31. Lemma awned from the middle or below; inflorescence compact,
     cylindrical, spikelike   **ALOPECURUS (p. 189)**
  31. Lemma awned from or near the tip
    32. Awn 2.5 mm or less long; strong perennial   **CINNA (p. 185)**
    32. Awn 6 mm or more long; short-lived annual
                                  LIMNODEA (p. 187)
29. Disarticulation above glumes
  33. Lemma indurate, awned, with well-developed callus at the base,
     permanently enclosing palea and caryopsis
    34. Awn of lemma 3-branched, lateral branches short or rudimentary
      in a few species   **ARISTIDA (p. 316)**
    34. Awn of lemma unbranched
      35. Awn straight or curved but not twisted, rarely more than 2 to 4
        times as long as body of lemma, early deciduous; body of lemma
        broad, usually subglobose, with a short, blunt callus
                                  ORYZOPSIS (p. 208)
      35. Awn twisted and geniculate, usually several to many times as
        long as body of lemma, persistent or finally disarticulating
        36. Palea brownish, large and broad, usually longer than the
          lemma and slightly protruding at the apex, grooved between
          the 2 heavy nerves; floret subglobose, with a short and
          abruptly pointed callus   **PIPTOCHAETIUM (p. 209)**
        36. Palea colorless, thin, small or rudimentary, much shorter
          than the lemma and not protruding from the apex, the 2
          nerves weak or absent; floret terete, slender, the callus with
          a long, sharp point   **STIPA (p. 206)**
  33. Lemma not indurate or permanently enclosing palea and caryopsis
    37. Glumes equal, broad, abruptly short-awned from an obtuse apex;
      lemma much shorter than glumes, awnless   **PHLEUM (p. 190)**
    37. Glumes not equal, or if nearly so, then not abruptly awned
      38. Second glume 4 to 5 times as long as lemma; annual with
        densely contracted, spikelike panicle
                                GASTRIDIUM (p. 191)
      38. Second glume shorter to slightly longer than lemma
        39. Glumes 1-nerved, pubescent, tapering to a plumose awn tip;
          panicles contracted, densely flowered, woolly, 2 to 3 cm long
          and about as broad   **LAGURUS (p. 192)**
        39. Glumes 1- to 3-nerved, glabrous or pubescent but without a
          plumose awn tip; panicles open or contracted, elongate and
          slender when contracted and densely flowered

40. Lemma awned from the back, base, or cleft apex; glumes equaling or exceeding lemma

    41. Lemma firm, awned from near the apex, awn straight or flexous, 3 to 4 times as long as the body; annuals

        APERA (p. 182)

    41. Lemma, thin, awned from the back or base, awn usually geniculate when long

        42. Floret with a tuft of hair at the base; palea present, usually well developed; rachilla prolonged above insertion of palea, perennials

            CALAMAGROSTIS (p. 181)

        42. Floret usually without hairs at the base; palea absent or poorly developed; rachilla not prolonged above insertion of palea; annuals and perennials

            AGROSTIS (p. 183)

40. Lemma awned from an entire or minutely cleft apex; glumes, at least the first, usually shorter than lemma

        MUHLENBERGIA (p. 279)

Group IV (Panicle with rebranched primary branches; fertile florets 2 or more)

1. Plants 2 to 6 meters tall

    2. Spikelets mostly 3 to 7 cm long and 7- to 13-flowered

        ARUNDINARIA (p. 319)

    2. Spikelets less than 2 cm long and with fewer than 7 florets

        3. Leaves mostly basal, the blades 0.5 to 1.5 cm broad; culms densely clumped, without creeping rhizomes   **CORTADERIA (p. 328)**

        3. Leaves evenly distributed on culm, blades 2 to 6 cm broad; culms with stout creeping rhizomes, forming large colonies

            4. Lemmas villous, rachilla glabrous   **ARUNDO (p. 326)**

            4. Lemmas glabrous, rachilla villous   **PHRAGMITES (p. 327)**

1. Plants less than 2 meters tall

    5. Lemmas with 3 nerves, these usually conspicuous   **E (p. 148)**

    5. Lemmas, at least some, 5- to 15-nerved, nerves conspicuous or obscure

        EE (p. 149)

E (Lemmas 3-nerved)

6. Nerves of lemma pubescent or puberulent, or base of lemma long-hairy

    7. Plants with rhizomes; panicles open, with long, spreading branches; lemmas glabrous on nerves, with tuft of hair at base

        REDFIELDIA (p. 274)

    7. Plants without rhizomes; panicles open or contracted; lemmas pubescent or puberulent on nerves, at least below

8. Palea densely long-ciliate on the upper half

                      TRIPLASIS (p. 268)

8. Palea not densely long-ciliate on the upper half
  9. Panicles 1 to 8 cm long, contracted, ovoid or oblong; lemmas conspicuously long-hairy on nerves, at least below

                      ERIONEURON (p. 267)

  9. Panicles open or contracted, 4 to 30 cm or more long; when less than 10 cm long, then lemmas inconspicuously puberulent on nerves   **TRIDENS (p. 267)**

6. Nerves of lemma not pubescent or puberulent, base of lemma not long-hairy
  10. Lemmas 3-awned
    11. Awns 4 to 10 cm long   **SCLEROPOGON (p. 275)**
    11. Awns less than 1 cm long   **BLEPHARIDACHNE (p. 275)**
  10. Lemmas awnless
    12. Lemmas 6 to 10 mm long; second glume 3- to 5-nerved; caryopsis large, turgid, beaked   **DIARRHENA (p. 211)**
    12. Lemmas less than 6 mm long, or if this long, then second glume 1-nerved; caryopsis not large, turgid or beaked
      13. Glumes much longer than lowermost floret; spikelets mostly 2-flowered   **DISSANTHELIUM (p. 181)**
      13. Glumes about equaling or shorter than lowermost floret
        14. Glumes flat, the first and often the second nerveless, irregularly toothed at the broad apex; spikelets usually with 2 florets; perennial of stream banks or marshy habitats

                      CATABROSA (p. 205)

        14. Glumes rounded or keeled, 1- to 3-nerved, tapering to an entire, acute apex
          15. Spikelets 2- to 4-flowered; lemmas 3 to 6 mm long; Maine to Pennsylvania   **MOLINIA (p. 328)**
          15. Spikelets 3- (rarely 2-) to many-flowered; lemmas less than 3 mm long, or if longer, then florets 6 or more

                      ERAGROSTIS (p. 265)

EE (Lemmas 5- to 13-nerved)
16. Lemmas awned   **F (p. 149)**
16. Lemmas awnless   **FF (p. 152)**

F (Lemmas awned)
17. Lemmas with 5 or more awns or awnlike lobes
  18. Lemmas with 5 awns or awnlike lobes   **ORCUTTIA (p. 316)**
  18. Lemmas with more than 5 awns

19. Glumes 1-nerved; florets falling together

PAPPOPHORUM (p. 314)

19. Glumes 5- to many-nerved; florets falling separately

COTTEA (p. 315)

17. Lemmas with a single awn

20. Culms woody, perennial; spikelets mostly 3 to 7 cm long

ARUNDINARIA (p. 319)

20. Culms not woody or perennial; spikelets rarely as much as 3 cm long

21. Glumes 2 cm or more long; lemmas 1.5 cm or more long; introduced annuals **AVENA (p. 179)**

21. Glumes less than 2 cm long, or if longer, then lemmas less than 1.5 cm long

22. Lemmas awned from back or base

23. Spikelets 2-flowered

24. Rachilla prolonged beyond insertion of upper floret

25. Awn clavate and jointed near the middle, joint with tuft of hair; annual **CORYNEPHORUS (p. 176)**

25. Awn not clavate or jointed; annuals and perennials

DESCHAMPSIA (p. 176)

24. Rachilla not prolonged beyond insertion of upper floret; delicate annuals **AIRA (p. 176)**

23. Spikelets 3- to several-flowered; perennials

HELICTOTRICHON (p. 179)

22. Lemmas awned from a bifid or entire apex

26. First glume longer than lowermost floret; lemmas awned from a bifid apex

27. Awns 5 to 15 mm or more long, usually flattened below

DANTHONIA (p. 329)

27. Awns 2 mm or less long, not flattened below

28. First glume 1- or 3-nerved; lemmas 5 to 6 mm long

SIEGLINGIA (p. 331)

28. First glume 5- or 7-nerved; lemmas 2 to 3 mm long

SCHISMUS (p. 331)

26. First glume about as long as or shorter than lowermost floret

29. Spikelets of 2 kinds, fertile and sterile; fertile spikelets sessile and nearly covered by sterile spikelets in a spicate or subcapitate panicle **CYNOSURUS (p. 172)**

29. Spikelets not as above

30. First glume with 3 or 5 distinct nerves; glumes and lemmas rounded on the back; lemmas 8 to 12 mm long, excluding awns

31. Lemmas long-pilose on callus; margins of leaf sheath
free to base   **SCHIZACHNE (p. 206)**
31. Lemmas not long-pilose on callus; margins of leaf sheath
connate, at least below
    32. Plants perennial; palea not adherent to caryopsis
MELICA (p. 204)
    32. Plants perennial or annual; palea adherent to caryopsis
BROMUS (p. 158)
30. First glume with 1 to 3 distinct or indistinct nerves; glumes
and lemmas keeled or rounded on the back
    33. Palea colorless
        34. Second glume obovate, broadest above the middle;
disarticulation below glumes
SPHENOPHOLIS (p. 173)
        34. Second glume broadest below the middle; disarticula-
tion above glumes   **KOELERIA (p. 173)**
    33. Palea colored green or brown, at least on nerves
        35. Spikelets 1.5 (infrequently 1.2) cm or more long;
lemma apex distinctly to minutely bifid
BROMUS (p. 158)
        35. Spikelets less than 1.2 cm long
            36. Lemmas awned from a distinctly bifid apex, awn
straight or geniculate; second glume equaling or
exceeding lowermost floret
                37. Plants perennial, or if annual, then spikelets dis-
articulating above glumes
TRISETUM (p. 174)
                37. Plants annual, spikelets disarticulating below
glumes   **SPHENOPHOLIS (p. 173)**
            36. Lemmas awned from an entire or minutely notched
apex, awn straight; second glume usually shorter
than lowermost floret
                38. Spikelets laterally compressed, more or less
asymmetrical, subsessile in dense clusters at
tips of stiff, erect, or spreading branches; glumes
and lemmas acute or irregularly short-awned;
perennial   **DACTYLIS (p. 171)**
                38. Spikelets not laterally compressed or asymmetri-
cal, not in dense clusters at branch tips
                    39. Plants annual   **VULPIA (p. 161)**
                    39. Plants perennial   **FESTUCA (p. 162)**

FF (Lemmas awnless)
40. Nerves of lemma strongly and uniformly developed and equally spaced
    41. Second glume 3-nerved, lemma 5-nerved; leaf sheath margins free
                                          PUCCINELLIA (p. 167)
    41. Second glume 1-nerved; lemma 7- to 9-nerved; leaf sheath margins connate, at least below   **GLYCERIA (p. 204)**
40. Nerves of lemma not strongly and uniformly developed, or if so, then not equally spaced
    42. Glumes and lemmas spreading at right angles to rachilla, inflated and papery, resembling the rattles of a rattlesnake; spikelets on slender pedicels   **BRIZA (p. 170)**
    42. Glumes and lemmas not as above
      43. Glumes and lemmas similar, large, thin, flabellate, 9- to 11-nerved; low annual; California   **NEOSTAPFIA (p. 316)**
      43. Glumes and lemmas not as above, or if so, then plant perennial
        44. First glume distinctly longer than lowermost lemma
          45. Glumes 7 mm or more long; lemmas 6 mm or more long
            46. First glume 7- to 11-nerved; lemmas 5- to 7-nerved, densely long-villous on the lower half; Inyo County, California
                              SWALLENIA (p. 312)
            46. First glume 1- to 3-nerved; lemmas 7- to 9-nerved, puberulent below   **SIEGLINGIA (p. 331)**
          45. Glumes less than 7 mm long; lemmas 2 to 3 mm long
                              SCHISMUS (p. 331)
        44. First glume about equaling or shorter than lowermost lemma
          47. Lowermost 1 to 3 florets reduced, sterile, about half as long as those above
            48. Disarticulation below glumes, spikelets falling entire; plants of coastal dunes   **UNIOLA (p. 312)**
            48. Disarticulation above glumes and between florets; plants of woodland sites   **CHASMANTHIUM (p. 331)**
          47. Lowermost florets not reduced, as large as those above
            49. Palea colorless; lateral nerves of lemma indistinct
              50. Second glume obovate, usually abruptly narrowing to an obtuse or broadly acute apex; disarticulation below glumes
                              SPHENOPHOLIS (p. 173)
              50. Second glume not broadened above the middle or only slightly so, acute at apex; disarticulation above glumes
                              KOELERIA (p. 173)
            49. Palea green or brown, at least on nerves
              51. Lemmas 7- to 13-nerved   **G (p. 153)**
              51. Lemmas 5-nerved (3 to 5 in *Diarrhena*)   **GG (p. 153)**

G (Lemmas 7- to 13-nerved)

52. Spikelets unisexual, the staminate and pistillate in separate inflorescences and usually on separate plants; glumes and lemmas thick, firm,
indistinctly nerved
 53. Plants without rhizomes but developing long, thick stolons
                   ALLOLEPIS (p. 310)
 53. Plants strongly rhizomatous, stolons usually not developed
                   DISTICHLIS (p. 308)
52. Spikelets perfect; glumes and lemmas relatively thin, lemmas mostly
with membranous margins
 54. Margins of leaf sheath united to or near apex; caryopsis oblong or
ovate, without persistent, hornlike styles
  55. Palea adhering to caryopsis   **BROMUS (p. 158)**
  55. Palea not adhering to caryopsis   **MELICA (p. 204)**
 54. Margins of leaf sheath free to base; caryopsis suborbicular, with persistent hornlike styles; Texas **VASEYOCHLOA (p. 272)**

GG (Lemmas 5-nerved, 3 to 5 in Diarrhena)

56. Lemmas thick, nerves converging at apex to form a stout beak
                 DIARRHENA (p. 211)
56. Lemmas thin or firm but not thick, nerves not converging to a beaked
apex
 57. Lemmas narrowly acute or attenuate at apex, not scarious on margins
  58. Spikelets unisexual, plants dioecious **LEUCOPOA (p. 165)**
  58. Spikelets perfect
   59. Lemmas villous on callus   **SCOLOCHLOA (p. 166)**
   59. Lemmas not villous on callus   **FESTUCA (p. 162)**
 57. Lemmas obtuse or broadly acute at apex, usually scarious on margins
above
  60. Lemmas moderately to strongly keeled, often with tuft of long,
kinky hair at base   **POA (p. 168)**
  60. Lemmas rounded on the back, glabrous or puberulent at base
                PUCCINELLIA (p. 167)

Group V (Panicle with unbranched primary branches)

1. Glumes with hooked spines; spikelets deciduous in burrlike clusters of 2
to 5   **TRAGUS (p. 308)**
1. Glumes without hooked spines
 2. Spikelets with 2 or more fertile florets   **H (p. 154)**
 2. Spikelets with 1 fertile floret, with or without reduced florets above
                HH (p. 154)

H (2 or more fertile florets)
3. Inflorescence branches paired, verticillate, or clustered at the culm apex
 4. Glumes and lemmas awnless   **ELEUSINE (p. 285)**
 4. Glumes or lemmas awned
  5. Lemmas mostly 3-awned, lateral awns short and sometimes lacking; tall, cespitose perennials, never developing stolons
                CHLORIS (p. 295)
  5. Lemmas 1-awned; plants annual, or if perennial, then with stout creeping stolons
   6. Second glume short-awned or mucronate; rachis projecting stiffly beyond terminal spikelet   **DACTYLOCTENIUM (p. 286)**
   6. Second glume not awned or mucronate; rachis not extended beyond terminal spikelet   **CHLORIS (p. 295)**
3. Inflorescence branches distributed along culm axis, seldom more than 1 at each rachis node
 7. Glumes 1 cm or more long, much longer than lower floret; lemmas long-ciliate on margins   **TRICHONEURA (p. 287)**
 7. Glumes less than 1 cm long, 1 or both shorter than lowermost floret; lemmas glabrous or puberulent on margins
  8. Lemmas 3-nerved
   9. Lemmas glabrous, acute and awnless at apex; spikelets widely spaced and not overlapping, on stiffly spreading branches
       ERAGROSTIS (*E. sessilispica*) **(p. 265)**
   9. Lemmas glabrous or puberulent on nerves or at base, usually awned or mucronate; spikelets closely spaced and overlapping when lemmas glabrous or unawned (See also *Tridens ambiguus* and *T. buckleyanus*)   **LEPTOCHLOA (p. 287)**
  8. Lemmas with 5 or more nerves
   10. Plants perennial, with tall culms and slender, flexuous inflorescence branches   **GLYCERIA (p. 204)**
   10. Plants annual, with short, tufted culms and short, stiff inflorescence branches
    11. Spikelets 3-flowered; lowermost lemma about 5 mm long; upper leaf sheaths broad and overlapping; disarticulation below glumes   **SCLEROCHLOA (p. 166)**
    11. Spikelets more than 3-flowered; lowermost lemma about 2.5 mm long; upper sheaths not large and overlapping; disarticulation above glumes   **CATAPODIUM (p. 167)**

HH (1 fertile floret)
12. Spikelets on main axis as well as on branches
 13. Glumes absent; lemmas firm, boat-shaped; spikelets strongly com-

pressed laterally, closely imbricated on branches   **LEERSIA (p. 322)**
13. Glumes, at least the second, present
   14. Lemmas awned; leaf blades mostly 8 to 12 mm or more broad
      GYMNOPOGON (p. 288)
   14. Lemmas awnless; leaf blades 5 mm or less broad
      15. Glumes stiff, the first narrowly acute or acuminate, strongly
         1-nerved   **SCHEDONNARDUS (p. 292)**
      15. Glumes soft, the first broad and irregularly lacerate or toothed at
         apex, nerveless   **WILLKOMMIA (p. 291)**
12. Spikelets all on branches and none on inflorescence axis, the latter sometimes terminating in a single branch
   16. Inflorescence branches 2 or more, digitate, clustered or in 2 or 3 ver-
      ticels at culm apex
      17. Rudimentary floret absent or represented by a minute scale; inflo-
         rescence branches slender, digitate, usually 2 to 6; spikelets awnless
         CYNODON (p. 293)
      17. Rudimentary floret or florets present above the fertile one; branches
         few to numerous; spikelets usually awned   **CHLORIS (p. 295)**
   16. Inflorescence branches 1 to several, not digitate, clustered or in
      verticels
      18. Inflorescence with a single, stout, curved, unilateral branch; second
         glume with a short, stout, dorsal awn   **CTENIUM (p. 305)**
      18. Inflorescence with 1 to several branches; second glume awnless
         19. Spikelets 1-flowered, without rudimentary florets; inflorescence
            branches erect-appressed or somewhat spreading, mostly 3 to 12
            cm long   **SPARTINA (p. 304)**
         19. Spikelets with 1 or more staminate or rudimentary florets above
            the fertile one; inflorescence branches spreading or reflexed, in-
            frequently over 4 cm long
            20. Spikelets in deciduous clusters of 3, middle (terminal) spikelet
               perfect, the lower 2 spikelets staminate or sterile; Southwest-
               ern United States, infrequent
               21. Plants perennial, stoloniferous; glumes acute to acuminate
                  CATHESTECUM (p. 302)
               21. Plants annual, delicate, tufted; glumes with a fine awn from a
                 notched tip   **AEGOPOGON (p. 302)**
            20. Spikelets not in deciduous clusters of 3, or if so, then lower 2
               not staminate or sterile   **BOUTELOUA (p. 297)**

Group VI (Spike or spicate raceme)

1. Spikelets in capitate clusters, these subsessile in leafy portion of plant;
   lemmas 3-nerved; low, tufted, or sod-forming grasses

2. Disarticulation below glumes, spikelets falling in burrlike clusters; inflorescence axis and outer (second) glumes of spikelet cluster becoming thick and indurate; spikelets pistillate; plant strongly stoloniferous
BUCHLOË (p. 301)
2. Disarticulation above glumes, spikelets not falling in clusters; glumes not becoming indurate; spikelets perfect
   3. Lemmas with 3 stout, ciliate awns **BLEPHARIDACHNE (p. 275)**
   3. Lemmas with a single awn
      4. Glumes much longer than lemmas; lemmas deeply bifid at apex
ERIONEURON (*E. pulchellum*) **(p. 267)**
      4. Glumes shorter than lemmas; lemmas acuminate at apex, not bifid
MUNROA (p. 271)
1. Spikelets not in capitate clusters, or if so, then these elevated well above basal clump of leaves
   5. Spikelets with a single floret
      6. Spikelets single at each node
         7. Plants annual, lacking stolons or rhizomes
            8. Lemmas bearing a delicate awn at the tip; West Coast
SCRIBNERIA (p. 178)
            8. Lemmas awnless
               9. Inflorescence a raceme; disarticulation above glumes
MIBORA (p. 184)
               9. Inflorescence a spike; each spikelet disarticulating with a section of rachis
                  10. First glume absent except on terminal spikelet
MONERMA (p. 211)
                  10. First glume present, 2 glumes paired in front of spikelet
PARAPHOLIS (p. 212)
         7. Plants perennial
            11. Margins of lemma ciliate; inflorescence a slender, curved, unilateral spike; southern Arizona, rare **MICROCHLOA (p. 294)**
            11. Margins of lemma glabrous
               12. Spikelets partially embedded in a thick, flattened rachis; leaf blades thick, flat, mostly 5 to 8 mm wide; plants with stout stolons **STENOTAPHRUM (p. 215)**
               12. Spikelets sessile or short-pediceled on a slender rachis; leaf blades not over 4 mm wide
                  13. Leaf blades involute, tightly inrolled and bristle-pointed; spikelets sessile, slender and elongate **NARDUS (p. 211)**
                  13. Leaf blades flat or folded, not bristle-pointed; spikelets short-pediceled, ovate **ZOYSIA (p. 307)**
      6. Spikelets 3 at each node, lateral spikelets pediceled and sterile (ex-

cept in *H. vulgare*)   **HORDEUM (p. 197)**

5. Spikelets with 2 or more florets
  14. Lemmas conspicuously 11- to 15-nerved, usually glandular-viscid in age; tufted annuals; California
    15. Lemmas flabellate, without definite lobes or awns; glumes absent
          NEOSTAPFIA (p. 316)
    15. Lemmas not flabellate, with 5 to 11 lobes, teeth, or awns; glumes present, about equal in length   **ORCUTTIA (p. 316)**
  14. Lemmas 1- to 7-nerved
    16. Spikelets oriented edgewise to rachis, first glume absent except on terminal spikelet   **LOLIUM (p. 163)**
    16. Spikelets not oriented edgewise to rachis, both glumes present on all spikelets
      17. Second glume bearing a stout, divergent awn on the back; inflorescence a short, densely flowered, unilateral, curved spike
          CTENIUM (p. 305)
      17. Second glume without dorsal awn
        18. Rachis mostly with 2 or more spikelets per node; perennials
          I (p. 157)
        18. Rachis mostly with 1 spikelet per node   **II (p. 157)**

I (Mostly 2 or more spikelets per node)
19. Spikelets disarticulating in clusters from a persistent rachis; plants mostly rhizomatous or stoloniferous   **HILARIA (p. 306)**
19. Spikelets disarticulating above glumes or with sections of rachis; plants not rhizomatous or stoloniferous
  20. Glumes minute or absent   **HYSTRIX (p. 196)**
  20. Glumes both well developed
    21. Spikelets with 1 fertile and 1 rudimentary floret; annual
          TAENIATHERUM (p. 196)
    21. Spikelets usually with 2 to several fertile florets; perennials
      22. Spikelets disarticulating above glumes, rachis persistent; glumes broadened at or above base, or if setaceous, then infrequently over 2.5 cm long   **ELYMUS (p. 193)**
      22. Spikelets falling with sections of a readily disarticulating rachis; glumes setaceous, mostly 2.5 to 5 cm long   **SITANION (p. 196)**

II (Mostly 1 spikelet per node)
23. Lemmas thin, awnless, distinctly 3-nerved
  24. Spikelets 1.5 to 3 cm long, staminate; stoloniferous perennial
          SCLEROPOGON (p. 275)
  24. Spikelets less than 1 cm long

25. Lemmas awnless; spikelets unisexual; stoloniferous annual
                                            NEERAGROSTIS (p. 265)
25. Lemmas with delicate awn; spikelets perfect; cespitose perennial,
    without stolons   **TRIPOGON (p. 289)**
23. Lemmas thick or thin, 5- to several-nerved, nerves often distinct
    26. Spikelets short-pediceled
        27. Lemmas awnless
            28. Disarticulation below glumes; upper leaf sheaths enlarged, par-
                tially enclosing inflorescence   **SCLEROCHLOA (p. 166)**
            28. Disarticulation above glumes; upper leaf sheaths not enlarged
                                            CATAPODIUM (p. 167)
        27. Lemmas awned; upper leaf sheths not enlarged, inflorescence well
            exserted
            29. Glumes much longer than lemmas; lemmas with a geniculate awn
                from a bifid apex   **DANTHONIA (p. 329)**
            29. Glumes not longer than lemmas; lemmas with a straight awn from
                an entire apex
                30. First glume 1-nerved, second glume 3-nerved; nerves of lemma
                    not converging at apex   **PLEUROPOGON (p. 205)**
                30. First glume usually 5-nerved, second glume usually 7-nerved;
                    nerves of lemma converging at apex
                                            BRACHYPODIUM (p. 160)
    26. Spikelets sessile
        31. Glumes narrow, rigid, setaceous; lemmas long-awned; cultivated
            annual   **SECALE (p. 203)**
        31. Glumes not setaceous, broadened at or above base
            32. Glumes thick, indurate; annuals   **TRITICUM (p. 201)**
            32. Glumes thick or thin; perennials   **AGROPYRON (p. 198)**

DESCRIPTION AND DISCUSSION OF GENERA

Subfamily I. Pooideae

Tribe 1. Poeae

1. *Bromus* L.

Annuals and perennials, a few rhizomatous. Leaves usually with closed
sheaths and broad, flat, thin blades. Inflorescence a panicle, infrequently a
raceme, of large, several-flowered spikelets, these 13 to 45 mm or more
long. Disarticulation above the glumes and between the florets. Glumes unequal, usually acute and awnless, 1- to 5-nerved. Lemmas 5- to several-
nerved, 1-awned from a slightly bifid apex or infrequently awnless. Palea adnate to the caryopsis. Basic chromosome number, $x = 7$.

A genus of about one hundred species, present in temperate and cool

regions of the world. Forty-two species, twenty-two of which are native, are
listed for the United States in Hitchcock's manual (1951). The native bromes
are referable to the sections *Ceratochloa* and *Pnigma*. Section *Ceratochloa*
includes annuals and perennials of both North and South America. In these
grasses the spikelets are strongly flattened laterally, and the glumes and lemmas are sharply keeled. Perhaps most important of the *Ceratochloa* bromes
in the United States is *Bromus carinatus* Hook. & Arn., CALIFORNIA
BROME. This tufted perennial is distributed from Montana, Arizona, and
New Mexico westward to the Pacific Coast ranges and south into Mexico.
Scarcely distinct and more abundant in the Rocky Mountain area is *B. mar-
ginatus* Nees, MOUNTAIN BROME. Frequent in the southern United
States as a cool-season grass of low elevations is *B. willdenowii* Kunth, RES-
CUEGRASS (Fig. 5-2). This is a South American annual or weak perennial

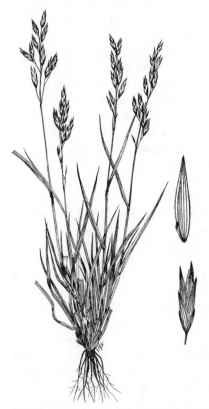

Fig. 5-2. *Bromus willldenowii*. Plant,
spikelet, and floret. (From Gould and
Box, 1965.)

that was introduced into the United States as a pasturegrass and now is common as a weed of roadsides and waste places. For many years rescuegrass went under the name *B. catharticus* Vahl, a name that has been shown to be invalid. *Bromus willldenowii* is closely related to another South American species, *B. unioloides* (Willd.) H.B.K. The two may not be specifically distinct, in which case the latter name would apply to the collective species.

Native perennial bromes with rounded rather than flattened and keeled spikelets are referred to the section *Pnigma*. Several species of this group are important forage grasses of the Rocky Mountain region. Included are *B. ciliatus* L.,* FRINGED BROME; *B. anomalus* Rupr. ex Fourn.; *B. frondosus* (Shear) Woot. & Standl.; and *B. purgans* L.,* CANADA BROME. Also included in section *Pnigma* is *B. inermis* Leyss, SMOOTH BROME, a rhizomatous sodgrass with thin, rounded, awnless, or minutely mucronate lemmas. Smooth brome was introduced into the western United States at an early period as a forage grass and as a soil binder along earthen dams, road banks, and other disturbed sites. It has been an exceptionally successful introduction and still is frequently seeded in western rangelands.

The numerous weedy annual bromes of the United States are mainly adventive European species referable to sections *Bromium* and *Eubromus*. Most of these grasses are relatively harmless weeds, but *B. diandrus* Roth (*B. rigidus* Am. auct.), RIPGUT BROME, and *B. rubens* L., FOXTAIL BROME, have stout-awned spikelets that cause mechanical injury to the eyes and mouthparts of grazing animals. *Bromus tectorum* L., DOWNY BROME or CHEATGRASS, is widespread throughout the United States except in the Southeast. Reference: Wagnon, 1952.

2. *Brachypodium* Beauv.

Annuals and perennials with short-pediceled or subsessile spikelets borne singly at the nodes of a stiffly erect, spicate raceme. Inflorescence sometimes reduced to a single, terminal spikelet. Spikelets large, several- to many-flowered. Disarticulation above the glumes and between the florets. Glumes unequal, stiffly pointed, 5- to 7-nerved. Lemmas firm, rounded or flattened on the back, 7-nerved, the midnerve extending into a short or long awn from an entire apex. Palea about as long as the lemma, 2-keeled, often with stiff, pectinate-ciliate hairs on the nerves. Basic chromosome number, $x = 7$.

A genus of about fifteen species, these mostly in the temperate regions of the world, two native to mountainous regions of Mexico. The weedy Eurasian annual *Brachypodium distachyon* (L.) Beauv. (Fig. 5-3) has been reported as adventive at several widely scattered localities in the United States.

---

*Baum (1967) presented evidence that the correct name for plants now referred to *B. ciliatus* L. is *B. canadensis* Michx. and that the correct name for *B. purgans* L. of Hitchcock's manual (1951) is *B. pubescens* Muhl.

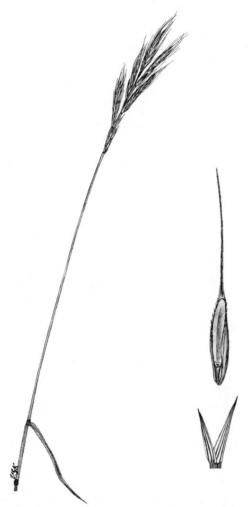

Fig. 5-3. *Brachypodium distachyon*. Inflorescence, pair of glumes, and 1 floret.

3. Vulpia K. C. Gmel.

Tufted annuals with narrow blades, usually contracted, spikelike panicles,
and 3- to many-flowered spikelets. Disarticulation above the glumes and between the florets. Glumes narrow, lanceolate or acuminate, 1- to 3-nerved,
the first often very short. Lemmas rounded on the back, inconspicuously 5-
nerved, tapering to a fine awn or merely acuminate. Caryopsis cylindrical
and elongate. Anthers usually 1, infrequently 3, per flower. Basic chromosome number, $x = 7$.

A small genus of weedy annuals, widely distributed in temperate regions of Europe and North and South America. Hitchcock (1935, 1936, 1951) treated *Vulpia* as a section of *Festuca*, and most United States systematists have followed him in this disposition. Fernald, in Gray's manual (1950), gave *Vulpia* generic status, as do most European workers. *Vulpia octoflora* (Walt.) Rydb., SIXWEEKS FESCUE (Fig. 5-4), is widely distributed throughout the United States. In the southern states it is frequently associated with *Hordeum pusillum* and annual species of *Aristida* around the perimeter of anthills, whereas in California it is characteristically found in recently burned areas at low elevations. Reference: Lonard and Gould, 1974.

Fig. 5-4. *Vulpia octoflora*. Plant and spikelet. (From Gould, 1951, as *Festuca octoflora*.)

4. *Festuca* L.

Tufted perennials with usually thin, flat or narrow and involute blades, and 3- to several-flowered spikelets in open or contracted panicles. Disarticulation above the glumes and between the florets. Glumes narrow, unequal, acute or acuminate, 1- to 3-nerved. Lemmas thin or firm, rounded on the back, usually 5- to 7-nerved, awned from a narrow, entire or minutely bifid apex, or awnless. Palea free from the caryopsis. Stamens 3. Basic chromosome number, $x = 7$.

A genus of more than one hundred species, in temperate and cool regions of the world. Of the some twenty species native to the United States, *Festuca ovina* L., SHEEP FESCUE, is perhaps the most important. Widespread in the cool and cold regions of Asia, Europe, and North America, *F. ovina* is an important range forage species of the Rocky Mountain region. It is also native in the Great Lakes region and has been rather widely introduced in the eastern states. Closely related and probably only varieties of *F. ovina* are *F. idahoensis* Elmer, IDAHO FESCUE, and *F. arizonica* Vasey, ARIZONA FESCUE (Fig. 5-5). *Festuca*, which is found throughout the cool and cold regions of the Northern Hemisphere, is native to the mountainous areas of the western United States and has been introduced into the eastern states as a forage grass. *Festuca pratensis* Huds., MEADOW FESCUE, and the closely related *F. arundinacea* Schreb., TALL FESCUE, REED FESCUE, or ALTA FESCUE, are Old World species that have become widespread in the United States from pasture seedlings. In some localities they are established as weeds of roadsides and other moist, disturbed sites. In the United States, meadow fescue has commonly gone under the name *F. elatior* L., but as explained by Terrell (1967), the correct name is *F. pratensis*.

5. *Lolium* L.

Annuals and short-lived perennials with usually succulent culms and flat or folded blades. Inflorescence a spike of several-flowered spikelets, these borne solitary and oriented edgewise at the nodes of a continuous rachis. First glume absent except on the terminal spikelet. Second glume usually large, broad, several-nerved, awnless. Lemmas 5- to 9-nerved, rounded on the back, awnless or awned from a usually broad apex. Palea large. Basic chromosome number, $x = 7$.

A group of probably ten species or less, with natural distribution in the temperate regions of Europe and Asia. *Lolium perenne* L., PERENNIAL RYEGRASS or ENGLISH RYEGRASS (Fig. 5-6), was perhaps the first grass to be cultivated as a pasture grass in Europe, and records of its use in England date back to 1681. Thomas Jefferson reported the grass to be a good producer in Virginia as early as 1782. Numerous strains and agricultural varieties of *L. perenne* have been developed, and ITALIAN RYEGRASS, commonly recognized as a distinct species (*L. multiflorum* Lam.), is one of the most important of these. In addition to *L. perenne*, which has been widely introduced throughout the United States, five adventive annuals have become established in this country, the most common of which is *L. temulentum* L., DARNEL. The grains of *L. temulentum* are frequently infected with

Fig. 5-5. *Festuca arizonica*. Inflorescence and spikelet. (From Gould and Box, 1965.)

a fungus which causes a poisonous alkaloid to develop. Darnel supposedly is the plant referred to as "the tares," an injurious weed sown by the enemy in the parable of the Bible.

Lolium was included in the tribe Triticeae (Hordeae) by Hitchcock

Fig. 5-6. *Lolium perenne*. Inflorescence. (From Gould and Box, 1965.)

(1936, 1951), but has been shown to be closely related to *Festuca*. Numerous natural hybrids between *Lolium* and European species of *Festuca* have been reported.

6. *Leucopoa* Griseb.

Our species a dioecious, rhizomatous perennial with culms in large, dense clumps. Blades firm, narrow, flat or loosely involute. Inflorescence a contracted panicle with short, mostly appressed branches, these spikelet-bearing to the base. Spikelets 3- to 5-flowered, 7 to 12 mm long, the staminate spikelets somewhat larger than the pistillate. Disarticulation above the glumes and between the florets. Glumes subequal, thin, lanceolate, acute, the first 1-nerved, the second 3-nerved. Lemmas awnless, acute or acuminate, 5-nerved, rounded on the back. Palea as long as the lemma, 2-nerved and 2-keeled, scabrous-ciliate on the keels. Caryopsis beaked, bidentate at the apex. Basic chromosome number, $x = 7$.

The single North American species of the genus, *Leucopoa kingii* (S. Wats.) Weber, ranges from western Nebraska to Montana, Oregon, and southern California on dry, open slopes at low elevations. In the first edition of Hitchcock's manual (1935), this grass was referred to *Festuca*, and in the second edition (1951) it was treated as *Hesperochloa kingii* (S. Wats.) Rydb. Weber (1966) found the monotypic genus *Hesperochloa* to be inseparable from the European *Leucopoa*. He reported that according to the descriptions and key in *Flora of the U.S.S.R.* (Komarov, 1934), *L. kingii* is most closely allied to *L. caucasica* (Hack.) V. Krecz. & Bobr.

7. *Scolochloa* Link

Stout perennials with thick rhizomes and erect culms 1 to 1.5 meters tall. Blades flat, scabrous, elongate. Inflorescence an open panicle of 3- to 4-flowered spikelets. Disarticulation above the glumes and between the florets. Glumes large, lanceolate, subequal, awnless, the first 3-nerved, the second 5-nerved. Lemmas rounded on the back, villous on the callus, faintly 7-nerved, the apex narrow, membranous, and lacerate. Palea narrow, about as long as the lemma. Ovary hairy at the apex. Basic chromosome number, $x = 7$.

A genus of two species, one in eastern Asia and the other in northern Eurasia and northern North America. The single North American species, *Scolochloa festucacea* (Willd.) Link, grows in marshes and shallow water from Canada to northern Iowa, Nebraska, and eastern Oregon.

8. *Sclerochloa* Beauv.

Low, tufted annual with culms mostly less than 10 cm long and inflorescences partially included in greatly enlarged upper leaf sheaths. Leaves scarcely differentiated into sheath and blade, without a ligule or thickened area at the junction. Inflorescence a spicate raceme or contracted panicle, 1 to 2 cm long. Spikelets awnless, 3-flowered, subsessile, and crowded on the main inflorescence axis and on short branches. Disarticulation below the glumes, the spikelet falling entire. Glumes broad, rather firm, the first short, 3-nerved, the second about half as long as the spikelet, 7-nerved. Lemmas broad, rounded on the back and obtuse at the apex, membranous on the margin above, 5-nerved. Palea hyaline. Basic chromosome number, $x = 7$.

A genus of one species, *Sclerochloa dura* (L.) Beauv., this native to southern Europe and adventive at several scattered localities in the United States.

9. *Catapodium* Link

Low, tufted annuals with narrow, contracted panicles or spicate racemes. Leaves glabrous, the blades flat or involute. Ligules membranous, lacerate at the apex. Spikelets several-flowered, awnless, short-pediceled, disarticulating above the glumes and between the florets. Glumes unequal or about equal, the first lanceolate, 1- to 3-nerved, the second elliptic or oblong, 3-nerved. Lemmas broad, nearly terete, longer than the glumes, 5-nerved, rounded on the back, blunt at the apex. Palea slightly shorter than the lemma, usually scabrous on the nerves. Lemma and palea indurate at maturity, tightly enclosing the oblong caryopsis. Basic chromosome number, $x = 7$.

A small genus of tufted annuals, these with natural distribution in western and southern Europe, northern Africa, and western Asia. *Catapodium rigidum* (L.) C. E. Hubb. (Fig. 5-7) is adventive in North and South America. The grass was treated by Hitchcock (1936, 1951) as *Scleropoa rigida* (L.) Griseb., but Hubbard (1954) considers the monotypic genus *Scleropoa* to be inseparable from *Catapodium*.

10. *Puccinellia* Parl.

Low, tufted annuals or, less frequently, moderately tall perennials with culms decumbent or rhizomatous at the base. Leaf sheaths free, the margins overlapping. Spikelets small, several-flowered, awnless, in open or contracted panicles. Disarticulation above the glumes and between the florets. Glumes unequal, short, firm, the first 1-nerved (occasionally 3-nerved), the second 3-nerved. Lemmas firm, rounded on the back, acute or obtuse and often erose at the apex. Nerves of the lemma 5, weak or strong, parallel and not converging at the apex, occasionally with an additional pair of short, faint lateral nerves. Palea as long as, or slightly shorter than, the lemma. Lodicules distinct, elongate, and hyaline. Stamens 3. Caryopsis ovoid to oblong, usually adhering to the palea. Basic chromosome number, $x = 7$.

A genus of about thirty species, these mostly in moist or marshy, often alkaline, sites in the cooler portions of North America. A few species in Europe and Asia. Included in the genus as recognized here are four species that were referred to *Glyceria* by Hitchcock (1935, 1951). These were grouped together in a new genus, *Torreyochloa*, by Church (1949) and then moved to *Puccinellia* by Clausen (1952) and Munz (1958). Species that have been transferred are *P. pallida* (Torr.) R. T. Clausen, *P. pauciflora* (Presl) Munz, *P. erecta* (Hitchc.) Munz, and *P. californica* (Beetle) Munz. *Glyceria*, to which these species were assigned previously, has been shown to differ from *Puc-*

Fig. 5-7. *Catapodium rigidum*. Plant and spikelet.
(From Gould and Box, 1965, as *Scleropogon rigidum*.)

cinellia in many characters, including a basic chromosome number of $x = 10$.
Currently *Glyceria* is referred to the tribe Meliceae.

Most species of *Puccinellia* are of little economic significance. *Puccinellia airoides* (Nutt.) Wats. & Coulter, NUTTALL ALKALIGRASS, which is widely distributed in the central and western United States, does provide considerable forage in some areas.

11. *Poa* L.

Low to moderately tall annuals and perennials, many with rhizomes. Blades mostly flat or folded, with boat-shaped tips. Inflorescence an open or contracted panicle or occasionally reduced to a raceme. Spikelets mostly small, 2- to 7-flowered, awnless. Disarticulation above the glumes and between the florets. Glumes relatively broad, 1- to 3-nerved. Lemmas thin, broad, usually keeled and with a membranous border, obtuse or broadly acute at the apex, 5-nerved, the nerves often puberulent below. Base of the lemma glabrous or with long, kinky, cottony hairs. Palea glabrous. Basic chromosome number, $x = 7$.

Species about 250, in the temperate and cold regions of the world, extending into the subtropics and tropics as cool-season and as montane grasses. Fifty-nine species are listed for the United States in Hitchcock's manual (1951), 5 of which are annuals and the remainder perennials. The bluegrasses rank high both as forage plants of the western mountain ranges and as cultivated pasturegrasses. *Poa pratensis* L., KENTUCKY BLUEGRASS (Fig. 5-8), has been called "king of the pasturelands" and with the possible exception of *Phleum pratense*, TIMOTHY, is the most important perennial pasturegrass cultivated in temperate North America. Grasses of the *P. pratensis* complex that appear to be native in Canada and the Rocky Mountains

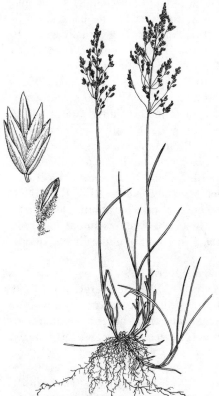

Fig. 5-8. *Poa pratensis*. Plant, spikelet, and floret. (From Gould, 1951.)

as far south as New Mexico were recently named *P. agassizensis* Boivin & D. Löve (Boivin and Löve, 1960). Both *P. pratensis* and *P. compressa* L., CANADA BLUEGRASS, are common constituents of lawn-grass mixtures in the cooler climates. *Poa fendleriana* (Steud.) Vasey, MUTTONGRASS or FENDLER BLUEGRASS, is the outstanding native Rocky Mountain bluegrass and is considered one of the twenty most valuable range grasses of the Rocky Mountain region (United States Forest Service, 1937). This is a vigorous perennial that is resistant to drought and able to hold up well under moderately heavy grazing.

A fine taxonomic treatment of the western species of *Poa* has been published by Keck (1965).

12. *Briza* L.

Tufted annuals and perennials with usually showy open panicles of several-flowered, awnless spikelets. In our species, the spikelets are pendulant or drooping on slender pedicels. Florets crowded and horizontally spreading. Glumes subequal, broad, thin and papery, rounded on the back, 3- to 9-nerved. Lemmas similar to the glumes, broader than long, usually with 7 to 9 nerves, these distinct or indistinct, and a broadly rounded apex. Palea short to nearly as long as the lemma. Basic chromosome number, $x = 7$.

Species about twenty, three native to Europe, one in Mexico and Central America, and the remainder in South America. All three of the European species now occur in scattered locations in the United States. *Briza minor* L., LITTLE QUAKINGGRASS (Fig. 5-9), an annual with a delicate panicle of spikelets 2 to 5 mm long, is locally common in coastal California, eastern Texas, and the eastern states. *Briza media* L., a perennial with slightly larger spikelets, has been reported from Ontario to Connecticut, Michigan, and California. The third species, *B. maxima* L., BIG QUAKING-GRASS, an annual with a showy inflorescence of spikelets as much as 20 to 25 mm long and 10 mm wide, is occasionally grown as an ornamental and has been reported as an escape in California, Texas, and sporadically elsewhere.

13. *Phippsia* (Trin.) R. Br.

Low, tufted, alpine perennials with contracted, few-flowered panicles of small, awnless spikelets. Spikelets 1-flowered, disarticulating above the glumes. Glumes unequal, minute, the first sometimes absent. Lemma thin, somewhat keeled, 3-nerved, abruptly acute. Palea slightly shorter than the lemma. Basic chromosome number, $x = 7$.

A genus of two species, one of which, *Phippsia algida* (Phipps) R. Br., is widespread in arctic and alpine regions of both hemispheres. This grass is known in the United States only from alpine peaks of Colorado (Weber, 1952).

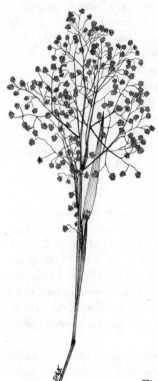

Fig. 5-9. *Briza minor*. Inflorescence.

14. *Coleanthus* Seidel

Dwarf annual with spreading culms that form little mats. Spikelets small, 1-flowered, in contracted, few-flowered panicles that are partially enclosed in the inflated upper sheaths. Glumes absent. Lemma thin, 1-nerved, about 1 mm long, abruptly tapered to a short awn tip. Palea well developed, 2-toothed, about as long as the lemma.

A genus of a single species, *Coleanthus subtilis* (Tratt.) Seidel, this native to Eurasia and occurring, probably adventive, on mud flats along the lower Columbia River in Oregon and Washington.

15. *Dactylis* L.

Erect perennials with usually densely clumped culms, keeled leaf sheaths, broad blades, and well-exserted panicles. Spikelets 2- to 5-flowered, laterally flattened, subsessile and crowded in dense asymmetrical clusters at the tips of stiff, erect, or spreading inflorescence branches. Glumes unequal to subequal, keeled, hispid-ciliate on the keel, 1- to 3-nerved, acute to acuminate or terminating in a short awn. Lemmas lanceolate-acuminate, keeled, awnless or short-awned, 5-nerved, hispid-ciliate on the keel. Palea well developed, short-ciliate on the keels. Basic chromosome number, $x = 7$.

A genus of about three species, these widely distributed in the temperate and cold portions of Europe and Asia. *Dactylis glomerata* L., ORCHARD-GRASS, has long been grown in the cooler portions of the United States as a cultivated pasture grass and in native grassland associations of mountainous areas. It is present in high mountain meadows as far south as Mexico and Central America. Reference: Stebbins and Zohary, 1959.

16. *Cynosurus* L.

Tufted annuals and perennials, with narrow, flat leaf blades and subcapitate or spicate panicles. Spikelets of two kinds, fertile and sterile, these together in dense clusters, the fertile ones sessile, the sterile ones short-pediceled, almost concealing the fertile ones. Fertile spikelets 1- to 5-flowered, disarticulating above the narrow, 1-nerved glumes. Lemmas of the fertile floret rounded on the back, inconspicuously 5-nerved, terminating in a short awn. Sterile spikelets with 2 narrow, 1-nerved glumes and several narrow, 1-nerved lemmas on a stiff, continuous rachis. Basic chromosome number, $x = 7$.

Species about four, native to the Mediterranean region. One perennial, *Cynosurus cristatus* L., and one annual, *C. echinatus* L., are now widespread in Europe and introduced or adventive at several localities in the United States. *Cynosurus cristatus*, CRESTED DOGTAIL, has been used in pasture-grass mixtures but has few of the qualities of a good forage grass.

17. *Lamarckia* Moench.

Low, short-lived annual with weak culms, soft, flat blades, and contracted panicles of fascicled spikelets. Terminal spikelet of each fascicle fertile, those below sterile, the fascicles falling entire. Fertile spikelet with a single perfect floret on a slender rachilla and a rudimentary floret borne above on a long bristle-like stipe. Lemmas of both the fertile and reduced florets with a delicate awn 5 to 10 mm long. Sterile spikelets mostly with 3 to 6 empty florets, the lemmas well developed, broad, awnless, scarious above. Glumes of the fertile and sterile spikelets similar, narrow, pointed. Basic chromosome number, $x = 7$.

A genus of a single species, *Lamarckia aurea* (L.) Moench., this adventive as a cool-season weed of disturbed soils in Texas, Arizona, southern California, and northern Mexico.

Tribe 2. Aveneae

18. *Koeleria* Pers.

Tufted annuals and perennials with narrow blades and shining, contracted, usually spikelike panicles of 2- to 4-flowered, flattened spikelets. Rachilla disarticulating above the glumes and between the florets, extended as a bristle above the uppermost floret. Glumes large, thin, acute, the first 1-nerved, the second broader, often slightly longer, indistinctly 3- to 5-nerved. Lemmas thin, shining, the lowermost usually slightly longer than the glumes, awnless or awned from a minutely bifid apex. Palea large, scarious and colorless. Basic chromosome number, $x = 7$.

Species about twenty, in the temperate regions of both hemispheres. Two species in the United States, one native perennial, *Koeleria pyramidata* (Lam.) Beauv. (*K. cristata* [L.] Pers.), JUNEGRASS, and one introduced annual, *K. phleoides* (Vill.) Pers. *Koeleria pyramidata* is frequent in prairie and woodland associations throughout temperate North America and is also widely distributed in Europe, Asia, and Africa. Hubbard (1954) used the name *K. gracilis* Pers. for this species, but Shinners (1956) suggested that the earliest valid name is probably *K. macrantha* (Ledeb.) Spreng.

Showing close affinities with both *Poa*, of the tribe Poeae and *Sphenopholis* and *Trisetum*, of the tribe Aveneae, *Koeleria* was referred to the former by Hackel (1887) and to the latter by Hitchcock (1951). *Koeleria pyramidata* bears considerable resemblance to *P. fendleriana*, from which it can be distinguished by its generally smaller spikelets, entirely colorless paleas, and puberulent panicle axis.

19. *Sphenopholis* Scribn.

Low, tufted annuals or short-lived perennials with soft, flat blades and usually contracted panicles of 2- (rarely 1-) to 3-flowered spikelets, these disarticulating below the glumes. First glume narrow, acute, 1-nerved (rarely 3-nerved), second glume broad, obovate, 3- to 5-nerved, obtuse or broadly acute at the apex, usually slightly shorter than the lowermost lemma. Lemmas firm, faintly 5-nerved or the nerves not visible, rounded on the smooth or rugose back, awnless or less frequently with an awn from just below the apex. Palea thin, membranous, colorless. Basic chromosome number, $x = 7$.

Species four, with one, *Sphenopholis obtusata* (Michx.) Scribn., PRAIRIE WEDGESCALE (Fig. 5-10), widespread in North America from Canada to northern Mexico and the Caribbean and the other three present in the central and eastern United States. In a monograph of the genus, Erdman (1965) reduced *S. intermedia* (Rydb.) Rydb. and *S. longiflora* (Vasey) Hitchc. to variental status under *S. obtusata* (var. *major* [Torr.] Erdman) and

Fig. 5-10. *Sphenopholis obtusata*. Plant and spikelet. (From Gould and Box, 1965.)

discussed evidence for considering *S. pallens* (Spreng.) Scribn. as comprising a series of hybrids between *S. obtusata* and *S. pennsylvanica* (L.) Hitchc.

Along with *Koeleria*, the genus *Sphenopholis* occupies a somewhat intermediate or transitional position between the tribes Poeae and Aveneae.
The most widespread species, *S. obtusata*, has been placed variously in 9
different genera, including *Koeleria*, *Trisetum*, *Poa*, *Festuca*, and *Agrostis*.
Also of interest is the imposing series of thirty-five specific and varietal synonyms listed for this species in Hitchcock's manual (1951).

20. *Trisetum* Pers.

Tufted perennials and a few annuals, with slender culms, flat blades, and
usually narrow panicles of 2-flowered (rarely 3- to 4-flowered) spikelets.
Rachilla usually villous, prolonged above the uppermost floret, disarticulating above the glumes and between the florets or, in a few species, below the

glumes. Glumes thin, nearly equal to very unequal, 1- to 3-nerved, acute, awnless, one or both usually equaling or exceeding the florets in length. Lemmas 5-nerved, bifid at the membranous apex, mostly with a straight or bent awn from the base of the notch but occasionally awnless, usually puberulent at the base. Basic chromosome number, $x = 7$.

Species about seventy-five, in the temperate and cold regions of both hemispheres. Nine species are reported for the United States, one of these, *Trisetum flavescens* (L.) Beauv., introduced from Europe. *Trisetum interruptum* Buckl. (Fig. 5-11) is a cool-season annual of the Southwest. *Trisetum spicatum* (L.) Richt., SPIKE TRISETUM, is a circumboreal species comprising a complicated polyploid series of populations that have been subdivided into several species by some authors on the basis of polyploid levels. It is frequent on the higher slopes in the Rocky Mountains, where it seldom occurs below 6,000 feet. On the mountains of the Pacific Coast, it is found as

Fig. 5-11. *Trisetum interruptum.* Inflorescence and spikelet. (From Gould and Box, 1965.)

low as 2,500 feet. Although not of uniform palatability and growth characteristics, this grass, with its adaptation to a wide range of habitats from 2,500 feet altitude to over 13,000 feet, ranks among the most valuable of the western range forage species. The genus *Trisetum* appears closely related to *Sphenopholis* and probably also has close affinities with *Koeleria*. Reference: Koch, 1979.

21. *Corynephorus* Beauv.

Low, tufted perennial with slender, densely clumped culms and filiform blades. Inflorescence a contracted panicle, this well exserted above the leaves. Spikelets 2-flowered, disarticulating above the glumes and between the florets. Glumes large, thin, shiny, lanceolate, 1-nerved, about equal. Lemma thin, obscurely nerved, much shorter than the glumes, with a tuft of hair on the callus, awned from near the base, minutely notched at the apex. Awn of lemma expanded and club-shaped at the apex, with a minute hairy joint near the middle. Palea about as large as the lemma, 2-nerved, toothed at the apex. Basic chromosome number, $x = 7$.

A small European genus, closely related to, and originally included in, *Aira*. The type species, *Corynephorus canescens* (L.) Beauv. (*Aira canescens* L.), has been collected at a few locations along the coast of the northeastern United States and in British Columbia.

22. *Aira* L.

Delicate, tufted annuals with thin, subfiliform, mostly basal leaves. Inflorescence an open or contracted panicle of small, 2-flowered spikelets. Rachis disarticulating above the glumes and between the florets, not prolonged behind the upper floret. Glumes about equal, longer than the lemmas, thin, lanceolate, 1- or obscurely 3-nerved. Lemmas firm, rounded on the back, awned from below the middle with a usually geniculate and twisted, hairlike awn, tapering to 2 slender teeth or setae at the tip. Awn of lower lemma sometimes wanting or reduced. Palea thin, shorter than the lemma. Basic chromosome number, $x = 7$.

A genus of about ten species, native to southern Europe but now widely distributed elsewhere. Three species have been introduced in the United States: *Aira caryophyllea* L., SILVER HAIRGRASS; *A. praecox* L.; and *A. elegans* Willd. ex Gaudin (Fig. 5-12), these often growing as weeds of disturbed sites in the eastern and southeastern states and also in the Far West.

23. *Deschampsia* Beauv.

Mostly perennials, a few annuals, with slender culms and narrow blades. Inflorescence an open or contracted panicle, commonly with slender, wiry branches and pedicels. Spikelets small, 2-flowered, disarticulating above the

Fig. 5-12. *Aira elegans*. Plant and spikelet.

glumes and between the florets. Rachilla usually hairy, prolonged beyond the terminal floret and sometimes bearing a rudiment at its tip. Glumes 1- to 3-nerved, lanceolate, nearly equal and exceeding the lower-most floret. Lemmas shiny, rounded on the back, mostly 5- to 7-nerved but the nerves often obscure, pubescent on the callus, deeply cleft at the apex, bearing a slender, usually twisted and geniculate awn from the middle or base. Palea 2-keeled, nearly as long as the lemma, often scabrous on the keels. Basic chromosome number, $x = 7$.

Species about forty, in temperate and cool regions of both hemispheres, growing at high altitudes in the warmer latitudes. These slender, tufted, mostly perennial grasses are closely related to *Aira* and at one time were included in that genus. As now interpreted, *Aira* is a genus of delicate annuals. Six of the seven species of *Deschampsia* in the United States are perennial. The only annual, *D. danthonioides* (Trin.) Munro ex Benth., ANNUAL HAIRGRASS, is widespread in the western states. It ranges from Alaska to Lower California and also occurs in Chile. *Deschampsia caespitosa* (L.) Beauv., TUFTED HAIRGRASS, is frequent in temperate and arctic regions and also occurs at high altitudes in the tropics. This "bunchgrass" is common in the mountain meadows of the western United States and is also present in the north central and eastern states. Of special interest is the presence of *D. antarctica* (Hook. f.) E. Desv. on the Antarctic continent. It is one of two flowering plants reported from that frigid landmass. Reference: Kawano, 1963; Koch, 1979.

24. *Scribneria* Hack.

Low, tufted annual with a slender, cylindrical spicate inflorescence, this usually one-third to one-half the entire length of the plant. Spikelets 1-flowered, solitary, and appressed flatwise against the nodes of a continuous rachis. Rachilla disarticulating above the glumes, extended back of the floret as a minute, hairy bristle. Glumes narrow, firm, awnless, about equal, the first 2-nerved, the second 4-nerved. Lemma shorter than the glumes, membranous, faintly nerved, minutely notched, and usually bidentate at the apex, with the midnerve extended as a short, straight awn. Palea about as long as the lemma. Flowers with a single stamen. Only one chromosome record reported, this $2n = 26$.

A monotypic genus, the single species, *Scribneria bolanderi* (Thurb.) Hack., occasional in sandy soil from Washington south to central California. *Scribneria* is treated in Hitchcock's manual (1951) as a member of the tribe Triticeae (Hordeae). In shifting the genus to the Aveneae, Stebbins and Crampton (1961) noted the relatively long glumes, bidentate and awned lemmas, and hairy lemma base, which suggested a relationship with *Deschampsia* and *Calamagrostis*.

25. *Avena* L.

Annuals with moderately tall, weak culms, broad, flat blades, and panicles of large, pendulous, usually 2- to 3-flowered spikelets on slender pedicels. The inflorescence of depauperate plants may be reduced to a raceme of a few spikelets or even a single spikelet. Disarticulation above the glumes and between the florets. Glumes about equal, broad, thin, several-nerved, longer than the lower floret and often exceeding the upper. Lemmas rounded on the back, 5- to 7-nerved, tough and indurate, often hairy on the callus, usually bearing a stout, geniculate awn from below the notch of a bifid apex. In the cultivated oats, *Avena fatua* var. *sativa*, the awn is absent or reduced. Palea 2-keeled, shorter than the lemma. Basic chromosome number, $x = 7$.

A genus of about ten to fifteen species, these native to temperate Europe and Asia but widely cultivated for human and animal food. *Avena fatua* L. var. *fatua*, WILD OATS, is common and weedy in California, where it was introduced at an early date. It is occasional on roadsides and in waste places throughout the country. *Avena fatua* var. *sativa* (L.) Haussk. (*A. sativa* L.), OATS, is grown throughout the central and western states and occasionally in the East. Although probably not persisting for long out of cultivation, it is frequent as a roadside weed in many areas. *Avena barbata* Brot., SLENDER OATS, is a common weed of fields and waste places in the Far West, Arizona to California, Oregon, and Washington.

26. *Ventenata* Koel.

Weak-stemmed annuals with narrow, flat blades. Inflorescence an open panicle, the branches bare on the lower half or two-thirds, bearing 2 to few spikelets near the tips. Spikelets with 2 to several florets, the upper one or ones perfect and awned, the lowermost staminate and awnless. Lemma awn geniculate, exserted from near the middle of the dorsal surface.

Species about six, native to Europe and Asia Minor. *Ventenata dubia* (Leers) Coss. & Dur. is adventive in Kootenai County, Idaho, and Spokane County, Washington. In reporting this grass from Idaho, Baker (1964) cited three collections from Kootenai County, stating that the grass apparently is becoming well established there. Basic chromosome number, $x = 7$.

27. *Helictotrichon* Besser

Tufted perennials with low to moderately tall culms, flat or involute blades, and usually contracted, few-flowered panicles of large, erect, 3- to 7-flowered spikelets. Rachilla villous, disarticulating about the glumes and between the florets. Glumes large, about equal or the first somewhat smaller and shorter, 1- to 5-nerved, thin and membranous at the acute apex. Lemmas rounded on the back, several-nerved, firm on the lower portion, membranous above and usually toothed at the apex, awned from about the middle, the awn stout, geniculate, twisted. Palea well developed, 2-keeled. Stamens 3, large. Basic chromosome number, $x = 7$.

A genus of thirty to forty species, closely related to *Avena* and included in that genus in the first edition of Hitchcock's manual (1935). Hubbard (1954) commented on the resemblance of *Helictotrichon* to *Arrhenatherum*, noting, however, that the latter has mostly 2-flowered spikelets, with the lower flower male, and the florets falling together at maturity. With distribution mostly in temperate Europe, Asia, and Africa, *Helictotrichon* is represented in the United States by two native and one introduced species. *Helictotrichon hookeri* (Scribn.) Henr., SPIKE OAT, with flat, glabrous leaf blades and glabrous sheaths, and *H. mortonianum* (Scribn.) Henr., ALPINE OAT, with flat, pubescent blades and pubescent lower sheaths, are present in the Rocky Mountain region, the former ranging eastward to Minnesota. *Helictotrichon pubescens* (Huds.) Pilger, with involute, pubescent blades, has been introduced from Europe in Connecticut and Vermont.

28. *Arrhenatherum* Beauv.

Moderately tall perennials with flat blades and narrow panicles. Spikelets usually 2-flowered, the upper floret perfect, the lower staminate. Rachilla continued back of the upper floret as a bristle. Disarticulation above the glumes, the 2 florets falling together. Glumes broad, thin, acute, the first 1-nerved, the second longer, 3-nerved. Lower floret staminate, larger than the upper floret and bearing a stout, geniculate awn below the middle. Upper floret perfect, the lemma 5- to 7-nerved, usually awnless or with a short, straight, hairlike awn from or just below the tip but with a well-developed awn in one variety of *Arrhenatherum elatius*. Palea 2-keeled, scabrous or hairy on the keels. Caryopsis narrowly oblong, pubescent (in *A. elatius*). Basic chromosome number, $x = 7$.

Species about six, native to the temperate regions of Europe and Asia. *Arrhenatherum elatius* (L.) Presl, TALL OATGRASS, has long been grown as a pasture grass on sandy soils in many parts of the United States and has frequently escaped from cultivation.

29. *Holcus* L.

Low to moderately tall perennials with velvety pubescent herbage. Culms weak, rather succulent. Blades broad, flat. Inflorescence a contracted panicle. Spikelets 2-flowered, the lower floret perfect, the upper smaller, staminate or sterile. Disarticulating below the glumes. Glumes large, thin, about equal, longer than the florets, the first 1-nerved, the second 3-nerved. Lemma of lower floret faintly 3- to 5-nerved, awnless, blunt at the apex. Lemma of the upper floret with a short, often hooked awn near the apex.

Palea as long as, or slightly shorter than, the lemma. Basic chromosome number, $x = 7$.

About eight species, these native to Europe and Africa. Two species, *Holcus lanatus* L., COMMON VELVETGRASS, and *H. mollis* L., have been introduced into the United States, the former now widespread and weedy in many localities.

30. *Dissanthelium* Trin.

Annuals and perennials, for the most part tufted alpine grasses 10 cm or less tall. Inflorescence a panicle, this usually contracted. Spikelets 2-flowered or less frequently 3-flowered. Disarticulation above the glumes and between the florets. Glumes glabrous or scaberulous on the keel toward the apex, the first 1-nerved, the second 3-nerved. Lemmas membranous or subcoriaceous, broad, awnless, in our species about 2 mm long and slightly pubescent, 3-nerved, the lateral nerves close to the margins. Palea membranous, slightly shorter than the lemma. Chromosome number unreported.

A genus of seventeen species, mostly dwarf perennials restricted to the high mountains of Peru and Bolivia. One annual, *Dissanthelium californicum* (Nutt.) Benth., is known only from a few collections made on Santa Catalina and San Clemente Islands, California, and on Guadalupe, Lower California. This species possibly is now extinct and it has not been collected in any locality for over sixty years (Peter Raven, personal communication). A brief monographic report on *Dissanthelium* in which ten new species were described was presented by Swallen and Tovar (1965).

31. *Calamagrostis* Adans.

Moderately large, mostly rhizomatous perennials with open or more often contracted panicles of 1-flowered spikelets. Rachilla disarticulating above the glumes and in many species extended back of the floret as a slender, hairy bristle. Glumes about equal, longer than the floret, 1- or 3-nerved, acute or acuminate. Lemma 3- or 5-nerved, tapering to a narrow, notched or toothed apex, with a slender, often geniculate awn from the middle or below, usually long-hairy on the callus, the hairs sometimes as long as the lemma. Palea about equaling the lemma or slightly shorter. Basic chromosome number, $x = 7$.

A large genus, with over one hundred species in the cool and temperate regions of both hemispheres. Hitchcock (1951) listed twenty-nine species for the United States, all native except *Calamagrostis epigeios* (L.) Roth. This grass was introduced from Eurasia and now is well established in the northeastern and north central states. While many of the United States species of *Calamagrostis* are of local distribution, a few, such as *C. canadensis* (Michx.)

Beauv., BLUEJOINT; *C. purpurascens* R. Br., PURPLE REEDGRASS; *C. inexpansa* A. Gray, NORTHERN REEDGRASS; *C. neglecta* (Ehrh.) Gaertn., Mey. & Schreb.; and *C. epigeios* have rather wide distribution. *Calamagrostis canadensis* grows in wet or moist habitats throughout the northern two-thirds of the United States and northward to Labrador on the east and Alaska on the west. It is present at elevations over 12,000 feet near the southern limits of its range in New Mexico. This grass furnishes much forage for cattle and horses, but tends to be tough and unpalatable when the herbage is mature.

32. *Ammophila* Host

Coarse, usually large perennials with stout, creeping rhizomes, tough, usually involute blades, and contracted, densely flowered panicles. Spikelets large, 1-flowered, awnless, disarticulating above the glumes. Rachilla hairy, at least above, extended behind the floret as a hairy bristle. Glumes firm, narrowly rounded or pointed at the apex, the first usually 1-nerved, the second 3-nerved. Lemma firm, lanceolate, rounded at the tip, 5- or 7-nerved, with a tuft of long hair on the base. Palea about as long as the lemma. Mature grain oblong or obovate, enclosed by the hardened lemma and palea. Basic chromosome number, $x = 7$.

A genus of four species: one, *Ammophila breviligulata* Fern., AMERICAN BEACHGRASS, along the coast of North America from Newfoundland to North Carolina and in the Great Lakes region; a second, *A. champlainensis* F. C. Seymour, in New York and Vermont; and two native to temperate Europe. *Ammophila arenaria* (L.) Link, EUROPEAN BEACHGRASS, has become well established on sand dunes of the Pacific Coast from Washington to the vicinity of San Francisco, California, where it was introduced as a sand binder-stabilizer. *Ammophila* is closely related to *Calamagrostis*, differing mainly in the awnless rather than awned lemma. Sterile hybrids between *A. arenaria* and *C. epigeios* collected on the coast of Great Britain have been named *Ammocalamagrostis baltica* (Fl.) P. Fourn. (Hubbard, 1954).

33. *Apera* Adans.

Tufted annuals with weak culms, flat blades, and open or contracted, densely flowered panicles. Spikelets small, 1-flowered, disarticulating above the glumes. Rachilla prolonged back of the floret as a naked bristle. Glumes membranous, lanceolate, acuminate at the tip, the first 1-nerved and usually slightly shorter than the 3-nerved second glume. Lemma firm, rounded on the back, indistinctly 5-nerved, puberulent at the base, awned from near the tip with a straight or flexuous awn. Palea well developed, 2-nerved, as long as the lemma or slightly shorter. Mature caryopsis narrowly oblong, tightly enclosed between the firm lemma and palea. Basic chromosome number, $x = 7$.

A small group of annuals, closely related to *Agrostis*, native to the cooler parts of Europe and Asia. Two adventive species, *Apera spica-venti* (L.) Beauv. and *A. interrupta* (L.) Beauv., are established at scattered localities in the United States. The species of *Apera* were included in *Agrostis* in the first edition of Hitchcock's manual (1935).

34. *Agrostis* L.

Low to moderately tall annuals and perennials with slender culms, flat or involute blades, and open or contracted panicles. Spikelets small, 1-flowered, disarticulating above the glumes. Rachilla usually not hairy or extended beyond the insertion of the floret. Glumes thin, lanceolate, acute to acumi-

Fig. 5-13. *Agrostis hiemalis.* Plant and spikelet. (From Gould and Box, 1965.)

nate, nearly equal, the first usually 1-nerved, the second 1- or 3-nerved. Lemma thin, broad, 3- or 5-nerved, acute to obtuse or truncate at the apex, glabrous or with a tuft of hair at the base, awnless or awned from the middle or below. Palea hyaline, usually small or absent but well developed in a few species. Mature grain narrowly oblong, loosely enclosed by the lemma or by the lemma and palea. Basic chromosome number, $x = 7$.

A genus of about 125 species, widespread in temperate and cold regions of the world and present at high altitudes in the tropics and subtropics. About 40 species are reported for the United States, these about equally represented in the eastern and western portions of the country. Several are important forage grasses, both on native rangelands and in improved pastures. Selections of *Agrostis stolonifera* L. (*A. alba* L.), RED TOP; *A. tenuis* Sibth., COLONIAL BENTGRASS; and *A. palustris* Huds., CREEPING BENTGRASS, are extensively used as lawngrasses and turfgrasses. *Agrostis hiemalis* (Walt.) B.S.P., WINTER BENTGRASS or TICKLEGRASS (Fig. 5-13), is frequent on open, disturbed sites throughout the eastern half of the United States. *Agrostis diegoensis* Vasey is one of the most common grasses at low elevations throughout nearly all of California.

35. *Polypogon* Desf.

Low to moderately tall annuals and perennials, with weak, decumbent-erect culms, these often rooting at the lower nodes. Blades thin and flat. Inflorescence a dense, contracted panicle of small, 1-flowered spikelets, these disarticulating below the glumes and falling entire. Glumes about equal, 1-nerved, abruptly awned from an entire or notched apex. Lemma broad, smooth, and shining, mostly 5-nerved, much shorter than the glumes, awnless or with a short, delicate awn from the broad, often minutely toothed apex. Palea slightly shorter than the lemma. Stamens 1 to 3. Basic chromosome number, $x = 7$.

Species about ten, throughout the temperate regions of the world but mostly in Europe and Asia. Apparently only one species, *Polypogon elongatus* H.B.K., is native to North America, and this is known in the United States only from southern Arizona and California. It ranges southward through Mexico to Argentina. Four European species of *Polypogon* have been introduced or are adventive in the United States. Most common and widespread is *P. monspeliensis* (L.) Desf., RABBITFOOT GRASS (Fig. 5-14), present in moist areas along streams, ditches, spring seeps, and lakes almost throughout the country.

36. *Mibora* Desv.

Low, tufted annual, 2 to 15 cm tall, with numerous delicate, unbranched culms, narrow basal leaves, and slender, 1-sided spicate racemes. Spikelets

1-flowered, single and subsessile at the rachis nodes. Disarticulation above the glumes, the rachis continuous, and the glumes persistent. Glumes membranous, 1-nerved, glabrous, awnless, broadly oblong, rounded on the back, blunt at the apex, about equal in length and longer than the lemma. Lemma broad, thin, finely 5-nerved, densely puberulent, truncate or minutely

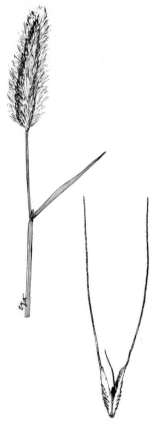

Fig. 5-14. *Polypogon monspeliensis.* Inflorescence and spikelet. (From Gould and Box, 1965.)

toothed at the apex. Palea as long as the lemma but narrower, puberulent. Caryopsis ovate or oblong, loosely enclosed by the lemma and palea. Basic chromosome number, $x = 7$.

A genus of a single species, *Mibora minima* (L.) Desv., this native to southern and southwestern Europe and northwestern Africa (Hubbard, 1954). Reported in the United States only at Plymouth, Massachusetts.

37. *Cinna* L.

Tall perennials with slender culms, thin, flat blades, and open or contracted panicles. Spikelets 1-flowered, disarticulating below the glumes, the rachilla usually prolonged behind the palea as a small, glabrous or scabrous stub or bristle. Glumes about equal or the first somewhat shorter, lanceolate, 1- to 3-nerved, acute at the apex. Lemma similar to the glumes, 3-nerved to 5-nerved, the mid-nerve usually protecting as a short, straight awn just below the acute or narrowly rounded and notched tip. Palea slightly shorter than the lemma, 2- or apparently 1-nerved, with a single keel and with the nerves close together. Stamens 1 or 2. Basic chromosome number, $x = 7$.

A genus of four species: one, *Cinna latifolia* (Trevir.) Griseb., DROOP-ING WOODREED, in northern Eurasia and North America; a second, *C. arundinacea* L., STOUT WOODREED (Fig. 5-15), in the eastern United

Fig. 5-15. *Cinna arundinacea.* Inflorescence and spikelet.

States; a third, *C. poaformis* (H.B.K.) Scribn. & Merr., ranging from Mexico to Venezuela and Peru on high mountain slopes, and a fourth, *C. bolanderi* Scribn., endemic to central montane California. *Cinna latifolia* is to be found in moist, shaded sites throughout the cooler portions of the United States, but is never locally abundant (D. M. Brandenberg, personal communication).

38. *Limnodea* L. H. Dewey

Low, short-lived annual, with soft culms; broad, thin, flat blades; and a loosely contracted panicle. Spikelets 1-flowered, disarticulating below the glumes, the spikelet falling entire. Rachilla extended behind the palea as a fine bristle. Glumes about equal, lanceolate, firm, gradually narrowing to an

Fig. 5-16. *Limnodea arkansana*. Inflorescence and spikelet. (From Gould and Box, 1965.)

acute tip. Lemma membranous, glabrous, nerveless, rounded on the back, 2-toothed at the apex, and bearing a slender, bent awn between the teeth. Palea membranous, shorter than the lemma, 2-keeled. Basic chromosome number, $x = 7$.

A North American genus of a single species, *Limnodea arkansana* (Nutt.) L. H. Dewey, OZARKGRASS (Fig. 5-16), this a cool-season grass with distribution from Florida to Arkansas, Texas, and northern Mexico. It is present in disturbed sites on many soil types and is commonly associated with the annuals *Vulpia octoflora* and *Hordeum pusillum*.

39. *Anthoxanthum* L.

Moderately tall annuals and perennials with fragrant herbage due to the presence of coumarin. Leaf blades thin, flat, and broad. Inflorescence a contracted, often spikelike panicle. Spikelets with 1 large, fertile floret and 2 sterile, reduced florets below. Disarticulation above the glumes, the fertile and reduced florets falling together. Glumes broad, thin, acute or mucronate, the first 1-nerved and short, the second 3-nerved, longer than the florets. Lemmas of the reduced florets hairy, awned from the back, notched or toothed at the apex. Lemma of the fertile floret broad, rounded, firm; shiny, brown, awnless. Palea present and apparently 1-nerved in the fertile floret, absent in the reduced florets. Caryopsis tightly enclosed in the firm lemma and palea. Stamens 2. Basic chromosome number, $x = 5$.

A genus of about four species, these native to Europe, North Africa, and Asia. Two species, the perennial *Anthoxanthum odoratum*, SWEET VERNALGRASS, and the annual *A. aristatum* Boiss., have been introduced into the United States and now are well established throughout the eastern states and on the Pacific Coast.

40. *Hierochloë* R. Br.

Low to moderately tall cespitose or rhizomatous perennials, with thin, flat blades and open or contracted, mostly few-flowered panicles. Herbage fragrant because of the presence of coumarin. Spikelets 3-flowered, the upper floret perfect, the lower 2 staminate. Disarticulation above the glumes, the 3 florets falling together. Glumes equal, broad, thin, membranous, shiny, 1- to 3-nerved, awnless, as long as the florets or slightly shorter. Lemmas of the staminate florets broadly elliptic, rounded on the back, 5-nerved, awnless or with a short awn from a notched apex. Palea of the staminate florets 2-nerved, enclosing 3 stamens. Lemma of the perfect floret about as long as the lemmas of the staminate florets, rounded on the back, 3- to 5-nerved, awnless, short-hairy at the apex, becoming indurate. Palea apparently 1- or 3-nerved, rounded on the back. Perfect floret with 2 stamens. Basic chromosome number, $x = 7$.

A genus of about twenty species, these in the temperate and cool regions of the world, mostly in the Northern Hemisphere. Three species are reported for the United States. *Hierochloë alpina* (Swartz) Roem. & Schult. is occasional in the northern states from Maine, New Hampshire, and New York to Montana. *Hierochloë odorata* (L.) Beauv., SWEETGRASS, is present in moist, cool meadows, shaded stream banks, and cool mountain canyons throughout the southwestern and western states. The third species, *H. occidentalis* Buckl., CALIFORNIA SWEETGRASS, grows only in a few forested areas of Washington, Oregon, and California.

41. *Phalaris* L.

Annuals and perennials, the perennials rather coarse, densely cespitose or in small clusters from rhizomatous bases. Leaves mostly glabrous, with membranous ligules and flat blades. Inflorescence a contracted, usually spikelike panicle. Spikelets with 1 terminal perfect floret and 1 or 2 reduced florets below, the latter reduced to scales. Disarticulation above the glumes. Glumes about equal, large, awnless, usually laterally flattened and dorsally keeled, the keel often with a thin, membranous wing. Lemma of the fertile floret awnless, coriaceous and glossy, shorter and firmer than the glumes, often more or less hairy, permanently enclosing the faintly 2-nerved palea and plump caryopsis. Basic chromosome numbers, $x = 6$ and 7.

Species fifteen (Anderson, 1961), in the temperate regions of Europe, Asia, Africa, and the Americas. Nine species are reported for the United States, five native and four introduced. *Phalaris arundinacea* L., REED CANARYGRASS, a tall rhizomatous perennial, is distributed throughout the northern half of the United States. In many areas it is an important hay grass of moist meadows. *Phalaris caroliniana* Walt., CAROLINA CANARYGRASS (Fig. 5-17), is frequent and weedy on disturbed soils throughout the southern half of the country. It grows mainly as a cool-season grass. *Phalaris canariensis* L., CANARYGRASS, the caryopses of which provide the "canary seed" of commerce, has been reported as a casual weed at many locations. This grass probably could not persist in the United States, however, without the continued seed source in bird food. A comprehensive taxonomic treatment of the genus *Phalaris* has been presented by Anderson (1961).

42. *Alopecurus* L.

Tufted annuals and perennials, a few rhizomatous, with flat blades and dense, cylindrical, spikelike panicles of 1-flowered spikelets. Disarticulation below the glumes, the spikelets falling entire. Glumes equal, awnless, usually united at the margins below, ciliate on the keels. Lemma about as long as the glumes, firm, 5-nerved, obtuse, awned from below the middle. Margins of the lemma usually united near the base. Palea typically absent. Basic chromosome number, $x = 7$.

About twenty-five species, distributed in the temperate regions of the Northern Hemisphere, six native to the United States and four introduced or adventive. The perennial *Alopecurus pratensis* L., MEADOW FOXTAIL, introduced from Europe as a cultivated pasture grass, is well established in fields and other disturbed sites throughout the northern states.

Fig. 5-17. *Phalaris caroliniana.* Plant, spikelet, and fertile floret with scalelike rudimentary florets at base. (From Gould and Box,
1965.)

The native annual *A. carolinianus* Walt. (Fig. 5-18) is widely distributed
throughout the United States. The united glumes, united margins of the
lemma, and absence of the palea in the spikelets of *Alopecurus* are features
of special interest.

43. *Phleum* L.

Tufted annuals and perennials. Leaves with flat blades and membranous
ligules to 6 mm or more long. Inflorescence a short, cylindrical, tightly
contracted panicle. Spikelets 1-flowered, disarticulating above or occasionally below the glumes. Glumes equal, laterally flattened, mostly 3-nerved,
broad, abruptly narrowed at the apex to a mucro or short, stout awn. Lemma

membranous, 3- to 7-nerved, broad and blunt, awned, much shorter than the glumes. Palea membranous, narrow, about as long as the lemma. Stamens 3. Basic chromosome number, $x = 7$.

Species about ten, in the temperate regions of both hemispheres. Four species are present in the United States, but only one, *Phleum alpinum* L., MOUNTAIN TIMOTHY, is native. *Phleum pratense* L., TIMOTHY, introduced from Europe, is perhaps the most important hay grass of the northeastern United States. Both *P. alpinum* and *P. pratense* are valuable forage plants in mountain meadows of the western states.

44. *Gastridium* Beauv.

Low annuals with weak culms, flat blades, membranous ligules, and cylindrical, tightly contracted panicles. Spikelets 1-flowered, disarticulating

Fig. 5-18. *Alopecurus carolinianus*. Inflorescence and spikelet.

above the glumes. Rachilla prolonged behind the palea as a minute bristle. Glumes long, somewhat unequal, lanceolate or acuminate, 1-nerved, firm and shiny, awnless. Lemma much shorter than the glumes, broad and blunt, rounded on the back, 5-nerved, awnless or with a delicate awn from below the minutely toothed apex. Palea about as long as the lemma. Basic chromosome number, $x = 7$.

A genus of two species, both native to the Mediterranean region. *Gastridium ventricosum* (Gouan) Schinz & Thell., NITGRASS, is adventive and weedy in the United States. It is now well established in Oregon and California and is also known from a few collections from the East Coast and Texas.

45. *Lagurus* L.

Low, tufted annual with a dense, hairy, ovate or oblong panicle of 1-flowered spikelets. Herbage pubescent. Leaf blades flat and soft. Ligule a short, truncate membrane. Spikelet disarticulating above the glumes. Glumes about equal, 1-nerved, thin, hairy, tapering to a slender tip. Lemma thin, glabrous, shorter than the glumes, rounded on the back, cleft at the apex into 2 slender, awn-tipped teeth, and bearing a slender, geniculate awn on the back below the cleft. Rachilla extended as a fine bristle behind the floret. Basic chromosome number, $x = 7$.

A genus of a single species, *Lagurus ovatus* L., HARE'S-TAIL, this native to the Mediterranean region and grown as an ornamental in the United States. *Lagurus ovatus* has been reported as an escape in California, North Carolina, and New Jersey.

46. *Milium* L.

Our species a tufted, succulent perennial with usually glabrous herbage and lax, flat blades. Ligule membranous, as much as 10 mm long, rounded at the apex. Inflorescence an open panicle, the spikelets sparsely distributed on slender branches. Spikelets small, 1-flowered, awnless, dorsoventrally flattened, disarticulating above the glumes. Glumes equal, ovate, slightly longer than the lemma, 3-nerved. Lemma lanceolate to elliptic, rounded on the back, obscurely 5-nerved, becoming indurate, dark brown, smooth and shining. Palea narrow, about as long as the lemma and similar in texture. Caryopsis obovate, tightly enclosed by the lemma and palea. Apparently there are three basic chromosome numbers, $x = 4$, 7, and 9.

Species about six, mostly in cool temperate regions of Europe and Asia. The genus is represented in the United States by *Milium effusum* L., a tall, slender grass of woodland sites. It is distributed from the northeastern states westward to Minnesota and northern Illinois and is also present in Europe and Asia.

47. *Beckmannia* Host

Coarse annuals with moderately thick culms and broad, flat blades. Inflorescence with 1-flowered spikelets crowded on the short, spicate branches of a narrow, loosely contracted, simple or sparingly rebranched, interrupted panicle. Spikelets suborbicular, flattened, nearly sessile, closely imbricated in two rows on a slender rachis. Disarticulation below the glumes, the spikelet falling entire. Glumes equal, broad, firm, 3-nerved, keeled, abruptly apiculate at the apex. Lemma about as long as the glumes, narrow, 5-nerved, tapering to a slender, pointed tip. Palea narrow, slightly shorter than the lemma. Basic chromosome number, $x = 7$.

A genus of two species: one, *Beckmannia erucaeformis* (L.) Host, in northern Europe and Asia, and the other, *B. syzigachne* (Steud.) Fern., AMERICAN SLOUGHGRASS, widespread in the cooler parts of Eurasia and North America. *Beckmannia syzigachne* is present in marshes and along ditches throughout the northwestern and north central states and is occasional in the Northeast. *Beckmannia* was included in the tribe Chlorideae by Hitchcock (1935, 1951), but differs from that group in many fundamental characteristics, as has been noted by Reeder (1953).

Tribe 3. Triticeae

48. *Elymus* L.

Cespitose or rhizomatous perennials with usually flat blades and membranous ligules. Inflorescence typically a spike with spikelets in pairs or threes at the nodes. A few species with mostly 1 spikelet at each node and others with 4 to 6 or more spikelets at each node. Disarticulation (in our species) above the glumes and between the florets. Glumes equal or nearly so, firm, narrow, and subulate to broad and soft, 1- to several-nerved, awned or awnless. Lemmas about equaling the second glume or longer, mostly 5- to 7-nerved, rounded on the back, tapering to a short or long awn, less frequently awnless. Palea well developed. Basic chromosome number, $x = 7$.

In the "classical" concept, this genus comprises about fifty species, with some twenty-four in the United States. It is well represented throughout the cool and temperate regions of the Northern Hemisphere. Widespread and variable in the United States are *Elymus canadensis* L., CANADA WILD-RYE (Fig. 5-19), and *E. virginicus* L., VIRGINIA WILDRYE (Fig. 5-20). These species frequently grow together, and hybrid swarms of intermediate plants are not uncommon (Brown and Pratt, 1960; Pohl, 1959). Perhaps most important as forage species in the western states are the bunchgrasses *E. glaucus* Buckl., BLUE WILDRYE; *E. cinereus* Scribn. & Merr.; and the

Fig. 5-19. *Elymus canadensis.* Plant, pair of glumes, and floret. (From Gould, 1951.)

rhizomatous *E. triticoides* Buckl., CREEPING WILDRYE or BEARDLESS WILDRYE.

The separation of the genera *Elymus* and *Agropyron* on the basis of the number of spikelets per node has the obvious advantage of simplicity in respect to the mechanical grouping of species. From a phylogenetic viewpoint, however, this separation is quite unsatisfactory. Investigations by Dr. G. L. Stebbins and his students (Stebbins and Singh, 1950; Stebbins, Valencia, and Valencia, 1946a, 1946b) supported the belief that a number of species now assigned to *Agropyron* are more closely related to species of *Elymus* than they are to one another. Parallel series of cespitose plants; rhizomatous plants of dry, open areas; rhizomatous plants of woodland sites; and plants of coastal sands are apparent. Even as presently recognized, the number of spikelets per node in *Elymus* is highly variable. *Elymus simplex* Scribn. & Merr. and *E. salinus* Jones have inflorescences with predominantly or entirely 1 spikelet per node. Plants of *E. glaucus* growing at an Arizona location were observed to have mostly 1 spikelet at each rachis node in a shaded site, whereas in an exposed, sunny habitat there were predominantly 2 spikelets at each node. *Elymus triticoides* and *E. pacificus* Gould typically have 1 spikelet at the terminal and basal nodes of the rachis and 2 spikelets at the other nodes. On the other hand, *E. cinereus* has a spike or spicate raceme with usually 4 to 6 spikelets per node, and *E. condensatus* Presl, the giant wildrye of the southern and central California coastal area, has a contracted, densely flowered panicle with as many as 40 spikelets crowded on short branches at the lower rachis nodes. References: Bowden, 1964; Gould, 1947.

Fig. 5-20. *Elymus virginicus.* Inflorescence. (From Gould and Box, 1965.)

49. *Sitanion* Raf.

Short-lived cespitose perennials. Inflorescence a spike, with 2 (infrequently 1) spikelets at each node of a readily disarticulating rachis. Spikelets usually 2- to 4-flowered. Glumes narrow, usually setaceous, 1- to 3-nerved, the midnerve extended into a long, stout, flexuous awn. Palea well developed. Basic chromosome number, $x = 7$.

According to the taxonomic treatment by Wilson (1963), this is a genus of four species: *Sitanion jubatum* J. G. Smith, BIG SQUIRRELTAIL; *S. hystrix* (Nutt.) J. G. Smith, SQUIRRELTAIL; *S. longifolium* J. G. Smith; and *S. hordeoides* Suksdorf. The first three are rather widespread in the western United States, and the fourth is restricted to the far Northwest. For the most part, these are grasses of open, rocky slopes. Much variation has been noted in plants referred to *Sitanion*, and more than twenty-five species have been recognized by various authors. It is now known that *S. hystrix* hybridizes freely with species of the *Elymus-Agropyron* complex. *Sitanion hanseni* (Scribn.) J. G. Smith, recognized by Hitchcock (1951) as a valid species, is now considered to consist of a series of hybrids between *Sitanion* species and *E. glaucus* (Stebbins, Valencia, and Valencia, 1946b). Gould (1947, 1951) followed earlier workers in treating *Sitanion* and *Elymus* as cogeneric, referring *S. hystrix* to *E. elymoides* (Raf.) Swezey and *S. jubatum* to *E. multisetus* (J. G. Smith) Jones. Reference: Dewey and Holmgren, 1962.

50. *Taeniatherum* Nevski

Short-lived annuals with culms mostly 20 to 60 cm tall and narrow, usually flat leaf blades. Inflorescence a short, dense, bristly spike, the spikelets closely placed in pairs at the nodes of a continuous rachis. Spikelets with a single perfect floret and usually a rudimentary second floret above. Disarticulation above the glumes. Glumes subulate, indurate and fused together below, tapering into a slender, stiff awn. Lemma lanceolate, scabrous, with a long, divergent, flattened awn.

A small genus of weedy, annual grasses with natural distribution in temperate regions of Europe and Asia. *Taeniatherum caput-medusae* (L.) Nevski, MEDUSAHEAD, has become a serious weed of rangelands in Idaho and the Pacific Coast area from Washington to northern California. Hitchcock (1951) followed the original placement of this species in *Elymus* (*E. caput-medusae* L.), and it has also been referred to *Hordeum* (*H. caput-medusae* [L.] Coss. & Dur.).

51. *Hystrix* Moench

Tufted perennials with flat blades and bristly spikes. Spikelets 2- to 4-flowered, 1 to 4 at each node of a continuous rachis. Glumes greatly reduced or absent, setaceous when present. Lemmas firm, faintly 5-nerved, rounded on the back, tapering into a long, flexuous awn. Palea membranous, about as long as the lemma. Basic chromosome number, $x = 7$.

Four species are generally assigned to this genus, two in the United States, one in New Zealand, and one in the Himalayas. *Hystrix patula* Moench (*Elymus hystrix* L.), BOTTLEBRUSH, is widespread but scattered and infrequent in woodland sites of the eastern half of the United States and in southeastern Canada. In this species the spikelets are extremely bristly, widely spaced, and widely spreading. Quite different in general aspect is *H. californica* (Boland.) Kuntze (*Elymus californicus* [Boland.] Gould), of woodland sites and shaded ravines in the California Coast ranges. In this grass the spikelets are less bristly, closely placed on the rachis, and erect rather than spreading. As in the case of *Sitanion*, the American *Hystrix* taxa are closely related to species of *Elymus* and have been assigned to that genus by some workers (Bowden, 1964; Gould, 1947).

52. *Hordeum* L.

Low to moderately tall annuals and perennials, without rhizomes. Blades flat, mostly broad and lax. Ligules membranous. Inflorescence a dense spicate raceme with 3 spikelets at a node, the central one usually fertile and sessile, the lateral ones pediceled and usually staminate or sterile (all sessile and fertile in cultivated barley, *Hordeum vulgare*). Rachis of inflorescence usually fragmenting at maturity, disarticulating above each node, the short internodes falling attached to the triad of spikelets. Central spikelet 1-flowered, the rachilla terminating in a bristle which occasionally bears a rudimentary floret. Lateral spikelets often represented by the glumes only. Glumes narrow, usually subulate or awned and rigid. Lemmas firm, rounded on the back and dorsally flattened, 5-nerved, the nerves usually obscure, tapering into an awn. Palea slightly shorter than the lemma, usually adnate to the caryopsis. Basic chromosome number, $x = 7$.

A genus of about twenty-five species, distributed in the temperate regions of both hemispheres. Twelve species are reported as occurring in the United States, about eight native to this country (Covas, 1949). Except for the cultivated barley, *H. vulgare* L. (Fig. 5-21), this group is of little economic importance, and most of the species occur as weeds. *Hordeum brachyantherum* Nevski is a good forage plant of the higher or cooler ranges. Most widespread of the species in the United States is the native annual *H. pusillum* Nutt., LITTLE BARLEY, which has been reported from all the states except Nevada. *Hordeum leporinum* Link (Fig. 5-22) is frequent in the western states and occasional in the East. The weedy barleys, while furnishing some forage when young and green, at maturity may cause mechanical injury to nose and mouth membranes of grazing animals. The sharp-pointed rachis and rachilla joints and the stiff awns of the mature spikelets are capable of penetrating both hide and flesh. The name × *Elymordeum* has been applied to hybrids between species of *Elymus* and *Hordeum*. Eight such hybrids were listed by Bowden (1958), and Pohl (1966) reported on an Iowa colony of plants presumed to be hybrids between *E. villosus* Muhl. ex Willd. and *H. jubatum* L. Reference: Bowden, 1962.

Fig. 5-21. *Hordeum vulgare.* Inflorescence and spikelet cluster. (From Gould and Box, 1965.)

53. *Agropyron* Gaertn.

Perennials of diverse habit, many with creeping rhizomes. Inflorescence a bilateral spike with spikelets solitary, infrequently paired, at the nodes of a continuous or disarticulating rachis. Glumes broad or narrow, few- to several-nerved (rarely 1-nerved), awn-tipped or awnless. Lemmas firm, faintly 5- or 7-nerved, rounded on the back, awned or awnless. Palea membranous, about as long as the lemma, usually adhering to the caryopsis. Basic chromosome number, $x = 7$.

Fig. 5-22. *Hordeum leporinum.* Inflorescence and 3
spikelets at node of rachis. (From Gould and Box,
1965.)

As delimited in Hitchcock's manual (1951), a genus of some sixty species, with thirty-one listed for the United States. The group is widely distributed in cool and temperate regions of both hemispheres. Several species
of *Agropyron* are important forage grasses on western rangelands; outstanding among them are *A. spicatum* (Pursh) Scribn. & Smith, BLUEBUNCH
WHEATGRASS, and *A. smithii* Rydb., WESTERN WHEATGRASS. *Agropyron repens* (L.) Beauv., QUACKGRASS, is a rhizomatous European species that has become well established throughout the northern portion of
the United States as a vigorous, persistent weed. *Agropyron cristatum* (L.)
Gaertn., CRESTED WHEATGRASS (Fig. 5-23) has been widely seeded in
the western states as soil binder and as forage grass.

In recent years, considerable attention has been given to natural hybrids within the genus *Agropyron* and between species of this genus and

Fig. 5-23. *Agropyron cristatum*. Inflorescence and spikelet.

Elymus. As has been noted, investigations by Stebbins and his students (1946a, 1946b, 1950) led him to conclude that several species of *Agropyron* are more closely related to species of *Elymus* than they are to one another. Similar views have been expressed by Parodi (1940) and Gould (1947) based on morphological characteristics. Numerous studies of the *Agropyron* complex of North America have been made (Bowden, 1962, 1965; Boyle and Holmgren, 1955; Dewey, 1961, 1963, 1964a, 1964b, 1965a, 1965b, 1967a, 1967b; Pohl, 1962), but additional genetical and taxonomic investigations of the entire Triticeae complex are needed to facilitate a more satisfactory taxonomic treatment of this group. Reference: Beetle, 1955b; Estes et al., 1982.

54. *Triticum* L.

Annuals with broad, flat blades. Inflorescence a thick, bilateral spike. Spikelets solitary at the nodes of a continuous or disarticulating rachis. Spikelets 2- to 5-flowered, laterally flattened and oriented flatwise to the rachis or rounded and closely fitted into the rachis joints. Glumes thick and firm, 3- to several-nerved, toothed, mucronate or with 1 to several awns at the apex. Lemmas similar to the glumes in texture, keeled or rounded on the back, asymmetric, many-nerved, awnless or with 1 to 3 awns. Basic chromosome number, $x = 7$.

Fig. 5-24. *Triticum aestivum*. Inflorescence. (From Gould and Box, 1965.)

Fig. 5-25. *Triticum cylindricum.* Inflorescence.

A genus of about thirty species native to cool and temperate regions of southern Europe and western Asia. Numerous varieties of cultivated wheat, *Triticum aestivum* L. (Fig. 5-24), are grown in Canada and throughout the temperate regions of the United States. Further south, as in Oklahoma and Texas, wheat is an important cool-season crop. *Triticum* is also represented in the United States by three adventive species previously referred to *Aegilops*, a genus now combined with *Triticum* by most workers. *Triticum cylindricum* (Host) Ces. (*Aegilops cylindrica* Host), JOINTED GOAT-GRASS (Fig. 5-25), is widespread and locally abundant as a weed of wheatfields, roadsides, and other disturbed areas in the central and southern states. *Triticum triunciale* (L.) Rasp. (*A. triuncialis* L.), a potentially bad weed of grazing areas, is well established in California and has also been reported from a few other United States localities. *Triticum ovatum* (L.) Raspail (*A. ovata* L.) is a weed of fields and roadsides in California and Virginia.

55. *Secale* L.

Annuals and perennials with flat blades and membranous ligules. Inflorescence a dense bilateral spike with spikelets solitary at the nodes of a continuous or disarticulating rachis. In the cultivated rye, disarticulation above the glumes and between the florets. Spikelets usually 2-flowered, the rachilla

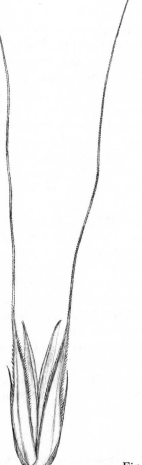

Fig. 5-26. *Secale cereale*. Spikelet.

continued behind the upper floret as a minute bristle. Glumes narrow, rigid, subulate. Lemmas broad, 5-nerved, sharply keeled, ciliate on the keel and the exposed margins, tapering to a long, stout awn. Basic chromosome number, $x = 7$.

Species five, widely distributed in temperate Europe and Asia; one species, *Secale cereale* L., RYE (Fig. 5-26), is cultivated extensively in Europe and to a lesser extent in the United States. Rye, like other cultivated cereals, frequently grows on roadsides and in waste places from chance seedings, but never becomes established outside cultivation.

Tribe 4. Meliceae

56. *Melica* L.

Moderately tall perennials with closed leaf sheaths and often swollen, cormlike culm bases. Ligules membranous, frequently decurrent. Blades flat, moderately broad. Inflorescence a panicle, this simple and contracted in most species but large and open in some. Spikelets relatively large, several-flowered, the fertile florets 2 or more (except *Melica imperfecta*, with 1 perfect floret). Upper florets reduced to a knoblike cluster of closely placed empty lemmas. Disarticulation above the glumes and between the florets in most species but below the glumes in some. Glumes subequal, large, broad, thin, 3- or 5-nerved, scarious-margined, awnless. Lemmas thin or firm, scarious margined, usually 7-nerved, awnless or with a straight awn from a slightly bifid apex. Callus not bearded. Basic chromosome number for North American species, $x = 9$.

Species about sixty, in the temperate and cooler parts of both hemispheres. Seventeen species and four varieties are reported for the United States (Boyle, 1945), these mostly in the Far West. Most species of *Melica* are rated high as range forage grasses but seldom occur in sufficient abundance to be of great significance. *Melica subulate* (Griseb.) Scribn., ALASKA ONION-GRASS, is frequent on shaded slopes and in mountain meadows from Alaska to central California. *Melica porteri* Scribn., PORTER MELIC, is the most common species of the Southwest. *Melica imperfecta* Trin., CALIFORNIA MELIC, is restricted to grasslands of central and southern California (and Baja California), where it may be locally abundant. *Melica mutica* Walt., TWO-FLOWERED MELIC, and *M. nitens* (Scribn.) Nutt. ex Piper, THREE-FLOWERED MELIC, are the only species with wide distribution in eastern and southeastern United States.

57. *Glyceria* R. Br.

Usually tall aquatic or marsh perennials, mostly rhizomatous or with culms decumbent at the base and rooting at the lower nodes. Sheaths partially or totally closed. Blades flat, usually broad and glabrous. Inflorescence with both sessile and pediceled spikelets in an open or contracted panicle, this occasionally reduced to a raceme. Spikelets awnless, several-flowered, usually linear, disarticulating above the glumes and between the florets. Glumes unequal, short, broad, mostly 1-nerved. Lemmas broad, rounded on the back, with 5 to 9 but usually 7 strong, parallel nerves, these not converging above. Apex of lemma acute, obtuse, or truncate. Palea large, broad, often slightly longer than the lemma. Stamens 3 or 2. Basic chromosome number, $x = 10$.

Species about thirty-five, in temperate regions of both hemispheres. The genus is represented in the United States by about fifteen species, present in marshy or moist woodland habitats. For the most part, these grasses are highly palatable to livestock, but few are of sufficient abundance to be of significance as forage plants. *Glyceria striata* (Lam.) Hitchc., FOWL MANNAGRASS, is distributed from Newfoundland to British Columbia, south to Florida, California, and Mexico. It ranges from near sea level to about 10,000 feet. This tall, tufted perennial grass is highly palatable to all types of livestock, but for the most part is limited to scattered stands along the margins of watercourses, swamps, and lakes.

As has been discussed in the section on *Puccinellia*, a number of species referred to *Glyceria* by Hitchcock (1951) were grouped together in a new genus, *Torreyochloa*, by Church (1949) and then moved to *Puccinellia* by Clausen (1952) and Munz (1958).

58. *Catabrosa* Beauv.

Aquatic perennials with decumbent-erect culms that root freely at the lower nodes. Leaves with closed or partially closed sheaths and soft, flat blades. Inflorescence an open panicle of small, awnless, mostly 2-flowered spikelets. Disarticulation above the glumes and between the florets. Glumes short, obtuse, unequal, nerveless or 1-nerved. Lemmas rounded on the back, prominently 3-nerved, the parallel nerves not converging at the broad, scarious apex. Palea broad, 2-keeled, about as long as the lemma. Caryopsis flat, oval, brownish, loosely enclosed in the lemma and palea. Basic chromosome number, $x = 10$.

Species seven, mostly in northern Eurasia; one species, *Catabrosa aquatica* (L.) Beauv., BROOKGRASS, in temperate and cold regions of Europe, Asia, Africa, and North America. In the United States, brookgrass is occasional along streams and spring areas from Wisconsin, Montana, and Oregon south to the high mountains of northern Arizona.

59. *Pleuropogon* R. Br.

Tufted annuals and perennials with erect or decumbent culms, flat leaf blades, and membranous ligules. Inflorescence a raceme of short-pediceled spikelets, these mostly large, many-flowered, and linear. Rachilla disarticulating above the glumes and between the florets. Glumes unequal, short, broad, awnless, the first 1-nerved, the second obscurely 3-nerved. Lemmas 7-nerved, short- to long-awned from an entire or slightly bifid apex. Palea strongly 2-keeled, the keels winged on the lower half. In *Pleuropogon sabinii* R. Br. and *P. oreganus* Chase, the palea keels are awned from below the middle. Basic chromosome numbers, apparently $x = 7$ and 8.

Species six, one in the Arctic and five along the Pacific Coast of North America. Three species, *P. californicus* (Nees) Benth. ex Vasey, *P. hooverianus* (Benson) J. T. Howell, and *P. davyi* Benson, are restricted to northwestern and central California.

60. *Schizachne* Hack.

Moderately tall cespitose perennials with erect, unbranched culms. Inflorescence a loose raceme or simple panicle of few, large, several-flowered spikelets borne on slender pedicels. Leaf sheaths closed. Leaf blades long, flat, and narrow. Spikelets 2.0 to 2.5 cm long, with short, awnless, 2- and 5-nerved glumes and strongly 7-nerved lemmas. Lemmas lanceolate, bearing a straight awn from between the teeth of a bifid apex. Disarticulation above the glumes and between the florets. Basic chromosome number, $x = 10$.

A genus of two species; one, *Schizachne purpurascens* (Torr.) Swallen, FALSE MELIC, widely distributed in the cooler parts of North America and also in Siberia and Japan. In the United States this grass occurs in open woodlands throughout the northeastern and north-central states and as far south as New Mexico.

Tribe 5. Stipeae

61. *Stipa* L.

Cespitose perennials with long, narrow, mostly involute blades, these usually in a basal clump. Inflorescence a large or small, usually contracted, often drooping panicle of 1-flowered spikelets. Spikelets variable in size but relatively large in most species. Glumes thin, mostly 1- to 5-nerved, acute, acuminate or infrequently aristate, longer than the body of the lemma. Lemma firm or indurate, relatively slender, terete or angular, awned, tightly enclosing the palea and caryopsis. Awn of the lemma usually stout, geniculate and twisted, scabrous or pubescent below in many species. Base of lemma and rachilla forming a sharp-pointed callus, this bearded with stiff hairs. Basic chromosome numbers, $x = 7$ and 11.

A genus of about 150 species, these widely distributed in temperate and tropical regions of the world. Hitchcock's manual (1951) lists 34 native U.S. species, all but a few of which are distributed in the western states. The large, coarse bunchgrass *Stipa spartea* Trin., PORCUPINEGRASS, is a common constituent of prairie vegetation from the northeastern states to Montana, Colorado, and northern New Mexico. *Stipa pulchra* Hitchc., PURPLE NEEDLEGRASS, is an important component of the native grasslands of the California Coast Ranges. In southern Oklahoma, Texas, and northern Mexico, *S. leucotricha* Trin. & Rupr., TEXAS WINTERGRASS (Fig. 5-27), is a valuable cool-season forage grass and one of the few native

Fig. 5-27. *Stipa leucotricha*. Inflorescence and spikelet. (From Gould and Box, 1965.)

cool-season perennials of the region. This species produces an abundance of basal cleistogamous spikelets under certain environmental influences (Brown, 1949, 1952). *Stipa robusta* (Vasey) Scribn., SLEEPY GRASS, of the southwestern United States and northern Mexico, is reported to have narcotic effects on animals that graze it, especially affecting horses.

Species of *Stipa* hybridize freely among themselves and also with the closely related *Oryzopsis hymenoides* (Johnson, 1945, 1960, 1962a, 1962b, 1963; Johnson and Rogler, 1943). A remarkable series of euploid and aneuploid chromosome numbers has been recorded for *Stipa*, with counts ranging from $2n = 22$ to $2n = 70$.

Fig. 5-28. *Oryzopsis hymenoides*. Plant and spikelet with glumes and floret separated. (From Gould, 1951.)

62. *Oryzopsis* Michx.

Cespitose perennials, closely related to *Stipa* and similar in general aspect. Inflorescence a narrow or open panicle of 1-flowered, awned spikelets. Disarticulation above the glumes. Glumes broad, obtuse to acuminate. Lemma broad, subterete, firm or indurate, about as long as the glumes, with a short, blunt callus at the base and a short, deciduous awn, either straight or twisted and bent. Palea completely enclosed by the lemma. Basic chromosome number, $x = 11$.

Species about twenty, distributed in the cool and temperate regions of both hemispheres. Twelve species reported for the United States by Hitchcock (1951). *Oryzopsis hymenoides* (Roem. & Schult.) Ricker, INDIAN RICEGRASS (Fig. 5-28), is an important range forage species of semiarid regions of the West. Johnson and Rogler (1943) gave evidence to indicate that the type specimen of *O. bloomeri* (Boland.) Ricker and several recognized "species" of *Stipa* are *Oryzopsis-Stipa* hybrids.

63. *Piptochaetium* Presl

Tufted perennials with slender, stiffly erect culms, filiform leaves, these mostly in a basal clump, and small panicles of plump, 1-flowered spikelets. Disarticulation above the glumes. Glumes slightly longer than the body of the lemma, broad, thin, acute or acuminate, 7-nerved. Lemma firm, rounded, slightly obovate, dark brown or gray at maturity, bearing a stout, twisted, and twice geniculate deciduous or persistent awn. Palea narrow, firm but with thin margins, the 2 nerves close together and slightly keeled. Palea enclosed by the lemma but slightly exserted at the apex as a minute projection. Basic chromosome number, $x = 11$.

Species about twenty, mostly in South America. One species, *Piptochaetium fimbriatum* (H.B.K.) Hitchc., PINYON RICEGRASS (Fig. 5-29), is frequent in the mountains of Colorado, New Mexico, Arizona, western Texas, and northern Mexico. It grows mainly in open pine forests and contributes significantly to the range forage in the areas where it occurs. *Piptochaetium avenaceum* (L.) Parodi (*Stipa avenacea* L.), BLACKSEED NEEDLEGRASS, is frequent and widely distributed throughout the open woodlands of the eastern and southeastern states, extending southwestward to eastern Texas.

Tribe 6. Brachyelytreae

64. *Brachyelytrum* Beauv.

Our species a slender, erect perennial with short, knotty rhizomes. Blades flat and broad. Inflorescence a short, contracted, few-flowered panicle. Spikelets with a single floret, disarticulating above the glumes. Rachilla extended beyond insertion of the floret as a bristle. First glume absent or minute. Second glume short, narrowly lanceolate, occasionally awned. Lemma about 1 cm long, firm, narrow, rounded on the back, 5-nerved, extending into an awn 1 to 3 cm long. Basic chromosome number, $x = 11$.

A genus of two species, one in Korea and Japan and the other, *Brachyelytrum erectum* (Schreb.) Beauv., occasional throughout the eastern half of the United States and in adjacent areas of Canada. Reference: Tateoka, 1957.

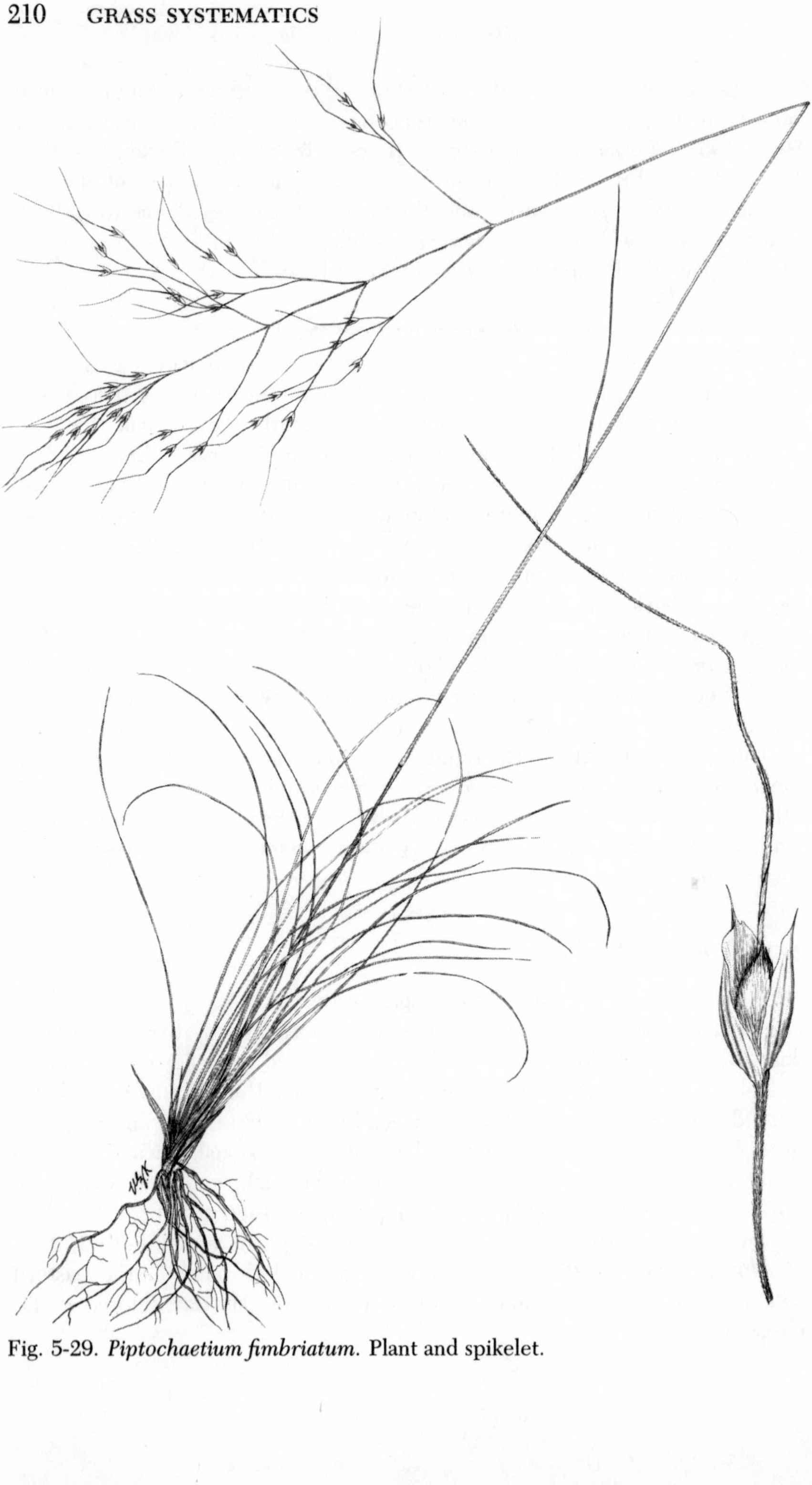

Fig. 5-29. *Piptochaetium fimbriatum*. Plant and spikelet.

Tribe 7. Diarrheneae

65. *Diarrhena* Beauv.

Slender, erect perennials. Leaf blades long, flat, 1 to 2 cm broad. Inflorescence a narrow, erect or drooping, few-flowered panicle. Spikelets mostly 3- to 5-flowered, disarticulating above the glumes and between the florets. Glumes unequal, shorter than the lemmas, the first 1-nerved, the second 3- or 5-nerved. Lemmas firm, tapering to a point, 3-nerved, the nerves converging at the tip. Palea broad, obtuse, strongly 2-nerved. Stamens 2 or 3, infrequently 1. Caryopsis large, hard, shiny, turgid, beaked above, at maturity conspicuously exserted from between the spreading lemma and palea. Chromosome numbers reported, $2n = 38$ and 60.

A genus of about five species, these mostly Asiatic. The single North American species, *Diarrhena americana* Beauv., is occasional in woodlands from Virginia to Michigan, South Dakota, Oklahoma, and Texas.

Tribe 8. Nardeae

66. *Nardus* L.

Low, tufted perennial with slender, densely crowded culms arising from short rhizomes. Leaf blades narrow, stiff, tightly inrolled, sharp-pointed. Ligules membranous, rounded. Inflorescence a short unilateral spike with slender, 1-flowered spikelets in two rows along one side of a narrow rachis. Disarticulation above the glumes. First glume absent, the second minute or absent. Lemma firm, narrowly lanceolate, 3-nerved, with an awn 1 to 3 mm long. Palea slightly shorter than the lemma. Stamens 3. Chromosome numbers reported, $2n = 26$ and 30.

A genus of a single species, this, *Nardus stricta* L., distributed throughout the temperate regions of Europe, Asia, northwestern Africa, Greenland, and Newfoundland. *Nardus stricta* has been sparingly introduced in dry, sandy sites in the northeastern United States and is also reported from Oregon (Chambers and Dennis, 1963).

Tribe 9. Monermeae

67. *Monerma* Beauv.

Tufted, branching annuals with weak, flat blades and slender, cylindrical spikes. Spikelets 1-flowered, solitary at the nodes of, and partially embedded in, the thickened, cylindrical rachis. Disarticulation in the rachis, each spikelet falling attached to a section of the rachis. First glume absent except on the terminal spikelet. Second glume large, firm or indurate, several-nerved, acute, longer than the rachis section to which it is attached, oriented with the dorsal surface away from the rachis. Lemma thin, hyaline, 3-nerved, awnless, shorter than the glume. Palea thin and hyaline. Chromosome numbers reported, $2n = 26$ and 52.

A genus of about three species, all native to the Old World. *Monerma cylindrica* (Willd.) Coss & Dur., THINTAIL, is adventive in salt marshes from the San Francisco Bay area south to Lower California. Reference: Tateoka, 1959.

68. *Parapholis* C. E. Hubb.

Low, tufted, much-branched annuals, with slender, curved, cylindrical spikes at each of numerous branch tips. Spikelets 1- or 2-flowered, solitary at the nodes of a thickened rachis, partially embedded in, and falling attached to, the rachis joint. Glumes large, subequal, coriaceous, 3- or 5-nerved, somewhat asymmetrical, tapering to a point, the two placed in front of the spikelet and appearing as halves of a single glume. Lemma thin, hyaline, 1-nerved, awnless, shorter than the glumes but exceeding the hyaline palea. Chromosome numbers reported, $2n = 14, 32, 36, 38,$ and 42.

A genus of about four species, all native to temperate regions of the Old World. *Parapholis incurva* (L.) C. E. Hubb., SICKLEGRASS (Fig. 5-30), is adventive and weedy in saline coastal sites on the Atlantic and Pacific Coasts and also in Texas along the Gulf of Mexico.

Fig. 5-30. *Parapholis incurva*. Plant and spikelet with section of the rachis. (From Gould and Box, 1965.)

Subfamily II. Panicoideae

Tribe 10. Paniceae

69. *Digitaria* Heister

Annuals and perennials, with erect or decumbent-spreading, stoloniferous culms. Blades mostly thin and flat. Inflorescence a panicle with few to numerous slender spikelike racemose branches, these unbranched or sparingly branched near the base. Spikelets slightly plano-convex; solitary, paired, or in groups of 3 to 5; subsessile or short-pediceled in two rows on a 3-angled, often winged rachis. Disarticulation below the glumes. First glume minute or absent. Second glume well developed but usually shorter than the lemma of the sterile floret. Nerves of the glumes and lemma of the sterile floret glabrous, puberulent, or long-ciliate. Lemma of the fertile floret relatively narrow, acute or acuminate, firm and cartilaginous, but not hard, the margins thin, flat, not inrolled over the palea. Basic chromosome number, $x = 9$.

Species about three hundred, in temperate and tropical regions throughout the world. Including the grasses referred to *Trichachne* by Hitchcock (1951), there are about twenty-four species of *Digitaria* native to the United States. Most of the American species are weedy and grow in moist, disturbed sites. *Digitaria californica* (Benth.) Henrard (*T. californica* [Benth.] Chase), ARIZONA COTTONTOP (Fig. 5-31), is a leafy perennial bunchgrass that contributes considerable range forage in the Southwest, from southern Colorado to Texas, Arizona, and northern Mexico. This grass makes rapid growth following winter rains and furnishes earlier forage than

Fig. 5-31. *Digitaria californica*. Plant and spikelet. (From Gould and Box, 1965, as *Trichachne californica*.)

most associated grass species. *Digitaria sanguinalis* (L.) Scop., HAIRY CRABGRASS, and the closely related *D. ciliaris* (Retz.) Koel. are widespread weedy annuals, the former in temperate climates almost throughout the United States, the latter from southern Virginia and southeastern Nebraska through Texas to Central and South America. Reference: Henrard, 1950.

70. *Leptoloma* Chase

Tufted perennials with large, open panicles that break off at the base and function as tumbleweeds. Base of plant usually with short, slender, "knotty" rhizomes. One species with stolons. Ligule a short, truncate membrane. Blades short, flat, crisped on the margins. Spikelets on spreading pedicels that much exceed the spikelets in length. Disarticulation below the glumes. First glume absent or vestigial. Second glume 3- or 5-nerved, about equal-

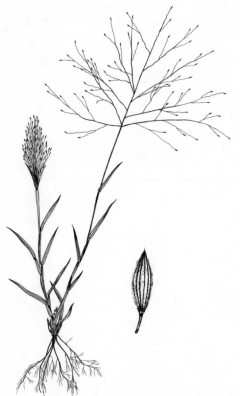

Fig. 5-32. *Leptoloma cognatum.* Plant and spikelet. (From Gould and Box, 1965.)

ing the 5- or 7-nerved lemma of the sterile floret. Glumes and lemma of sterile floret short- to long-pubescent between the nerves and on the margins. Lemma of the fertile floret dark brown, smooth, cartilaginous, narrow, acute at the apex, with thin, flat (not inrolled) margins. Palea like the lemma in texture. Basic chromosome number, $x = 9$.

Species about five, two in the United States and three in Australia. *Leptoloma cognatum* (Schult.) Chase, FALL WITCHGRASS (Fig. 5-32), is a low, weak perennial that is widely distributed throughout the eastern United States, ranging westward to Minnesota, Nebraska, Arizona, and northern Mexico. It flowers almost continuously throughout the growing season. Despite its abundance in many areas, relatively little herbage is developed, and the species is of little value as a forage grass. Henrard (1950) included the species of *Leptoloma* in the genus *Digitaria*. The spikelets are similar but the inflorescence of *Leptoloma*, with the long, spreading pedicels, is much more open than any of the North American *Digitaria* representatives.

71. *Anthaenantia* Beauv.
Slender, erect perennials with short rhizomes and firm, flat blades. Spikelets conspicuously villous, short-pediceled in small, loosely contracted panicles, disarticulating below the glume. First glume absent, the second glume 5-nerved, obovate, rounded at the apex, densely villous. Lemma of the sterile floret villous, 5-nerved, about as long as the second glume. Lemma of the fertile floret 3-nerved, cartilaginous, dark brown, much narrower than the second glume, with pale, thin, flat margins. Palea narrow, the 2 nerves nearly parallel. Chromosome number not reported.

Species two, occasional in pine woods along the Atlantic coastal plain, North Carolina to Florida and Texas. *Anthaenantia villosa* (Michx.) Beauv., PURPLE SILKYSCALE, and *A. rufa* (Ell.) Schult, GREEN SILKYSCALE (Fig. 5-33), provide some forage in the pine-woods grazing areas of the Southeast.

72. *Stenotaphrum* Trin.
Low, mat-forming, stoloniferous perennials with thick, succulent, flat blades and (in our species) spikelets partially embedded in the flattened and thickened rachis of a short, unilateral, spicate inflorescence. Disarticulation at the nodes of the rachis, the spikelets falling attached to the sections. (*Note*: The spikelets are actually borne on rudimentary branch axes that are extended as a point beyond the insertion of the 1 to 3 spikelets present at each node of the main rachis.) First glume short, irregularly rounded. Second glume and lemma of the sterile floret about equal, glabrous, pointed at the apex, faintly nerved. Reduced (lower) floret often staminate. Lemma of the fertile floret chartaceous. Basic chromosome number, $x = 9$.

Fig. 5-33. *Anthaenantia rufa*. Inflorescence and spikelet.

A genus of six species, in tropical and subtropical regions of Africa, the Americas, Hawaii and other Pacific islands, and Australia. *Stenotaphrum secundatum* (Walt.) Kuntze, ST. AUGUSTINE GRASS (Fig. 5-34), is native to the southeastern United States and the American tropics, where it grows naturally in moist, sandy sites. This and bermudagrass, *Cynodon dactylon*, are the two most common southern lawn grasses. St. Augustine grass is relatively coarse, but with water and fertilizer it makes a dense, uniform stand. It successfully competes with most lawn weeds. It does not develop rhizomes and thus is easily controlled in flower beds and gardens.

Fig. 5-34. *Stenotaphrum secundatum.* Plant and spikelet. (From Gould and Box, 1965.)

73. *Brachiaria* Griseb.

Annuals and perennials, with stoloniferous or decumbent culm bases and flat leaf blades. Inflorescence a small panicle with 2 to numerous short, usually simple, erect-appressed or spreading branches. Disarticulation below the glumes. First glume present, shorter than to nearly as long as the spikelet.

Fig. 5-35. *Brachiaria fasciculata*. Plant, spikelet, and fertile floret. (From Gould and Box, 1965.)

Second glume and lemma of the lower floret about equal, acute, 5- to 7-nerved. Lemma of the upper floret indurate, glabrous, obtuse or broadly acute at the apex, rarely short-awned, with the margins tightly inrolled over the palea. Lemma and palea of upper floret strongly transverse rugose. Basic chromosome number $x = 9$, or 7 in one species.

Species about twenty-five, in tropical and subtropical regions of both hemispheres. Nine species are native to the southern and southwestern United States since the transfer of the "Fasciculata" group of *Panicum* (*B. adspersa* [Trin.] Parodi, *B. arizonica* [S.&M.] S. T. Blake, *B. fasciculata* [Sw.] Parodi, *B. ramosa* Stapf., *B. reptans* [L.] Gard. & Hubb., and *B. texana*

Fig. 5-36. *Brachiaria platyphylla*. Plant. (From Gould and Box, 1965.)

[Buckl.] S. T. Blake) to the genus. BROWNTOP BRACHIARIA (*B. fasciculata*) (Fig. 5-35) is probably the most common species ranging from Florida to Arizona and southward at low elevations to northern South America. *Brachiaria platyphylla* (Griseb.) Nash, BROADLEAF SIGNALGRASS (Fig. 5-36), and *B. plantaginea* (Link) Hitchc., CREEPING SIGNALGRASS, are weedy annuals of loose, sandy, or loamy soils. *Brachiaria ciliatissima* (Buckl.) Chase, FRINGED SIGNALGRASS or SANDHILL GRASS (Fig. 5-37), is a low, stoloniferous perennial adapted to loose, sandy soils. It has considerable value as a soil binder of drifting sand. A chromosome number of $2n = 14$ has been reported for *Brachiaria reptans* (L.) Gard. & Hubb. This is the only New World panicoid grass known to have a basic chromosome of $x = 7$, a common characteristic of pooid grasses. References: Blake, 1958, 1969; Brown, 1977; Parodi, 1969.

Fig. 5-37. *Brachiaria ciliatissima*. Inflorescence. (From Gould and Box, 1965.)

74. *Axonopus* Beauv.

Stoloniferous or cespitose perennials (the U.S. species), with flat or folded, rounded or pointed blades. Inflorescence with 2 to several slender spicate branches, the spikelets solitary at the nodes and rather widely spaced in two rows on one side of a flattened rachis. Spikelets with the rounded back of the lemma of the fertile floret oriented away from the branch rachis. Disarticulation below the spikelet. First glume absent, the second glume and lemma of the sterile floret about equal, narrowly ovate or oblong, often pointed beyond the fertile lemma. Lemma and palea of the fertile floret indurate, oblong, glabrous, usually obtuse at the apex. Basic chromosome number, $x = 10$.

A genus of about seventy species, all native to the American tropics and subtropics. One species, *Axonopus compressus* (Swartz) Beauv., is found also in Africa, where it apparently is an introduction. Only three species are reported for the United States, these all in the southeastern states, from Virginia and Oklahoma south to Florida and Texas. *Axonopus affinis* Chase, COMMON CARPETGRASS (Fig. 5-38), is frequent on sandy soils in low, moist, woodland or open sites. It is a strongly stoloniferous, mat-forming perennial, adapted to relatively sterile soils and occasionally used as a lawn grass. Reference: Black, 1963.

75. *Reimarochloa* Hitchc.

Perennials, ours with stout stolons. Blades short, flat or folded. Inflorescence with the spikelets subsessile and solitary at the nodes of few to several

Fig. 5-38. *Axonopus affinis*. Plant and spikelet. (From Gould and Box, 1965.)

short, unbranched primary branches. Spikelets loosely imbricated in two rows along one side of a narrow, flattened rachis, the rounded back of the lemma of the fertile floret turned toward the branch rachis as in *Paspalum*. Disarticulation below the glumes. Both glumes absent or the second occasionally present in the terminal spikelet of a branch. Lemmas of sterile and fertile florets about equal, gradually tapering to a point. Lemma of the fertile floret relatively thin, inrolled over the palea only at the base, the upper portion of the palea free. Chromosome number unreported.

A genus of about four species, these in the American tropics, with one, *Reimarochloa oligostachya* (Munro) Hitchc., ranging northward into Florida. This species has no economic importance in our region.

76. *Eriochloa* H.B.K.

Cespitose annuals and perennials. Blades flat, mostly thin. Ligule a ciliated membrane. Inflorescence a loosely contracted panicle, the spikelets subsessile or short-pediceled on unbranched or sparingly rebranched primary branches. Disarticulation below the glumes. First glume reduced and fused with the rachis node to form a cup or disk. Second glume and lemma of the sterile floret about equal, usually scabrous, hispid or hirsute, acute or more commonly acuminate at the apex. Lemma of the fertile floret indurate, glabrous, finely rugose, with slightly inrolled margins, apiculate or short-awned at the apex. Basic chromosome number, $x = 9$.

A genus of about twenty-five species in the tropics and warm-temperate regions of both hemispheres. Of the seven species occurring in the United States, three are perennial and four annual. The annuals often grow as weeds of roadsides, ditch banks, low meadows, and similar moist or marshy habitats. *Eriochloa sericea* (Scheele) Vasey ex Munro, TEXAS CUPGRASS (Fig. 5-39), a perennial, is widespread in the central and southern Great Plains region. It is highly palatable and a good forage grass in pastures protected from heavy grazing. Reference: Gould, 1950; Shaw and Smeins, 1979, 1981.

Fig. 5-39. *Eriochloa sericea.* Plant, spikelet, and fertile floret. (From Gould and Box, 1965.)

77. *Paspalum* L.

Annuals and perennials, the United States species perennial, many with rhizomes or stolons. Blades usually flat, often thin and broad. Inflorescence with 1 to many unilateral spikelike branches, these scattered or, in a few species, paired at the culm apex. Spikelets subsessile or short-pediceled, solitary or in pairs on a flattened, occasionally broadly winged rachis, with the rounded back of the lemma of the fertile floret turned toward the rachis. Disarticulation at the base of the spikelet. First glume typically absent but irregularly present in a few species. Second glume and lemma of the sterile floret usually about equal, broad and rounded at the apex, infrequently acute. Lemma of the fertile floret firm or indurate, rounded on the back, usually obtuse, with inrolled margins. Palea broad, flat or slightly convex, the margins entirely enfolded by the lemma. Basic chromosome number, $x = 10$.

Fig. 5-40. *Paspalum floridanum*. Plant, spikelet, and fertile floret. (From Gould and Box, 1965.)

A genus of about four hundred species distributed throughout the warmer regions of the world. Hitchcock's manual (1951) reported forty-five native species and a number of introductions in the United States, these mostly in the Southeast. Despite the relatively large number of species, few are of special importance as forage grasses in the United States. Among the most widespread and locally abundant species are *Paspalum floridanum* Michx., FLORIDA PASPALUM (Fig. 5-40); *P. plicatulum* Michx., BROWN-SEED PASPALUM (Fig. 5-41); *P. setaceum* Michx., THIN PASPALUM;

Fig. 5-41. *Paspalum plicatulum.* inflorescence and pair of spikelets. (From Gould and Box, 1965.)

P. pubiflorum Rupr. ex Fourn., HAIRYSEED PASPALUM; *P. urvillei* Steud., VASEYGRASS; and *P. dilatatum* Poir., DALLISGRASS. A number of stoloniferous species, such as *P. distichum* L., KNOTGRASS, grow in wet, marshy, or shoreline habitats. Reference: Banks, 1966.

78. *Paspalidium* Stapf

Rhizomatous perennials. Culms to 1 meter tall, thick and succulent, often decumbent and stoloniferous at the base. Ligule a ring of hairs. Inflorescence a panicle of subsessile spikelets on short, unbranched primary branches, these widely spaced on the main rachis. Spikelets borne singly in two rows on the flattened branch rachis, oriented with the back of the lemma of the fertile floret turned toward the rachis as in *Paspalum*. Disarticulation below the glumes. Glumes both present, the first short, acute to broadly obtuse or truncate at the apex, the second slightly shorter than the pointed lemma of the sterile floret, obtuse to acute at the apex. Lemma of the fertile floret firm or hard, moderately rugose, broadly pointed at the tip. Palea flat or slightly convex on the back. Basic chromosome number, $x = 9$.

A genus of about four species, represented in the United States by *Paspalidium geminatum* (Forsk.) Stapf (Fig. 5-42) and *P. paludivagum* (Hitchc.

Fig. 5-42. *Paspalidium geminatum*. Inflorescence and spikelet. (From Gould and Box, 1965, as *Panicum geminatum*.)

& Chase) Parodi. These grasses are adapted to marshy or aquatic habitats, usually growing in shallow water along rivers, lake shores, and marshes. *Paspalidium paludivagum* is occasional from Florida to Texas and southward into Mexico and Central America. *Paspalidium geminatum* has approximately the same range but is more frequent. Although *Paspalum*-like in the general appearance of the inflorescence, the two species of *Paspalidium* were referred to *Panicum* by Hitchcock (1951).

79. *Panicum* L.

Annuals and perennials of extremely diverse habit. Ligule a membrane or a ring of hairs. Inflorescence an open or contracted panicle. Disarticulation below the glumes. Glumes usually both present, the first commonly short. Lowermost floret sterile or occasionally staminate, with a lemma similar to the glumes in texture and usually as long as, or slightly longer than, the second glume. Lemma of the upper floret shiny and glabrous in our species, smooth, firm or indurate, tightly clasping the palea with thick, inrolled margins. Palea of the upper floret like the lemma in texture. Basic chromosome number, $x = 9$ and 10.

Panicum is the largest of the grass genera, with some 475 species widely distributed throughout the warmer parts of the world. Hitchcock (1951) listed 170 species for the United States, grouping these in the three subgenera *Paurochaetium*, *Dichanthelium*, and *Eupanicum*. Following the monographic treatments of *Setaria* by Rominger (1962) and *Dichanthelium* by Gould and Clark (1978), the species included by Hitchcock (1951) in the subgenera *Paurochaetium* and *Dichanthelium* are transferred to *Setaria* and *Dichanthelium*, respectively. In addition, the species of *Panicum* in the group "Fasciculata" of Hitchcock and Chase (1910) are transferred to *Brachiaria* (Parodi, 1969; Blake, 1969). Also, the species in group "Gymnocarpa" is moved to the genus *Phanopyrum* (*P. gymnocarpon* Nash ex Small), and *P. hians* Ell. in the group "Laxa" is transferred to the genus *Steinchisma* (*S. hians* [Ell.] Nash ex Small).

Subgenus *Eupanicum* includes about 43 species of both annuals and perennials. For the most part, the annuals are plants of the semiarid Southwest or weedy grasses of gardens, roadsides, and other disturbed sites. Outstanding among the perennials is *P. virgatum* L., SWITCHGRASS (Fig. 5-43), which has been reported from all regions of the United States except the Northwest and California. In its most common form, *P. virgatum* is a tall, robust bunchgrass, adapted to low prairie sites, riverbanks, and swale areas. It is rated as a good forage grass when utilized in the earlier stages of growth and also makes good hay. *Panicum bulbosum* H.B.K., BULB PANICUM, is an important forage grass of the southwestern mountain ranges. This species

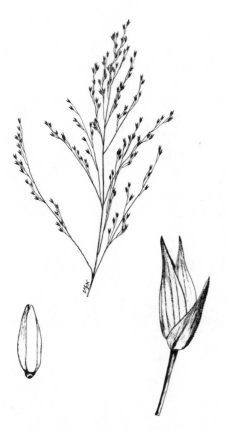

Fig. 5-43. *Panicum virgatum.* Inflorescence, spikelet, and fertile floret. (From
Gould and Box, 1965.)

develops large, cormlike swellings at the base of the culms (Fig. 5-44). At
low altitudes in the Southwest, the tufted perennial *P. hallii* Vasey, HALL'S
PANICUM, and the stoloniferous *P. obtusum* H.B.K., VINE MESQUITE
(Fig. 5-45), are valuable forage grasses, the first on well-drained, calcareous
soils and the second mostly in low, poorly drained situations. *Panicum hemitomon* Schult., MAIDENCANE, is present in moist sites along the Atlantic
coastal plain from New Jersey to Texas. This is a good forage species, and
because of the extensive rhizome system developed, it is important as a soil
binder in some areas. Many large, coarse *Panicum* species are important
tropical forage plants, and some have been introduced with considerable

Fig. 5-44. *Panicum bulbosum*. Plant, spikelet, and fertile floret. (From Gould, 1951.)

success in the southern states. Among these are *P. maximum* Jacq., GUINEA GRASS, and *P. purpurascens* Raddi, PARA GRASS. Reference: Fairbrothers, 1953; Hsu, 1965.

80. *Dichanthelium* (Hitchc. & Chase) Gould

Perennials, typically tufted and usually forming early in the growing season a basal rosette of shorter, broader blades than those of the culms. Ligule usually a ring of hairs, infrequently a short membrane or absent. Culm blades narrow or broad, usually flat. Secondary branching after first culm growth and elongation often resulting in densely clustered fascicles of much-reduced leaves and inflorescences. Inflorescences typically a small panicle with spreading or occasionally contracted branches, the spikelets short- or long-pediceled. Disarticulation below the glumes. Glumes both present, but the lower most often greatly reduced. Lower floret usually neuter but staminate in a few species. Upper floret perfect, with a shiny, glabrous, coriaceous lemma and palea. The lemma tightly inrolled over the palea. Basic chromosome number $x = 9$.

A genus of about 40 species predominantly found in the eastern United States and Canada, Mexico, the Antilles, and Central America, and with a few in South America (Gould 1980). Six species also have been reported from Hawaii (Clark and Gould, 1978). In their monograph of *Panicum* (1910), Hitchcock and Chase included 109 species within the subgenus *Dichanthelium*, and Hitchcock (1951) recognized 111 species for the United States. Gould and Clark (1978) reduced the number of taxa in *Dichanthelium* to 26 species and 19 varieties in the United States and Canada. Many of the species are widespread throughout the eastern half of the United States,

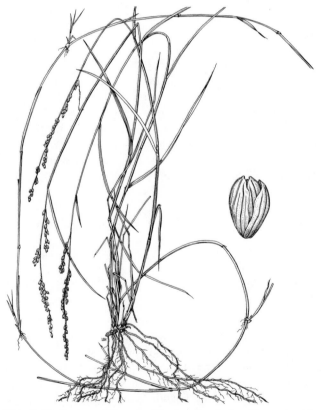

Fig. 5-45. *Panicum obtusum*. Plant and spikelet. (From Gould, 1951.)

and the center of diversity appears to be in southeastern North America. Perhaps the most common species is *D. oligosanthes* (Schult.) Gould which occurs almost throughout the United States. Also frequent in the eastern states is *D. sphaerocarpon* (Ell.) Gould, ROUNDSEED DICHANTHE-LIUM (Fig. 5-46).

Although closely related to *Panicum*, *Dichanthelium* comprises a closely interrelated series of taxa which possess numerous characteristics to delimit the group from *Panicum* (Gould and Clark, 1978). Cytologically, the genus is unique among the larger panicoid genera in the relatively low incidence of polyploidy. Of the twenty-six species in the United States, all but three for which counts have been made are diploid ($2n = 18$). The three polyploid taxa are all tetraploid ($2n = 36$). Characteristically, *Dichanthelium* species flower early in the spring and again in the summer or autumn. Little seed is set in the spring or vernal inflorescences. Most seed apparently is produced in cleistogamous spikelets of the later-developed and reduced lateral inflorescences. References: Brown and Smith, 1975; Clark and Gould, 1975; Spellenberg, 1975a, 1975b.

81. *Steinchisma* Raf.

Perennial with slender, tufted culms which are often freely branching. Ligule a fringed scale. Panicles small, few-flowered, variable in branching and general aspect, with spikelets clustered along erect-spreading primary and short secondary branches. Spikelets glabrous and at first broadly oblong but later widely gaping at the apex between the florets. Glumes both present, the lower most reduced. Lower floret neuter, the palea becoming inflated, obovate, apiculate, and much broader than the lemma. Lemma of the upper floret smooth but not shining. Basic chromosome number, $x = 10$.

A genus of about four species, one in the United States and the remainder in the Caribbean, Mexico, or Brazil. *Steinchisma hians* (Ell.) Nash ex Small, GAPING STEINCHISMA (Fig. 5-47), occurs throughout the eastern United States from Virginia to Texas and ranges into Mexico. Hitchcock (1951) included New Mexico in the distribution of this species (listed as *Panicum hians* Ell.).

Hitchcock and Chase (1910) included *Steinchisma hians* in the informal group "Laxa" along with twelve other *Panicum* species with an inflated lower palea. Brown (1977), however, warranted segregation of the genus from *Panicum* based on differences in inflorescence type, epidermal characters of the upper palea, and habitat. Brown (1977) and Brown and Brown (1975) reported that *S. hians* and *S. exiguiflorum* (Griseb.) Brown are intermediate between the C_3 and C_4 photosynthetic pathways in numerous characters and concluded that these taxa may be in some middle stage of evolving from non-Kranz to Kranz.

Fig. 5-46. *Dichanthelium sphaerocarpon* var. *sphaerocarpon.*
Plant and spikelet. (From Gould and Box, 1965, as *Panicum
sphaerocarpon.*)

Fig. 5-47. *Steinchisma hians*. Plant and spikelet. (From Gould and Box, 1965, as *Panicum hians*.)

82. *Phanopyrum* Nash

Stout perennial with erect culms from creeping stoloniferous bases, these rooting freely at the nodes. Ligule a membrane. Blades long, broad, and flat with auriculate or cordate bases. Panicles large with numerous stiffly erect-spreading primary branches, with small clusters of spikelets on short pedicels, many on short, appressed lateral branches. Glumes both present, the lowermost only slightly shorter than the second. These, along with the lemma of the lower floret, more or less dorsally compressed. Lower floret neuter with a palea shorter than the lemma. Lemma and palea of the upper floret narrowly oblong, much shorter than the glumes and lower lemma. Basic chromosome number $x = 10$ (R. D. Webster, personal communication).

A genus of a single species, this, *Phanopyrum gymnocarpon* (Ell.) Nash ex Small (Fig. 5-48), distributed in wet woodlands or forest habitats from South Carolina to Texas and southern Arkansas. Hitchcock and Chase (1910) placed this species in the informal grouping "Gymnocarpa" of *Panicum*; however, generic rank is based on differences in panicle type, frequent dorsal compression of the glumes and lemma of the lower floret, and habitat. This genus appears closely related to the genus *Ichnanthus* of South America.

83. *Lasiacis* (Griseb.) Hitchc.

Strong perennials with stout, woody culms, these frequently climbing over shrubs and low branches of trees. Inflorescence a small, few-flowered panicle. Spikelets relatively large, subglobose, oriented obliquely on the pedicels, disarticulating below the glumes. First glume present, short. Second glume and lemma of the sterile floret about equal, broad, thin, shiny, usually more or less beaked and often puberulent at the tip. Lemma and palea of the fertile floret indurate, obtuse, with a tuft of wooly hair from a slight depression at the tip. Basic chromosome number, $x = 9$.

Species about twenty-five, in the American tropics; one, *Lasiacis divaricata* (L.) Hitchc., extending northward into southern Florida, where it grows along the border of woodlands at low altitudes. Reference: Davidse, 1978.

84. *Oplismenus* Beauv.

Low, usually prostrate and creeping annuals and perennials with culms freely branching below and short, thin, flat leaf blades. Inflorescences small, few-flowered, with spikelets subsessile and crowded on 2 to several short, distant, spicate branches. Disarticulation below the glumes. Glumes about equal, awned from a minutely notched apex, the awn of the first glume considerably longer than that of the second. Lemma of sterile floret about as long as the second glume, mucronate or short-awned. Lemma of fertile flo-

Fig. 5-48. *Phanopyrum gymnocarpon*. Inflorescence, spikelet, and fertile floret. (From Gould, 1975, as *Panicum gymnocarpon*.)

ret smooth, indurate, elliptic, acute at the apex, the margins enclosing the palea. Basic chromosome number, $x = 9$.

A genus of about ten species, in the tropics of both hemispheres, none native to the United States. *Oplismenus setarius* (Lam.) Roem. & Schult. (Fig. 5-49) and the closely related *O. hirtellus* (L.) Beauv. have been introduced or are adventive in the United States. The former is reported from several localities in the Southeast, the latter from southern Texas.

Fig. 5-49. *Oplismenus setarius*. Plant and spikelet. (From Gould and Box, 1965.)

85. *Echinochloa* Beauv.

Coarse annuals and perennials, usually with weak, succulent culms and broad, flat blades. Ligule a ring of hairs or absent. Inflorescence a contracted or moderately open panicle, with few to numerous, simple or rebranched, densely flowered branches. Spikelets subsessile, in irregular fascicles or regular rows, disarticulating below the glumes. Glumes and lemma of sterile floret usually with stout spicules and long or short hairs, these often glandular-pustulate at the base. First glume well developed but much shorter than the second, acute or slightly awned. Second glume and lemma of the sterile floret about equal, acute, short-awned or with a long, flexuous awn. Lemma of the fertile floret indurate, smooth and shiny, with inrolled margins and usually an abruptly pointed apex. Palea of the fertile floret similar to the lemma in texture, broad but narrowing to a pointed tip that is free from the lemma margins. Basic chromosome number, $x = 9$.

Fig. 5-50. *Echinochloa crusgalli*. Inflorescence and spikelet. (From Gould and Box, 1965.)

Species about twenty, throughout the warmer regions of the world. Hitchcock's manual (1951) lists six species for the United States, but the number of species and varieties in this country is a controversial matter. Widespread and weedy on moist, disturbed sites throughout the United States are populations of the European species *Echinochloa crusgalli* (L.) Beauv., BARNYARD GRASS (Fig. 5-50), and the closely related native species *E. muricata* (Beauv.) Fern. The latter is similar to the former in most characteristics and was not recognized as taxonomically distinct by Hitchcock (1935, 1951). Common throughout the southern United States as a

weed of flower beds and other cultivated areas is the adventive species
E. colonum (L.) Link, JUNGLERICE. References: Ali, 1967; Fairbrothers,
1952; Gould et al., 1972; Shinners, 1954; Wiegand, 1921.

86. *Sacciolepis* Nash

Medium-sized to tall annuals and perennials, with rather weak culms and
thin, flat blades. Inflorescence a contracted and slender, usually long and
spikelike panicle. Spikelets slender, broad and gibbous at the base, disarticulating below the glumes. First glume short, acute, usually much narrower than the spikelet. Second glume broad, inflated-saccate at the base,

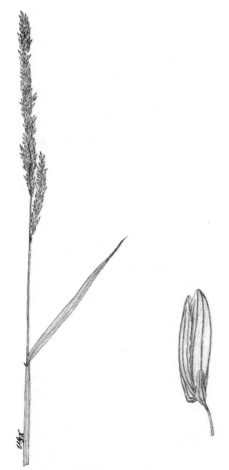

Fig. 5-51. *Sacciolepis striata.* Inflorescence and spikelet.

strongly several-nerved, awnless. Lemma of the sterile floret narrow, usually 3- or 5-nerved, awnless, about as long as the second glume. Lemma and palea of the fertile floret smooth, indurate, rounded at the apex, much shorter than the second glume and lemma of the sterile floret. Basic chromosome number, $x = 9$.

A genus of about thirty species in the tropics and subtropics of both hemispheres. Only one species, *Sacciolepis striata* (L.) Nash, AMERICAN CUPSCALE (Fig. 5-51), is native to the United States. This tall perennial is occasional in moist or marshy habitats throughout the Southeast, ranging westward to Texas and Oklahoma.

87. *Rhynchelytrum* Nees

Annuals and perennials, mostly with erect, moderately tall culms. Ligule a ring of hairs. Blades thin and flat. Inflorescence an open or contracted much-branched panicle, the spikelets on slender pedicels. Spikelets laterally compressed, disarticulating below the glumes. First glume minute to about one-third as long as the second. Second glume and lemma of the lower floret about equal, usually silky-villous with fine hairs, tapering to a notched, usually awned apex. Lower floret usually staminate, with a well-developed palea. Lemma of the fertile floret much shorter than the second glume and lemma of the lower floret, slender, membranous, glabrous, narrowly rounded at the apex, the margins thin and not inrolled over the palea. Basic chromosome number, $x = 9$.

A genus of thirty-five to forty species, mostly African. One species, *Rhynchelytrum repens* (Willd.) C. E. Hubb., NATAL GRASS (Fig. 5-52), a native of Africa, now is frequent in the American tropics and occasional as far north as Florida, Texas, and Arizona. This grass is a cespitose perennial with a showy panicle, the silky-villous spikelets first rosy purple and then fading to pink and white. Natal grass appears relatively unpalatable to livestock in Mexico, where it is frequently abundant on disturbed soils, especially along roadsides.

88. *Setaria* Beauv.

Cespitose annuals and perennials, with erect or geniculate culms, these often branching at the base. Leaf blades typically flat and thin, in subgenus *Ptychophyllum* the blades plicate and the leaves with a petiole-like constriction between the sheath and blade. Inflorescence a slender, usually contracted and densely flowered, bristly panicle, the spikelets subsessile on the main axis and on short branches. Some or all of the spikelets subtended by 1 to several persistent bristles (reduced branches), the spikelets disarticulating above the bristles. Glumes and lemma of sterile floret typically glabrous, prominently nerved, acute or obtuse, the first glume short, the second

Fig. 5-52. *Rhynchelytrum repens.* Inflorescence and spikelet.

glume and lemma of the sterile floret equal, or more frequently the second
glume one-half to two-thirds as long. Lemma and palea of the fertile floret
indurate, rounded at the apex, usually finely or coarsely transverse-rugose.
Basic chromosome number, $x = 9$.

A genus of 125 species, mostly in tropical Africa but some in the warmer
parts of all the continents. In a monographic treatment of the North American species, Rominger (1962) listed 43 species and four varieties, about half
of which occur in the Unitd States. Only 12 species are native to the United
States. Rominger grouped the North American taxa into three subgenera:

Ptychophyllum, with 6 species; *Paurochaetium*, with 10 species; and *Setaria*, with 27 species and four varieties. Species of subgenus *Paurochaetium*, including *S. reverchonii* (Vasey) Pilger (Fig. 5-53), were referred to *Panicum* by Hitchcock. Many of the U.S. species are weedy, but the perennial bunch-

Fig. 5-53. *Setaria reverchonii*. Plant and spikelet.
(From Gould and Box, 1965.)

grass *S. leucopila* (Scribn. & Merr.) K. Schum. (*S. macrostachya* H.B.K. of Hitchcock [1951] for the most part) is an important southwestern forage grass. *Setaria geniculata* (Lam.) Beauv., KNOTROOT BRISTLEGRASS (Fig. 5-54), a weedy perennial with short, knotty rhizomes, is common in moist soils throughout the southern and eastern portions of the country and southward through Mexico. References: Emery, 1957a, 1957b.

Fig. 5-54. *Setaria geniculata.* Inflorescence and spikelet. (From Gould and Box, 1965.)

89. *Pennisetum* L. Rich.

Annuals and perennials with usually tall, erect culms. One species with stout, creeping stolons. Leaf blades usually thin and flat. Inflorescence typically a dense, bristly, tightly contracted panicle, the spikelets solitary or in groups of 2 to several in involucres of numerous bristles. Spikelets falling together with the involucre. First glume small or vestigial. Second glume and lemma of the sterile floret about equal or the glume shorter. Lemma and palea of the fertile floret firm, smooth and shiny, the margins of the lemma thin and flat. Basic chromosome number, $x = 9$ (and 7?).

A genus of about eighty species in the tropics and subtropics of both hemispheres. *Pennisetum setosum* (Swartz) L. Rich., a large, coarse, cespitose perennial, ranges from southern Florida to Brazil and Bolivia. *Pennisetum purpureum* Schumach., NAPIER GRASS or ELEPHANT GRASS, a robust perennial with culms 2 to 4 meters tall, has been used in the South as a cultivated forage plant. This grass is native to Africa but is extensively grown in Central America for forage. *Pennisetum clandestinum* Hochst ex Chiov., KIKUYU GRASS, a low, sod-forming perennial with stout, creeping rhizomes, is an important pasture grass in Central America. Native to Africa, it is reported as a weed of orchards and gardens in southern California,

Fig. 5-55. *Pennisetum villosum*. Inflorescence and spikelet.

where it is planted for lawns. *Pennisetum villosum* R. Br., FEATHERTOP (Fig. 5-55), has been introduced into the United States as an ornamental and grows as an escape in widely scattered localities. The *Pennisetum* of most economic importance is *P. spicatum* (L.) Koenicke (*P. glaucum* [L.] R. Br.), PEARL MILLET, a robust annual that is extensively cultivated in the tropics for human food. Trial plantings of this grass have been made in southern United States.

90. *Cenchrus* L.

Annuals and perennials, mostly with weak, geniculate-decumbent culms and soft, flat blades but a few with tall, coarse, stiffly erect culms. Ligule a fringe of hairs. Spikelets enclosed in burrs, these sessile or subsessile on a short, stout rachis. Burrs of bristles and/or flattened spines (modified branches) fused together at least at the base. Bristles and spines usually retrorsely barbed. Spikelets 1 to 8 in each burr, this readily disarticulating at maturity. Glumes thin, membranous, unequal. First glume 1- or 3-nerved, the second 1- to 7-nerved. Lemma of the sterile floret thin, 1- to 7-nerved, equaling or exceeding the second glume. Palea of the sterile floret about equaling the lemma. Lemma of the fertile floret thin, membranous, 5- or 7-nerved, tapering to a slender, usually acuminate tip, the margins not inrolled. Caryopsis elliptic to ovoid, dorsally flattened. Basic chromosome number, $x = 9$.

In a monographic treatment of the genus, DeLisle (1963) recognized

Fig. 5-56. *Cenchrus incertus*. Plant, burr, and spikelet. (From Gould and Box, 1965.)

twenty species, these mostly American but a few in the warmer regions of Africa, southern Europe, Asia, Australia, and the South Pacific. All but one of the six species of *Cenchrus* native to the United States are weedy annuals or short-lived perennials of open, sandy, usually disturbed sites. They are commonly known as "sandburs." *Cenchrus incertus* M. A. Curtis (Fig. 5-56) is a widespread annual or biennial that is interpreted by DeLisle as including populations referred to *C. pauciflorus* Benth. by Hitchcock (1951). The distribution of this variable sandbur as stated by DeLisle is throughout the southern United States from North Carolina west to California and south through Mexico and Central and South America, and in the West Indies. *Cenchrus myosuroides* H.B.K., BIG SANDBUR (Fig. 5-57), is a cespitose perennial with stout culms to 1.5 meters tall. It is present on sandy soils along the eastern coastal plain from Georgia and Florida to southern Louisiana and southern Texas and ranges southward through Mexico into Central America and the Caribbean Islands, and in South America to central Argentina. Frequently seeded in southern Texas as a pasture grass and now well

Fig. 5-57. *Cenchrus myosuroides.* Inflorescence burr of 1 spikelet surrounded by bristles. (From Gould and Box, 1965.)

established throughout the area is *C. ciliaris* L., BUFFELGRASS (Fig. 5-58). This species was referred to *Pennisetum* (*P. ciliare* [L.] Link) in Hitchcock's manual (1951).

91. *Amphicarpum* Kunth

One annual and one perennial, both with moderately tall culms, flat blades, and narrow panicles. Spikelets of two kinds, those of the terminal panicle and large, subterranean, cleistogamous spikelets borne on slender, leafless rhizomes. The aerial spikelets reportedly do not mature seeds, but the cleistogamous spikelets are fruitful. First glume of the aerial spikelets small or absent. Second glume and lemma of the sterile floret about equal, firm, acute. Lemma of the fertile floret narrow, smooth, acute, with thin, flat margins. Palea like the lemma in texture. First glume of the subterranean spikelet absent, the second glume and lemma of the sterile floret subequal, firm, many-nerved. Lemma of the fertile floret indurate, longer than the second glume, abruptly tapering to a stout, beaked apex, the margins thin and flat. Stamens with small anthers on short filaments. Basic chromosome number, $x = 9$.

Fig. 5-58. *Cenchrus ciliaris*. Inflorescence, burr, and spikelet.
(From Gould and Box, 1965.)

A North American genus of two species, both in sandy soils of pine
woodlands along the Atlantic coastal plain. *Amphicarpum purshii* Kunth, an
annual with hirsute leaves, is distributed from New Jersey to Georgia, and
A. muhlenbergianum (Schult.) Hitchc., a perennial with essentially glabrous
culms and leaves, is restricted to South Carolina and Florida.

92. *Melinis* Beauv.

Annuals and perennials, many with slender, weak, branching culms and
glandular, hairy foliage. Leaf blades flat, usually soft. Ligule a fringe of hairs.
Inflorescence a contracted or open but narrow panicle. Spikelets on slender
pedicels, awned or awnless, glabrous or hairy. Disarticulation below the
glumes. First glume a minute scale or rim. Second glume and lemma of ster-

ile floret equal, the glume 5- or 7-nerved, notched or 2-lobed at the apex, awnless or short-awned, the lemma 3- or 5-nerved (rarely 7-nerved), narrow, deeply lobed, awnless or more frequently with a fine, hairlike awn at the apex. Lemma and palea of the fertile floret thin and membranous. Basic chromosome number, $x = 9$.

A genus of about twenty species, all but one restricted to Africa. *Melinis minutiflora* Beauv., MOLASSESGRASS, a cespitose perennial, appears to be native to tropical America as well as Africa. It is widely used in Central and South America as a pasture grass and also has been tried with some success in southern Florida. The common name "molassesgrass" comes from the strong-scented, glandular-viscid leaves. Although not especially palatable, this grass provides good forage when taken by livestock.

93. *Anthephora* Schreb.

Cespitose or rhizomatous annuals and perennials, with flat blades and well-developed membranous ligules. Inflorescence a spikelike raceme, the spikelets short-pediceled or subsessile, in deciduous clusters of 4 to 9. Lower glumes of a cluster indurate and fused together at the base to form an involucre. Second glume shorter than the first, thin, 1-nerved, often reduced to little more than the nerve. Lemma of the sterile floret 3- to 7-nerved, thin, membranous, hairy with pustulate-based hairs, awnless. Upper floret bisexual or staminate, the lemma firm but not indurate, awnless. Basic chromosome number, $x = 9$.

A genus of six species, five in Africa and one, *Anthephora hermaphrodita* (L.) Kuntze, in the American tropics. *Anthephora hermaphrodita* is a low, rather coarse, weedy annual. It has been introduced into southern Florida as an escape from experimental grass plots.

Tribe 11. Andropogoneae

94. *Imperata* Cyrillo

Coarse, mostly tall perennials with stout, creeping rhizomes. Inflorescence a dense, contracted, narrowly oblong or slender and spikelike panicle, the spikelets more or less obscured by silky hairs. Spikelets all alike, awnless, with a tuft of long, silky hair at the base, in unequally pediceled pairs on a continuous branch rachis. Disarticulation below the spikelet, the long-hairy callus falling with the spikelet. Glumes subequal, thin, several-nerved. Lemmas of the sterile and fertile florets fertile, and palea of the fertile floret membranous and hyaline. Basic chromosome number, $x = 10$.

Species about seven, in the warmer parts of both hemispheres. *Imperata brasiliensis* Trin. is present on low prairies and woods borders in Alabama and Florida; it also grows in Central and South America. *Imperata brevifolia* Vasey, SATINTAIL (Fig. 5-59), is a large cespitose perennial of semi-

Fig. 5-59. *Imperata brevifolia*. Inflorescence and spikelet.

arid regions from Utah and Nevada to western Texas, California, and northern Mexico. A third species, *I. cylindrica* (L.) Beauv., has been introduced in Florida and Oregon. This grass is widespread in the temperate and subtropical regions of the Old World.

95. *Miscanthus* Anderss.

Tall, coarse perennials, with long, flat blades and flabellate, silky-pubescent panicles of numerous, crowded, closely flowered branches. Spikelets all alike and perfect, unequally pediceled in pairs along a continuous rachis, with a tuft of long, silky hair at the base of the glumes and on the callus. Disarticulation below the callus. Glumes subequal, mostly thin but firm, awnless. Lemma of the sterile floret membranous, awnless, slightly shorter than the glumes. Lemma of the fertile floret membranous, hyaline, awnless or with a geniculate and twisted awn from a bifid, toothed apex. Palea of the fertile floret small, membranous. Basic chromosome number, $x = 9$ (most records $2n = 38$).

Species about eight, mostly native to southeastern Asia. Three species, *M. sinensis* Anderss., *M. sacchariflorus* (Maxim.) Hack., and *M. nepalensis* (Trin.) Hack., are occasionally cultivated in the United States as ornamentals.

96. *Saccharum* L.

Tall, coarse perennials, mostly with culms in dense clumps from short, stout rhizomes. Leaves large, the blades flat and broad. Inflorescence a large, densely flowered, plumose panicle as much as 60 cm in length. Spikelets all alike and perfect, in pairs of 1 sessile and 1 pediceled, with a tuft of long, silky hair at the base, disarticulating below the glumes or the sessile spikelet falling attached to the pedicel and a section of the rachis. Glumes large and firm. Lemma of the sterile floret and lemma and palea of the fertile floret hyaline, sometimes absent. Basic chromosome number, $x = 10$.

Species about twelve, mainly in the Old World tropics. *Saccharum officinarum* L., SUGAR CANE, is cultivated in Louisiana and other southern states for the production of molasses and sugar. This species and its hybrid derivatives supply about two-thirds of the world's commercial sugar.

97. *Erianthus* Michx.

Stout perennials with culms mostly 1 to 4 meters tall. Leaves large, the blades flat and broad. Inflorescence usually a large, dense panicle. Spikelets all alike and perfect, in pairs of 1 sessile and 1 pediceled, the rachis section and pedicel falling attached to the sessile spikelet. Glumes large, firm, subequal, usually with a tuft of long hair at the base. Lemma of the sterile floret and lemma and palea of the fertile floret hyaline. Lemma of the fertile floret with a long, straight or loosely twisted awn. Basic chromosome number, $x = 10$.

Species about twenty, in the temperate-subtropical regions of the world; six in the United States. Our species are restricted mainly to the eastern and southeastern states, where they grow in moist or marshy soils. *Erianthus giganteus* (Walt.) Muhl., SUGARCANE PLUMEGRASS, ranges from New York to Florida and Texas. With about the same ranges are *E. contortus* Baldw. ex Ell., BENTAWN PLUMEGRASS; *E. alopecuroides* (L.) Ell., SILVER PLUMEGRASS; *E. strictus* Baldw., NARROW PLUME-GRASS; and *E. coarctatus* Fern. *Erianthus brevibarbis* Michx. is a relatively rare species reported from Illinois and Arkansas.

98. *Sorghum* Moench

Annuals and perennials, many with tall, stout culms. Blades long, flat, narrow or broad. Inflorescence a large, open or contracted panicle, the spikelets clustered on short racemose branchlets. Spikelets in threes at the branchlet tips, 1 sessile and fertile and 2 pediceled and staminate or neuter, and below the tips in pairs of 1 sessile and perfect and 1 pediceled and reduced. Disarticulation below the sessile spikelet, the rachis section and the pedicel or pedicels falling attached to the sessile spikelet. Glumes coriaceous, awnless, about equal in length. Lemma of the sterile floret and lemma and palea of the fertile floret membranous, the lemma of the fertile floret usually with a geniculate and twisted awn, this readily deciduous in *Sorghum halepense*. Basic chromosome number, $x = 5$.

A genus of about thirty-five species, mostly in the warmer parts of Africa, two species native to Mexico and Central America. *Sorghum* is represented in the United States by two introduced species, the cultivated annual crop plant, *S. vulgare* Pers., SORGHUM, and the rhizomatous perennial forage grass, *S. halepense* (L.) Pers., JOHNSONGRASS (Fig. 5-60). *Sorghum vulgare* has been cultivated from prehistoric times, and the list of important varieties and races includes hegari, sorgo, kafir, milo, broomcorn, and durra. *Sorghum halepense* has become widely established as a pasture grass and weed of roadsides and cultivated fields throughout the southern and eastern states. It has been reported from as far north as Maine, Michigan, Iowa, and Wyoming, but probably does not persist in the cooler regions. Although widely grown for forage, *S. halepense* develops sufficient cyanogenetic compounds under certain conditions of growth to cause prussic acid poisoning in grazing animals. Reference: de Wet, 1978.

99. *Sorghastrum* Nash

Slender, rather tall perennials with flat blades and narrow, contracted panicles. Spikelets basically in pairs of 1 sessile and perfect and 1 pediceled and rudimentary, but the pediceled spikelets usually completely reduced and represented only by the slender, hairy pedicel. Disarticulation in the rachis, the sessile spikelet falling attached to a section of the rachis and pedicel or,

Fig. 5-60. *Sorghum halepense*. Plant in natural habitat.

at the branch tips, to 2 pedicels. Glumes firm, subequal, hirsute, awnless. Lemma of the sterile floret absent or membranous and rudimentary. Lemma of the fertile floret membranous, with a stout, geniculate, and twisted awn. Basic chromosome number, $x = 10$.

A genus of about fifteen species, in the warmer regions of Africa and the Americas. *Sorghastrum* is represented in the United States by three species: *S. nutans* (L.) Nash,* INDIANGRASS (Fig. 5-61); *S. elliottii* (Mohr) Nash; and *S. secundum* (Ell.) Nash. *Sorghastrum nutans* is widespread in the United States east of the Rocky Mountains. It is an important constituent of the western tall-grass prairies and is a good range forage plant. The other two U.S. species of *Sorghastrum* are present only in the southeastern states, growing for the most part in sandy woods openings.

100. *Andropogon* L.

Cespitose perennials with usually stiffly erect culms, rounded or flattened and keeled sheaths, and flat or folded blades. Ligule membranous. Flower-

---

*Baum (1967) presented evidence that the correct name for this species is *S. avenaceum* (Michx.) Nash.

ing culm much-branched and "broomlike" in some species, unbranched or little-branched above the base in others. Each culm or branch terminated by an inflorescence of 2 to several racemose branches. Sessile spikelet of a pair fertile. Pediceled spikelet well developed, rudimentary, or absent. Disarticulation in the rachis, the sessile spikelets falling attached to the associated pedicel and section of the rachis. Glumes large, firm, awnless. Lemmas of the sterile and fertile florets membranous, the lemma of the fertile floret awned or awnless. Basic chromosome numbers, $x = 9$ and 10.

Fig. 5-61. *Sorghastrum nutans*. Plant and spikelet, the spikelet falling attached to the hairy pedicel of the vestigial or entirely reduced pediceled spikelet. (From Gould and Box, 1965.)

A relatively large group of grasses, represented in the warmer regions of all continents. Some seventeen species occur in the United States, these mostly in the Southeast. *Andropogon gerardii* Vitman, BIG BLUESTEM (Fig. 5-62), ranges from central Mexico to Canada and is one of the most

Fig. 5-62. *Andropogon gerardii.* Inflorescence, spikelet pair, and spikelet. (From Gould, 1951.)

important forage species of the tall-grass prairie regions. *Andropogon virginicus* L., BROOMSEDGE BLUESTEM, and the closely related *A. glomeratus* (Walt.) B.S.P., BUSHY BEARDGRASS (Fig. 5-63), are frequent and commonly associated together on sterile, sandy soils of the southern and southeastern states. Both are coarse bunchgrasses of low palatability for livestock.

As interpreted by Hitchcock (1935, 1951), this genus includes the grasses herein referred to *Dichanthium, Bothriochloa,* and *Schizachyrium.* Despite the exclusion of these groups from *Andropogon,* the genus remains a heterogeneous aggregation of not too closely related species groups (Gould, 1967). *Andropogon distachyus* L., the type species, is from the Mediterranean region. This and related Mediterranean-African species are somewhat similar in general aspect to *A. gerardii,* but are very unlike the *A. virginicus–A. glomeratus* group represented in the United States by a dozen or more species. In most, perhaps all, of the latter group, the stamens are reduced to one per spikelet. *Andropogon ternarius* Michx., SPLIT-BEARD BLUESTEM, of open woodland sites in the southern and southeastern states, represents a third series of U.S. *Andropogon* species. The closest relationships of this and similar species are probably with the *A. virginicus*–*A. glomeratus* group.

Fig. 5-63. *Andropogon glomeratus*. Portion of plant showing the densely congested terminal flowering branches, each ultimate branch representing a reduced lateral shoot.

101. *Arthraxon* Beauv.

Low, creeping or decumbent annuals with broad, flat, mostly short, cordate-clasping blades. Inflorescence of few (rarely 1) to several short, sparingly rebranched, spicate or racemose branches, these scattered or digitate at the culm apex. Pediceled spikelet of a pair completely reduced and absent, the pedicels also usually absent except near the base of the inflorescence. Rachis disarticulating at the base of the sessile spikelet. Glumes subequal, the first firm, acute, the second acute or acuminate with membranous margins. Lemma of the fertile floret hyaline, entire or minutely notched, usually awned from the back. Basic chromosome number, $x = 9$.

A genus of about twenty species, these native to the Old World tropics and with a few introduced or adventive in the Americas. *Arthraxon hispidus* (Thunb.) Makino, native to Asia, has been reported as adventive at several locations in the eastern states from Maryland and Pennsylvania to Florida and Louisiana, and also in Washington.

102. *Microstegium* Nees

Annuals with weak, usually decumbent-spreading culms. Inflorescence of few to several short, spreading, racemose branches, these widely spaced or subdigitate on the main culm. Spikelets all alike and fertile, in pairs of 1 sessile and 1 pediceled or both unequally pediceled. Disarticulation first in the nodes of the rachis and secondarily at the base of the spikelets. Glumes equal, usually firm but thin. Lemma of the sterile floret reduced or absent.

Lemma of the fertile floret hyaline, awnless or awned from an entire or notched apex. Palea of fertile floret small or absent. Lodicules 2, cuneate. Stamens 3 or 2, the anthers small. Basic chromosome number, $x = 10$.

A small genus with most species in Asia and the East Indies and a few in Africa. *Microstegium vimineum* (Trin.) A. Camus is adventive in the United States and has been reported from Ohio, Virginia, North Carolina, Kentucky, Tennessee, and Alabama.

103. *Dichanthium* Willemet

Low to moderately tall perennials, mostly cespitose but some with extensive creeping stolons. Inflorescence as in *Bothriochloa*, but pedicels and internodes of the rachis flat or rounded, without a groove or membranous central area, and lower pair of spikelets of the inflorescence branches usually both sterile and awnless. Basic chromosome number, $x = 10$.

Fig. 5-64. *Dichanthium annulatum*. Inflorescence and spikelet pair. (From Gould and Box, 1965.)

A small genus of Asiatic, Australian, and African species, closely related to *Bothriochloa* and differing in the characters mentioned above. Three species, *Dichanthium annulatum* (Forsk.) Stapf, KLEBERG BLUESTEM (Fig. 5-64); *D. aristatum* (Poir.) C. E. Hubb., ANGLETON BLUESTEM; and *D. sericeum* (R. Br.) Camus, SILKY BLUESTEM, have been introduced as pasture grasses in Texas and elsewhere in the South. All three occur as casual weeds along roads and ditches in southern Texas. References: Celarier et al., 1961; Swallen, 1950.

104. *Bothriochloa* Kuntze

Cespitose perennials with erect or decumbent-spreading culms and flat blades. Inflorescence a terminal panicle, the spikelets on few to several spicate primary branches, these sparingly rebranched in a few species. Pedicels and upper rachis internodes with a central groove or membranous area. Sessile spikelets of a pair fertile and awned, more or less triangular in outline, the first glume dorsally flattened, the second glume with a median keel. Pediceled spikelet staminate or sterile, usually well developed. Disarticulation in the rachis, the sessile spikelet falling attached to a pedicel and section of the rachis. Basic chromosome number, $x = 10$.

A genus of about thirty species, these distributed throughout the warmer regions of all continents. Gould (1957) listed eight native and three introduced species for the United States, treating them as *Andropogon*. *Bothriochloa barbinodis* (Lag.) Herter, CANE BLUESTEM, and *B. saccharoides* (Swartz) Rydb., SILVER BLUESTEM (Fig. 5-65), are important range forage species of the Southwest. *Bothriochloa ischaemum* (L.) Keng var. *songaricus* (Rupr.) Celarier & Harlan, KING RANCH BLUESTEM, has been introduced in the South in native and cultivated pasture seedings. This grass has become well established in many areas, especially in Texas, where it is present almost throughout the state in pastures and on roadsides.

About half of the species of *Bothriochloa* regularly or irregularly have a depressed glandular area (pit) in the middle or upper portion of the first glume of the sessile spikelet. In some of the Old World species the glumes have multiple pits, and the glume of the pediceled spikelet is also pitted. The occurrence of the glandular glume pit as correlated with polyploidy and speciation in this group has been discussed by Gould (1959). References: Gould, 1953; Swallen, 1950.

105. *Chrysopogon* Trin.

Cespitose annuals and perennials with flat blades and few-flowered panicles of large spikelets borne at the tips of slender, spreading branches. Spikelets usually in threes, 1 sessile and perfect and 2 pediceled and reduced. Glumes of the fertile spikelet firm and hard, rounded on the back, awnless, with a hard, sharp-pointed, hairy callus at the base. Lemmas of the sterile and fer-

Fig. 5-65. *Bothriochloa saccharoides* var. *torreyanus* (Steud.) Gould. Inflorescence and spikelet pair. (From Gould and Box, 1965, as *Andropogon*.)

tile florets thin, membranous, the latter with a stout, usually geniculate awn. Basic chromosome number, $x = 10$.

A genus of about twenty species, all in the warmer regions of the Old World except *Chrysopogon pauciflorus* (Chapm.) Benth. ex Vasey, which is native to Florida and Cuba. *Chrysopogon pauciflorus* is a moderately tall, tufted annual with spikelets about 1.5 cm long and stout, geniculate awns to 15 cm long. It grows in sandy woods openings and open fields.

106. *Hyparrhenia* Anderss. ex Stapf

Perennials, mostly tall, coarse, and robust. Flowering culms terminating in a series of short branches, each bearing a panicle of 2 flowering branchlets, these separate or on a common peduncle and subtended by an expanded, spathelike leaf sheath. Spikelets in pairs of 1 sessile and 1 pediceled, the lowermost pairs sessile, awnless, staminate or neuter, the uppermost pairs as in *Andropogon*, with the sessile spikelet bisexual and awned and the pediceled spikelet staminate or neuter and usually awnless. In some species only one of the sessile spikelets of each branch is fertile. Disarticulation usually in the rachis, the spikelets falling singly or in clusters. Glumes of the sessile spikelet large, firm, flat or rounded, the first exceeding the second in length and thickness. Lemma of the sterile floret and lemma and palea of the fertile floret thin and membranous, the palea minute or absent. Lemma of the fertile floret usually with a stout, geniculate awn. Basic chromosome number, $x = 10$.

Species about seventy, native to the tropics of the Old World, many in Africa. *Hyparrhenia rufa* (Nees) Stapf and *H. hirta* (L.) Stapf have been introduced as forage grasses in the American tropics and also have been grown in experimental gardens in Florida, Louisiana, and Texas. Hitchcock (1951) reported that *H. rufa* grows as an escape in Florida.

Fig. 5-66. *Schizachyrium scoparium* var. *frequens*. Plant and spikelet pair. (From Gould and Box, 1965, as *Andropogon scoparius* var. *frequens*.)

107. *Schizachyrium* Nees

Cespitose perennials, some with short, stout rhizomes. Leaves with rounded or compressed and keeled sheaths, and flat or folded, infrequently terete, blades. Flowering culms much-branched, each leafy branch terminating in a single pedunculate raceme. Spikelets appressed to the rachis or somewhat divergent at maturity. Disarticulation in the rachis, the sessile spikelet falling attached to a pedicel and a section of the rachis. Sessile spikelet fertile, awned, the pediceled spikelet reduced but usually present. Glumes of the sessile spikelet large and firm. Lemmas of the sterile and fertile florets thin and membranous, the latter usually with a geniculate arm. Basic chromosome number, $x = 10$.

A relatively large genus, well represented in subtropical-tropical grasslands of the world, many in the Americas and about twelve species in the warmer portions of the United States. Hitchcock (1935, 1951) did not recognize this genus as distinct from *Andropogon*. *Schizachyrium scoparium* (Michx.) Nash, LITTLE BLUESTEM (Fig. 5-66), ranges throughout North America from Canada to Mexico and is absent in the United States only from Nevada and the Pacific coastal states. This is one of the most important of the tall-grass prairie forage species and with *Andropogon gerardii* and *Sorghastrum nutans* is dominant over much of the prairie region. *Schizachyrium scoparium* exhibits considerable variation throughout its wide range. It is most widely represented by var. *frequens* (Hubb.) Gould. The varieties *S. scoparium* var. *divergens* (Hack.) Gould, EASTERN LITTLE BLUE-STEM, and *S. scoparium* var. *littoralis* (Hitchc.) Gould, SEACOAST BLUE-STEM, were recognized as distinct species (of *Andropogon*) by Hitchcock.

108. *Eremochloa* Buese

Perennials with compressed, spikelike racemes. Pediceled spikelet greatly reduced, in some species represented only by the stiffly erect pedicel. Sessile spikelets dorsally compressed, awnless, imbricated along one side of, but not sunken in, the slender rachis.

Species about ten, native to regions of tropical and temperate Asia. *Eremochloa ophiuroides* (Munro) Hack., CENTIPEDEGRASS, has been introduced with considerable success as a lawn grass from Florida to South Carolina and also in the Gulf states. It is a low, stoloniferous perennial that forms a dense turf.

109. *Trachypogon* Nees

Moderately tall, cespitose perennials with culms terminating in a spikelike raceme or less frequently 2 to few spicate branches. Spikelets in pairs on a continuous rachis, 1 subsessile and 1 with a slightly longer pedicel. Subsessile spikelet staminate, as large as the perfect spikelet but awnless and persistent on the rachis. Longer-pediceled spikelet perfect, with a firm, rounded, several-nerved first glume; a firm, few-nerved second glume; and a thin, narrow lemma (of the fertile floret) bearing a stout, twisted, geniculate or flexuous awn. Palea of fertile floret usually absent. Basic chromosome number, $x = 10$.

A genus of about fifteen species, in tropical-subtropical regions of Africa and the Americas, represented in the United States by the single species *Trachypogon secundus* (Presl) Scribn., CRINKLEAWN (Fig. 5-67). This

Fig. 5-67. *Trachypogon secundus*. Plant and spikelet. (From Gould and Box, 1965.)

moderately tall bunchgrass is present on dry, open, rocky slopes in southern Texas, New Mexico, and Arizona. It ranges southward into Mexico and also is present in Argentina.

110. *Elyonurus* Humb. & Bonpl. ex Willd.

Cespitose perennials with slender, moderately tall culms and narrow, flat or involute blades. Inflorescence a cylindrical, spikelike raceme, with disarticulation in the rachis. Spikelets awnless, paired, the pediceled spikelets staminate, similar in size and appearance to the sessile, perfect ones. First glume of the perfect spikelet firm and moderately coriaceous, the second glume thinner. Lemmas of the sterile and fertile florets thin and hyaline, the paleas absent. Basic chromosome number, $x = 10$.

A genus of about fifteen species, in the tropical-subtropical regions of both hemispheres, represented in the United States by two species, *Elyonurus tripsacoides* Humb. & Bonpl. ex Willd., WOOLSPIKE BAL-SAMSCALE (Fig. 5-68), and *E. barbiculmis* Hack., PAN AMERICAN BAL-SAMSCALE. *Elyonurus tripsacoides* is present in coastal prairies and sandy

Fig. 5-68. *Elyonurus tripsacoides.* Inflorescence and spikelet pair. (From Gould and Box, 1965.)

pine woodlands from Georgia and Florida to eastern and southern Texas. It is relatively unpalatable to livestock. *Elyonurus barbiculmis*, with densely wooly-pubescent inflorescences, has about the same distribution as *Trachypogon secundus*, from western Texas to southern New Mexico and southern Arizona through Mexico; also in southern South America. It is a good forage grass but grows mostly in scattered stands.

111. *Heteropogon* Pers.

Tufted annuals and perennials. Blades flat or keeled and folded on the midnerve. Culms usually branched at the upper nodes, terminating in unilateral spicate racemes. Spikelets in pairs, 1 sessile and the other short-pediceled. Sessile spikelets, except for the lowermost, fertile and long-awned, with a sharp-pointed, bearded callus. Both the sessile and pediceled spikelets staminate or sterile (and awnless) on the lower portion of the raceme. Disarticulation in the rachilla, at the base of the callus of the fertile spikelets. Glumes of the fertile spikelet about equal, the first thick, indurate, and dark-colored at maturity, enclosing the second glume. Glumes of the sterile or staminate spikelets thin, the first broad, green, faintly many-nerved. Lemmas of the sterile and fertile florets membranous, the latter with a stout geniculate and twisted awn. Mature fertile spikelet superficially similar to the floret of *Stipa* in appearance. Basic chromosome number, $x = 10$.

A genus of about eight species, distributed in the warmer parts of the world but relatively few in the Americas. *Heteropogon* is represented in the United States by two species, the perennial *H. contortus* (L.) Beauv., TANGLEHEAD, and the annual *H. melanocarpus* (Ell.) Benth., SWEET TANGLEHEAD. Both species are widely distributed in the tropics and subtropics of both hemispheres. *Heteropogon contortus* is occasional in mixed grasslands of the Southwest (Texas to Arizona) and is also locally abundant in grasslands of the lower Texas coast. It is rated as a good range forage grass. *Heteropogon melanocarpus* is similar to *H. contortus* in general aspect, but is annual and has a median row of large impressed glands on the upper leaf sheaths and the glumes of the sterile spikelets.

112. *Coelorachis* L.

Perennials with erect, usually much-branched culms and numerous cylindrical, spicate racemes. Spikelets awnless, in pairs on a thickened, readily disarticulating rachis. Sessile spikelet of a pair perfect, the pediceled spikelet staminate, both more or less sunken in the corky rachis. First glume of the sessile spikelet thick and firm, rounded at the apex, smooth or variously pitted, covering the remainder of the spikelet. Lemmas of the sterile and fertile florets membranous and hyaline. Pediceled spikelet reduced, sterile, the pedicel short and thick. Basic chromosome number, $x = 9$.

A genus of about twenty-five species, in the warmer parts of both hemispheres. Closely related to, and perhaps not distinct from, the tropical genus *Rottboellia*. Five species are listed (as *Manisuris*) by Hitchcock (1951) for the United States, all in the Southeast. *Coelorachis cylindrica* (Michx.) Nash, CAROLINA JOINTTAIL (Fig. 5-69), is rather frequent in prairies and open pine woods from North Carolina and Oklahoma to Florida and Texas.

Fig. 5-69. *Coelorachis cylindrica.* Inflorescence and spikelet pair. (From Gould and Box, 1965, as *Manisuris cylindrica.*)

113. *Hackelochloa* Kuntze

Much-branched annual with erect or decumbent-spreading culms. Blades flat and thin. Spikelets in short, spikelike racemes, these terminating the main culm axis and the numerous branches that eventually are developed at all but the lowermost culm nodes. Racemes subtended by, and often partially enclosed in, expanded leaf sheaths. Spikelets in pairs of 1 sessile and 1 pediceled, on a stout, short rachis. Disarticulation at the nodes of the rachis. Sessile spikelet fertile, awnless, globose, the first glume thick and rounded, coarsely rugose or alveolate. Lemmas of the sterile and fertile florets membranous, hyaline. Pediceled spikelets ovate-lanceolate, with thin, flat glumes. Pedicels fused on one side to the rachis. The only reported chromosome count is $2n = 14$, but this record needs confirmation.

A genus of a single species, *Hackelochloa granularis* (L.) Kuntze, this widely distributed and weedy in tropical regions of the world but apparently introduced in the Americas. *Hackelochloa granularis* has been reported from Georgia, Florida, Louisiana, New Mexico, and Arizona.

114. *Tripsacum* L.

Large cespitose perennials with stout, thick-based culms and usually broad, flat blades. Inflorescence a spikelike raceme or series of 2 to few spikelike racemose branches bearing staminate spikelets above and pistillate spikelets below. Staminate spikelets 2-flowered, in pairs on one side of a continuous rachis. Pistillate spikelets below the staminate and on the same rachis, single, sessile, and partially embedded on the rachis. Glumes of the staminate spikelet flat, several-nerved, relatively thin. Glumes of the pistillate spikelet hard and bony, fused with the rachis and tightly enclosing the rest of the spikelet. Lemmas of the sterile and fertile florets thin and membranous, awnless, often reduced. Staminate portion of the rachis deciduous as a whole, the pistillate portion breaking up at the nodes into beadlike units. Basic chromosome number, $x = 9$.

A genus of seven species, native to the warmer parts of the Americas (Cutler and Anderson, 1941). Three species occur in the United States: one, *Tripsacum floridanum* Porter ex Vasey, FLORIDA GAMAGRASS, restricted to southern Florida; one, *T. lanceolatum* Rupr., MEXICAN GAMAGRASS, ranging from southern Arizona south through Mexico to Guatemala; and *T. dactyloides* (L.) L., EASTERN GAMAGRASS (Fig. 5-70), widespread in

Fig. 5-70. *Tripsacum dactyloides*. Inflorescence. (From Gould and Box, 1965.)

the eastern half of the United States and also in the West Indies. *Tripsacum dactyloides* grows in large clumps, mostly in low prairie sites. It is highly palatable to livestock.

115. *Zea* L.

Monoecious plants with tall, thick, usually succulent culms and broad, flat blades. Staminate spikelets in unequally pediceled pairs on spikelike branches, these forming large panicles at the culm apex. Glumes of the staminate spikelet broad, thin, several-nerved. Lemma and palea hyaline. Pistillate spikelets sessile in pairs on a thickened woody or corky axis (cob). Glumes of the pistillate spikelets broad, thin, rounded at the apex, much shorter than the mature caryopsis. Lower floret sterile or occasionally fertile. Lemma of lower floret and lemma and palea of upper (fertile) floret membranous and hyaline. Basic chromosome number, $x = 10$.

The above description applies specifically to the annual crop plant *Zea mays* L., CORN or MAIZE. Also included in the genus *Zea* as interpreted by Mangelsdorf and Reeves (1942) are the two species of Mexican TEO-SINTE, the annual *Z. mexicana* (Schrad.) Reeves and Mangelsdorf and the perennial *Z. perennis* (Hitchc.) Reeves and Mangelsdorf. These were previously referred to the genus *Euchlaena* Schrad. As interpreted by Reeves and Mangelsdorf, the teosinte species arose from hybrids between *Z. mays* and species of *Tripsacum*. Much has been written about *Z. mays*, and hundreds of strains or races have been developed. From earliest historic times, it has been cultivated from central North America southward to Peru. The region of origin now is believed to be Mexico. Corn is grown throughout temperate and subtropical North America. In the United States, its cultivation is centered in the "corn belt" of Iowa and Illinois. Scattered plants are occasional on roadsides and in waste places adjacent to croplands, but corn is unable to reproduce itself for any length of time out of cultivation. References: Mangelsdorf and Reeves, 1939, 1959.

116. *Coix* L.

Tall, coarse grasses, with thick culms; broad, flat blades; and unisexual spikelets. Spikelets on spicate branches, the staminate in twos or threes on a continuous rachis above the pistillate, the latter enclosed in hard, bony, bead-like involucres of modified bracts. Staminate spikelets 2-flowered, with thin, obscurely nerved glumes and hyaline lemma and palea. Pistillate spikelets 3 in each involucre, 1 fertile and 2 sterile. Glumes of the fertile floret hyaline below, firmer at the pointed tip. Lemma and palea of fertile floret hyaline. Basic chromosome number, $x = 10$.

A genus of about four species, one, *Coix lacryma-jobi* L., JOB'S TEARS, widespread in tropical regions of the world, and the others in the East Indies. *Coix lacryma-jobi*, a tall, coarse annual, has been occasionally cultivated in the warmer parts of the United States and is reported to persist as an escape in Florida. Although frequent and often weedy in the American tropics, this grass is believed to be native only to the Old World. The hard, bony fruiting involucres have long been used as beads, especially for rosaries, in tropical countries.

Subfamily III. Chloridoideae

Tribe 12. Eragrosteae

117. *Eragrostis* Beauv.

Cespitose annuals and perennials, a few with rhizomes. Inflorescence an open (infrequently contracted) panicle. Spikelets 3- to many-flowered, awnless. Glumes unequal, 1- to 3-nerved, shorter than the lemmas. Lemmas 3-nerved, acute or acuminate at the apex, keeled or rounded on the back. Palea strongly 2-nerved and usually 2-keeled, often ciliolate on the keels, as long or nearly as long as the lemma. Glumes, lemmas, and mature caryopsis usually early deciduous, the paleas persistent on the rachilla. Caryopsis oblong or subelliptic, typically reddish brown and translucent. Basic chromosome number, $x = 10$.

Species about 250, in tropical and temperate regions of the world. Some 33 species are native to the United States, and an additional 15 are listed as introduced or adventive. The latter have come mainly from South Africa and the Mediterranean region. About half of the *Eragrostis* species are weedy annuals. The introduced annual *E. barrelieri* Daveau (Fig. 5-71) is becoming frequent as a roadside weed throughout the south central and southwestern United States. *Eragrostis intermedia* Hitchc., PLAINS LOVEGRASS (Fig. 5-72), is one of the few native species of significance as a range forage plant. This grass is widely distributed in the southern United States and northern Mexico. In the present concept, *E. intermedia* includes most of the United States plants referred to *E. lugens* Nees by Hitchcock (1951). Gould and Box (1965) grouped these plants together but referred them to *E. lugens*, a species that apparently does not occur in the United States. A number of vigorous perennial *Eragrostis* species have been introduced, mainly from Africa, as forage grasses for the South and Southwest. Most successful of these are *E. lehmanniana* Nees, LEHMANN LOVEGRASS; *E. curvula* (Schrad.) Nees, WEEPING LOVEGRASS; and *E. chloromelas* Steud., BOER LOVEGRASS. References: Koch, 1978; Witherspoon, 1977.

118. *Neeragrostis* Bush

Stoloniferous, dioecious, annual. Sheaths short, membranous on the margins below. Ligule a minute fringe of hairs. Inflorescence a simple panicle, raceme, or capitate cluster of subsessile spikelets. Spikelets linear, many-

Fig. 5-71. *Eragrostis barrelieri.* Plant and spikelet.
(From Gould and Box, 1965.)

flowered, awnless, the caryopses, glumes, and lemmas disarticulating from the rachilla as in *Eragrostis*. Paleas strongly 2-nerved, about as long as the lemma in the pistillate spikelets and half as long in the staminate florets. Basic chromosome number, $x = 10$.

A genus of a single species, *Neeragrostis reptans* (Michx.) Nicora, CREEPING LOVEGRASS. This low, mat-forming annual is distributed from Kentucky and South Dakota south to Florida, Texas, and northern Mexico. It is locally abundant on poorly drained sites with fine-textured soils. Frequently it forms solid stands in "hog wallows" and drying lake beds. Hitchcock treated the species as *E. reptans* (Michx.) Nees. Nicora (1962), in comparing the characteristics of this taxon and *Eragrostis*, noted differences in ovary and epidermal features as well as the dioecious nature of *Neeragrostis*. She also suggested probable close relationships between *Neeragrostis* and the genera *Distichlis* and *Monanthochloë*.

119. *Tridens* Roem. & Schult.

Moderately tall perennials with usually flat blades and open or contracted panicles of several-flowered spikelets. Spikelet disarticulation above the glumes and between the florets. Glumes mostly thin and membranous, subequal, the first 1-nerved, the second 1- to 3-nerved, rarely 5-nerved. Lem-

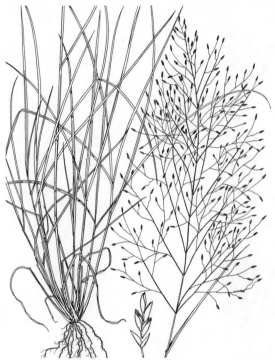

Fig. 5-72. *Eragrostis intermedia* Hitchc. Plant and spikelet. (From Gould, 1951.)

mas broad, thin, 3-nerved, short-hairy on the nerves below (except in *Tridens albescens*), rounded on the back, mostly bidentate at the apex, the midnerve and often the lateral nerves extended as minute mucros. Palea slightly shorter than the lemma, the 2 strong nerves usually glabrous (hairy below in *T. muticus*). Caryopsis dark brown, the embryo about two-fifths the length of the caryopsis. Basic chromosome number, $x = 10$.

With the removal of the grasses now grouped under *Erioneuron* (see discussion of *Erioneuron*), the genus *Tridens* comprises some sixteen species, these present for the most part in the eastern and southern United

States and northern Mexico. *Tridens flavus* (L.) Hitchc., PURPLETOP (Fig. 5-73), is frequent and often abundant in woodlands and on shaded slopes throughout the eastern half of the United States. *Tridens muticus* (Torr.) Nash, SLIM TRIDENS (Fig. 5-74), is widespread on dry, rocky slopes in the Southwest and adjacent Mexico.

Fig. 5-73. *Tridens flavus.* Inflorescence and spikelet. (From Gould and Box, 1965.)

120. *Triplasis* Beauv.

Low to moderately tall tufted grasses, one species annual, the other perennial. Culms with many nodes and short internodes, eventually breaking up at the nodes. Terminal inflorescence a short, few-flowered panicle or raceme. Clusters of cleistogamous spikelets regularly borne in the axils of the upper leaf sheaths. Spikelets of the terminal panicle mostly 2- to 4-flowered, those of the sheath axils usually 1-flowered. Glumes subequal, 1-nerved,

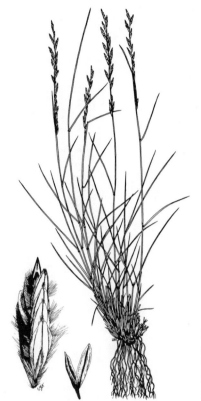

Fig. 5-74. *Tridens muticus*. Plant and spikelet with glumes separated from the florets. (From Gould, 1951, in part, and Gould and Box, 1965, in part.)

glabrous. Lemmas narrow, 3-nerved, the nerves ciliate or densely pubescent, the midnerve extended as a mucro or short awn from a notched apex. Palea narrow, strongly 2-nerved and 2-keeled, silky-villous on the keels. Basic chromosome number, $x = 10$.

A genus of two species, *Triplasis purpurea* (Walt.) Chapm. (Fig. 5-75) and *T. americana* Beauv., these growing on sandy soils, often in partial shade, through the eastern United States from Maine to the Great Lakes region and south to Florida and Texas.

121. *Erioneuron* Nash

Low, tufted, mostly stoloniferous perennials with narrow, often involute, cartilaginous-margined leaf blades. Inflorescence a short, usually capitate raceme or panicle of several-flowered spikelets, these disarticulating above the glumes and between the florets. Glumes large, membranous, subequal, 1-nerved. Lemmas broad, rounded on the back, 3-nerved, conspicuously

Fig. 5-75. *Triplasis purpurea.* Inflorescence and spikelet. (From Gould
and Box, 1965.)

long-hairy along the nerves at least below, bilobed at the apex (except in
Erioneuron pilosum), the midnerve short-awned, each lateral nerve often
prolonged as a short mucro. Palea slightly shorter than the lemma, ciliate on
the keels, long-hairy on the lower part between the nerves. Caryopsis
oblong, glossy and translucent, with an embryo more than half as long as the
caryopsis. Basic chromosome number, $x = 8$.

A group of five species, distributed in low rainfall areas of the southwestern United States and Mexico. None of these grasses is of economic importance, but *E. pilosum* (Buckl.) Nash is frequent on dry rangelands throughout the Southwest. *Erioneuron pulchellum* (H.B.K.) Tateoka, FLUFF-
GRASS (Fig. 5-76), is a characteristic grass of sand and "desert pavement"
sites in *Larrea tridentata* associations. Species herein included in *Erioneu-*

ron were included in *Tridens* by Hitchcock (1951). The transfer of four species to *Erioneuron* was made by Tateoka (1961) on the basis of numerous differences, including chromosome number, shape and color of the caryopsis, color of the stigma, shape and pubescence of the lemma, pubescence of the palea, shape of the bicellular microhairs of the epidermis, leaf anatomy, general plant habit, and nature of the panicle. Tateoka discussed similarities and possible close relationships between *Erioneuron* and the genus *Munroa*.

Fig. 5-76. *Erioneuron pulchellum*. Plant, showing stoloniferous habit. (From Gould, 1951, as *Tridens pulchellus*.)

122. *Munroa* Torr.

Low, tufted monoclinous or gymnomonoecious annuals with spreading, much-branched culms. Leaves short, fascicled, the blades stiff, pungent, and flat. Inflorescences of small, subsessile clusters of spikelets, these almost hidden in fascicles of leaves at the branch tips. Spikelets few-flowered, the lower ones of a cluster 3- or 4-flowered, the upper ones 2- or 3-flowered. Typically distal floret perfect, lower florets pistillate or rarely perfect. Disarticulation above the glumes and sometimes between the florets. Glumes of the lower spikelets subequal, narrow, acute, 1-nerved, slightly shorter than the lowermost lemma. Glumes of the upper spikelets unequal, the first reduced or even absent. Lemmas prominently 3-nerved, those of the lower spikelets coriaceous, usually with tufts of hairs on the margins near the middle, gradually narrowing above, abruptly mucronate or short-awned from a slender, spreading tip. Palea narrow, membraneous. Chromosome numbers reported, $2n = 14$ and 16.

A genus of five species, one, *Munroa squarrosa* (Nutt.) Torr. (Fig. 5-77), in western North America (Alberta to Texas, Arizona, and northern Mexico), and four in South America. *Munroa squarrosa* may be locally abundant but is relatively unpalatable to livestock. Stebbins and Crampton (1961) followed Hitchcock (1951) in referring *Munroa* to the tribe *Chlorideae*. The placement of this genus in the *Eragrosteae*, however, appears more satisfactory, especially in view of the comments of Tateoka (1961), who noted apparent close relationships of *Munroa* and *Erioneuron*. Reference: Anton and Hunziker, 1978.

Fig. 5-77. *Munroa squarrosa*. Plant and spikelet.

123. *Vaseyochloa* Hitchc.

Perennial with moderately tall, erect culms in small clumps. Slender creeping rhizomes frequently developed. Sheaths pilose externally at the apex. Ligule a short, lacerate, densely pilose membrane. Blades long, narrow, flat or folded. Inflorescence a panicle, with several-flowered spikelets on slender, erect or spreading branches. Disarticulation above the glumes and between the florets. Glumes firm, acute, shorter than the lemmas, the first narrow, 3- or 5-nerved, the second broader, 7- or 9-nerved. Lemmas broad, firm, rounded and hairy on the back, 7- or 9-nerved, tapering to a narrow, obtuse apex. Palea shorter than the lemma, broad, splitting down the middle at maturity. Caryopsis dark brown or black, oval, concave-convex, with 2 persistent hornlike style bases at the rounded apex. Basic chromosome number not known; counts of $2n = 60$ and $2n = 68$ reported.

A genus of a single species, *Vaseyochloa multinervosa* (Vasey) Hitchc. (Fig. 5-78), known only from southern Texas, in San Patricio and Kleberg counties, and on Padre Island, Nueces County. Relatively rare and restricted in its range, this unique and attractive grass is locally abundant on sandy sites. *Vaseyochloa* was placed in the tribe Aeluropodeae by Stebbins and Crampton (1961) and in the tribe Eragrosteae by Decker (1964). Its affinities are definitely chloridoid, but it does not fit especially well into any of the recognized tribes. As a member of the tribe Eragrosteae, it is atypical in having many-nerved rather than 3-nerved lemmas.

Fig. 5-78. *Vaseyochloa multinervosa*. Plant, spikelet, floret with palea split down the middle, and caryopsis. (From Gould and Box, 1965.)

124. *Redfieldia* Vasey

Plants large, coarse, perennial, with extensive rhizomes and firm, tough culms up to 1 meter tall. Blades long, firm, often involute. Panicle large and open, usually one-third to one-half the length of the culm, with 2- (infrequently 1-) to 6-flowered spikelets borne on flexuous pedicels. Disarticulating above the glumes and between the florets. Glumes narrow, lanceolate, 1-nerved. Lemmas lanceolate, distinctly 3-nerved, glabrous or scabrous,

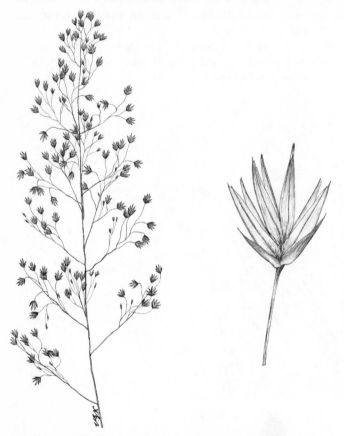

Fig. 5-79. *Redfieldia flexuosa*. Inflorescence and spikelet.

with a tuft of long, stiff hair at the base, occasionally mucronate at the apex. Caryopsis oblong, free from the palea. Basic chromosome number, $x = 10$ (only count reported is $2n = 25$). Reference: Reeder, 1976.

A genus of one or two species. *Redfieldia flexuosa* (Thurb.) Vasey, BLOWOUT GRASS (Fig. 5-79), is present on sandy hills and dune formations from South Dakota and Oklahoma to Colorado and northeastern Arizona. This is a tough, rhizomatous perennial of little value for forage but important as a soil binder in areas of drifting sand.

125. *Scleropogon* Philippi

Low, stoloniferous, mat-forming perennial with short, tufted leaves and flat, firm blades. Flowers mostly imperfect, the staminate and pistillate spikelets on different plants (dioecious) or less frequently on the same plant (monoecious). Perfect flowers occasionally produced. Spikelets large, in few-flowered spicate racemes or contracted panicles, the pistillate ones bristly with long awns. Staminate spikelets mostly with 5 to 10 widely spaced and persistent florets, the florets occasionally as many as 20. Glumes 1-nerved or obscurely 3-nerved, thin, pale, lanceolate, awnless, the first and second glumes separated by a short internode. Lemmas of the staminate spikelets similar to the glumes. Palea obtuse, shorter than the lemma. Pistillate spikelets mostly with 3 to 5 fertile florets and 1 to several rudimentary florets (reduced to awns) above. Rachilla disarticulating above the glumes, the florets falling together. Glumes of the pistillate spikelet unequal, acuminate, awnless, 3-nerved or with additional fine lateral nerves. Lemmas of the pistillate spikelet firm, rounded on the back, 3-nerved, the nerves extending into slender, spreading awns 5 to 10 cm long. Lowermost lemma with a bearded, sharp-pointed callus at the base. Palea narrow, the 2 nerves extended into short awns. Basic chromosome number, $x = 10$.

Species one, *Scleropogon brevifolius* Philippi, BURROGRASS (Fig. 5-80), present on dry plains, Colorado to Texas, Arizona, and central Mexico; also in Chile and Argentina. A coarse grass of low palatability, developing in large stands and apparently increasing on rangelands under heavy grazing pressure.

126. *Blepharidachne* Hack.

Low, tufted monoclinous or monoecious perennials or annuals. Culms with widely spaced internodes and spur shoots with short, congested internodes. Prophyll with two aristate appendages. Panicles short, congested, few-flowered, not or only slightly exserted above the subtending leaves. Sheaths open. Ligule of minute hairs or absent. Spikelets 4-flowered, first and second florets sterile or staminate, the third floret fertile (perfect or pistillate), and the fourth floret reduced to a 3-awned rudiment. Glumes about equal, thin, 1-nerved, acute or acuminate. Disarticulation above the glumes but not between the florets. Lemmas 3-nerved, 3-lobed and 3-awned at the apex, the awns conspicuously plumose. A chromosome number of $2n = 14$ has been reported, but this record needs confirmation.

A genus of four species, two in the southwestern United States and two

Fig. 5-80. *Scleropogon brevifolius*. Plant, staminate spikelet (awnless), and pistillate spikelet (awned).

in Argentina. *Blepharidachne kingii* (S. Wats.) Hack. has been reported from desert regions in Utah, Nevada, and southern California (Death Valley). *Blepharidachne bigelovii* (S. Wats.) Hack. (Fig. 5-81) is known only from dry, rocky slopes in Trans-Pecos Texas and adjoining Coahuila, Mexico. Reference: Hunziker and Anton, 1979.

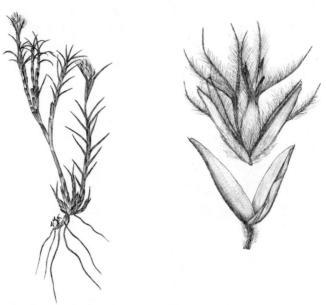

Fig. 5-81. *Blepharidachne bigelovii.* Plant and spikelet with glumes separated from the florets.

127. *Calamovilfa* Hack.

Tall, stout perennials, with short or widely spreading rhizomes and large, open or contracted panicles of small spikelets. Spikelets 1-flowered, the rachilla disarticulating above the glumes, not prolonged behind the floret. Glumes firm, unequal, 1-nerved, acute, the second nearly as long as the lemma, the first shorter. Lemma firm, 1-nerved, rounded on the back, glabrous or pubescent, bearded on the callus. Palea usually greatly reduced. Grain elongate. Basic chromosome number, $x = 10$.

A genus of four species, all North American. *Calamovilfa longifolia* (Hook.) Scribn. ranges on sandy soils from Alberta and Indiana west to Colorado and Idaho. *Calamovilfa gigantea* (Nutt.) Scribn. & Merr., BIG SANDREED, occurs on sandy hills and dunes from Kansas to Texas, Arizona, and Utah. *Calamovilfa brevipilis* (Torr.) Scribn. is present on the East Coast

from New Jersey to South Carolina, while *C. curtissii* (Vasey) Scribn. is
known only from the pine barrens of northern Florida.

Calamovilfa closely resembles *Calamagrostis* and *Ammophila* in gross
morphological characters, and was placed close to them in the tribe Agrostideae by Hitchcock (1935, 1951). Reeder and Ellington (1960), however,
pointed out that *Calamovilfa* differs from these genera in several important
embryo characters, type of lodicules, leaf anatomy, presence of epidermal
bicellular microhairs, and chromosome size and basic number. *Calamovilfa*
appears closely related to *Sporobolus*, and the two genera have many features in common, including the peculiar fruit character of having the seed
coat free from the ovary wall. Reference: Thieret, 1966.

128. *Lycurus* H.B.K.

Low to moderately tall, tufted perennial with dense, spikelike panicles of 1-
flowered spikelets. Leaves mostly in a basal clump, typically grayish-green

Fig. 5-82. *Lycurus phleoides*. Plant and spikelet with glumes
separated from the floret. (From Gould, 1951.)

in color. Sheaths laterally compressed and sharply keeled. Blades narrow, flat or folded, usually with a whitish midnerve and margins. Inflorescence mostly 3 to 8 cm long and 5 to 8 mm thick. Spikelets short-pediceled, deciduous in pairs together with the pedicels, the lowermost spikelet of the pair often sterile or staminate. Glumes shorter than the lemma, the first 2- or 3-nerved and with 2 or 3 short awns, the second similar but 1-nerved and 1-awned. Palea similar in texture to the lemma and about as long, enclosed by the lemma only at the base. Basic chromosome number, $x = 10$.

A genus of a single species, this, *Lycurus phleoides* H.B.K., WOLF-TAIL (Fig. 5-82), distributed in the southwestern United States, northern Mexico, and South America. *Lycurus phleoides* is frequent but seldom abundant on rocky, open slopes at medium altitudes in Oklahoma, Colorado, Utah, western Texas, New Mexico, and Arizona. It is similar to *Muhlenbergia wrightii* Vasey in general appearance and is often confused with that species.

129. *Muhlenbergia* Schreb.

Plants of diverse habits, from delicate, tufted annuals to large, coarse, cespitose perennials. Several species with creeping rhizomes. Culms simple to much-branched. Leaves various, usually with well-developed membranous ligules and narrow, flat or involute blades. Inflorescence an open or contracted panicle, spikelike in a few species. Spikelets typically 1-flowered, a second floret occasionally produced. Disarticulation above the glumes. Glumes usually 1-nerved or nerveless, occasionally 3-nerved, mostly shorter than the lemma, obtuse, acute, acuminate or short-awned. Lemma as firm as, or firmer than, the glumes, 3-nerved (the nerves indistinct in some species), with a short, usually bearded callus at the base and a single flexuous awn at the apex, less frequently mucronate or awnless. Palea well developed, shorter than or about equaling the lemma. Caryopsis elongate, cylindrical or slightly dorsally compressed, usually not falling free from the lemma and palea. Basic chromosome number, $x = 10$.

A large, diverse group of grasses which as presently interpreted includes over 125 species, these mostly in the Americas but a few in the Old World. Hitchcock (1951) reported some 70 species in the United States. *Muhlenbergia* is predominantly a genus of western grasses, but several species, mostly rhizomatous perennials, are to be found in the Midwest and the East. Typical of the eastern species is *M. schreberi* Gmel., NIMBLEWILL (Fig. 5-83), which ranges from the Atlantic Coast westward to Nebraska and Texas. The large, coarse bunchgrasses, *M. emersleyi* Vasey, BULLGRASS, and *M. longiligula* Hitchc., LONGTONGUE MUHLY, are frequent on rocky mountain slopes at medium altitudes in the Southwest. These species are representatives of the group that has been recognized as a distinct genus, *Epicampes* Presl. *Muhlenbergia porteri* Scribn., BUSH MUHLY, is widespread on dry mesas and rocky slopes from Colorado and Nevada to Texas, California, and northern Mexico. Originally it existed in extensive stands, but now it occurs for the most part in the protection of shrubs and subshrubs. It is highly palatable to livestock. Representative of the several western species with slender, tightly contracted panicles is *M. rigens* (Benth.)

Fig. 5-83. *Muhlenbergia schreberi*. Plant and spikelet. (From Gould and Box, 1965.)

Hitchc. (Fig. 5-84). *Muhlenbergia capillaris* (Lam.) Trin. (Fig. 5-85) is one of several closely related taxa with large, open, lacy panicles. This species ranges throughout the eastern states and westward to eastern Kansas and Texas. References: Soderstrom, 1967; Swallen, 1947.

130. *Sporobolus* R. Br.

Annuals and perennials, mostly cespitose but a few with creeping rhizomes. Leaves various, usually in a basal clump, with blades flat, folded or involute.

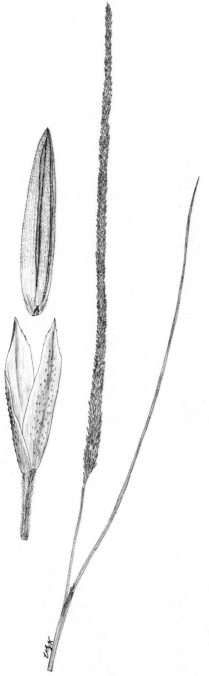

Fig. 5-84. *Muhlenbergia rigens*. Inflorescence and spikelet with glumes separated from the floret.

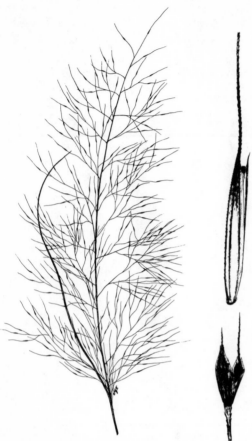

Fig. 5-85. *Muhlenbergia capillaris*. Inflorescence and spikelet with glumes separated from the floret. (From Gould and Box, 1965.)

Inflorescence an open or less frequently contracted panicle of small, awn-less, 1-flowered spikelets. Disarticulation above the glumes. Glumes 1-nerved, usually unequal and shorter than the lemma. Lemma 1-nerved, thin, awnless. Palea well developed, mostly as long as, or longer than, the lemma. Grain obovate, somewhat asymmetrical and flattened, usually falling free from the lemma and palea. Grain not a true caryopsis, as the seed coat (testa) is not fused to the ovary wall. Basic chromosome numbers, $x = 6$ and 9.

Species about one hundred, in temperate and tropical regions of both hemispheres. Thirty species were reported for the United States by Hitchcock (1951). From the closely related genus *Muhlenbergia*, these grasses differ in having thin, 1-nerved, consistently awnless lemmas; obovate, flattened fruits; and usually ciliate ligules. *Sporobolus cryptandrus* (Torr.) A. Gray, SAND DROPSEED (Fig. 5-86), is frequent on sandy soils, often on disturbed sites, throughout the United States except in the extreme South-

Fig. 5-86. *Sporobolus cryptandrus.* Plant and spikelet. (From Gould, 1951.)

east. *Sporobolus airoides* (Torr.) Torr., ALKALI SACATON, is a characteristic bunchgrass of alkaline areas in the western states. *Sporobolus indicus* (L.) R. Br., SMUTGRASS or RATTAIL SMUTGRASS, a tufted weedy perennial with coarse, tough herbage and contracted, spikelike, often smut-infested panicles, is frequent in the southeastern United States, where it is believed to be adventive. This grass is widespread throughout the tropics and subtropics of Asia and the Americas. Reference: Riggins, 1977.

131. *Blepharoneuron* Nash

Cespitose perennial, with culms 20 to 70 cm tall. Leaves mostly in a basal tuft. Ligule membranous, short, rounded, appearing as a continuation of the

Fig. 5-87. *Blepharoneuron tricholepis*. Plant and spikelets
with glumes separated from the floret.

sheath margins. Blades narrow, flat but soon becoming involute. Inflorescence a loosely contracted panicle. Spikelets 1-flowered, borne on slender pedicels, disarticulating above the glumes. Glumes broad, nearly equal, faintly-nerved, rounded dorsally, broadly acute or obtuse at the awnless apex. Lemma firmer than the glumes and slightly longer, rounded on the back, 3-nerved, the nerves usually densely ciliate-pubescent, the apex acute, occasionally mucronate. Palea slightly exceeding the lemma, pubescent on the 2 nerves. Basic chromosome number, $x = 8$.

A genus of a single species, *Blepharoneuron tricholepis* (Torr.) Nash, PINE DROPSEED (Fig. 5-87), this distributed from Colorado and Utah south to western Texas, Arizona, and northern Mexico. *Blepharoneuron tricholepis* occurs mainly in pine or spruce forest at high elevations, but has been reported from as low as 2,300 feet. It is palatable to all types of livestock and is considered a good range forage grass.

132. *Crypsis* Ait.

Low, tufted annuals with weak, spreading, many-noded culms and short, flat blades. Inflorescences of short, dense spicate panicles, these terminating the main shoots and also in the axils of the upper leaves. Panicles remaining partially enclosed by the expanded leaf sheaths. Spikelets 1-flowered, disarticulating below or above the glumes. Glumes subequal, narrow, acute. Lemma thin, broad, 1-nerved, acute, awnless. Palea broad and about as long as the lemma, splitting between the nerves at maturity. Grain with the seed coat free from the pericarp, the fruit thus not a true caryopsis. Basic chromosome number, $x = 8$ and 9.

A genus of about seven species, these native to the Mediterranean region (Lorch, 1962). Three species are adventive in the United States. *Crypsis alopecuroides* (Pill. & Mitterp.) Schrad. (*Heleochloa alopecuroides* [Pill. & Mitterp.] Host) has been reported from a few widely scattered localities across the country. *Crypsis vaginiflora* (Forsk.) Opiz is known from California and Idaho, and *C. schoenoides* (L.) Lam. has been collected in the northeastern states, Massachusetts to Wisconsin and south to Illinois and Iowa, and also California. In the 1951 revision of Hitchcock's manual, *C. alopecuroides* and *C. schoenoides* were referred to the genus *Heleochloa*. Reference: Hammel and Reeder, 1979.

133. *Eleusine* Gaertn.

Low, spreading annuals with thick, weak culms and soft, flat or folded, succulent leaf blades. Ligule a short, lacerate membrane. Spikelet sessile in two rows on two (occasionally one) to several branches digitately arranged at the culm apex. One or two branches frequently developed below the apical cluster. Spikelets 3- to several-flowered, disarticulating above the glumes and between the florets. Glumes firm, acute, unequal, the first short, 1-nerved, the second 3- to 7-nerved. Lemmas acute, awnless or mucronate, broadly keeled, 3-nerved, the lateral nerves very close to the midnerve. Palea shorter than the lemma. Grain plump, with a minutely transversely rugose seed loosely enclosed in a thin pericarp. Basic chromosome number, $x = 9$.

A genus of about six species, mostly of warmer regions of the Old World, one species native to South America and one adventive in the New

Fig. 5-88. *Eleusine indica*. Inflorescence and spikelet. (From Gould and Box, 1965.)

World. *Eleusine indica* (L.) Gaertn., GOOSEGRASS (Fig. 5-88), a common weed of the tropics and subtropics of both hemispheres, is frequent on disturbed soil in the southern United States and has been recorded throughout the country except in the Northwest.

134. *Dactyloctenium* Willd.

Annuals with thick, weak culms and soft, flat blades. Culms often decumbent below and rooting at the lower nodes. Inflorescence of 2 (occasionally 1) to several digitately arranged, unilateral spicate branches. Spikelets closely placed and pectinate in two rows on one side of a short, stout rachis, this projecting as a point beyond the insertion of the uppermost spikelet. Spikelets 2- to several-flowered, laterally compressed, with disarticulation often between the glumes, the first remaining on the rachis. Glumes keeled, subequal, 1-nerved, the first awnless, the second mucronate or with a short, stout awn. Lemmas firm, broad, 3-nerved, the lateral nerve indistinct, abruptly narrowing to a beaked, usually short-awned tip. Palea well developed, about as long as the lemma. Grain plump, usually subglobose, with a minutely ridged and rugose seed loosely enclosed in a thin pericarp. Basic chromosome number, $x = 10$.

A genus of three species, native to the warmer parts of the Eastern Hemisphere, one introduced in the Americas. *Dactyloctenium aegyptium* (L.) Beauv., CROWFOOT GRASS, is occasional as a weed of disturbed sites and coastal sands throughout the southern United States, from North Carolina and Florida to the West Coast and south into Mexico. The species of *Dactyloctenium* are closely related to those of *Eleusine* and are included in that genus by some systematists.

135. *Leptochloa* Beauv.

Cespitose annuals and perennials, apparently none with rhizomes or stolons. Culms usually leafy to well above the base, the leaves with flat, linear blades. Inflorescence a panicle, with few to numerous unbranched primary branches distributed along the upper portion of the culm or clustered near the tip. Spikelets 2- to several-flowered, overlapping and closely spaced on the branches, disarticulating above the glumes and between the florets. Glumes thin, 1-nerved or the second occasionally 3-nerved, acute, awnless or mucronate. Lemmas 3-nerved, frequently puberulent on the nerves. Apex of the lemma acute to obtuse or notched; awnless, mucronate, or awned. Palea well developed, occasionally puberulent on the nerves. Basic chromosome number, $x = 10$.

Species about seventy, in the warmer parts of both hemispheres; eleven species reported for the United States, all native. Most species of *Leptochloa* are plants of moist or marshy sites; many are weedy. *Leptochloa dubia* (H.B.K.) Nees, GREEN SPRANGLETOP (Fig. 5-89), is a good forage grass of the southwestern ranges but is seldom more than a minor element of the vegetation. Reference: McNeill, 1979.

136. *Trichoneura* Anderss.

Cespitose annuals and perennials with spikelets borne on few to several spicate primary branches, these widely scattered on the upper portion of the culm or clustered near its apex. Spikelets 3- to 9-flowered, disarticulating above the glumes and between the florets. Glumes about equal, longer than

Fig. 5-89. *Leptochloa dubia*. Plant and spikelet. (From Gould, 1951.)

the lemmas, thin, 1-nerved, acuminate or short-awned. Lemmas 3-nerved, the lateral nerves conspicuously long-ciliate, the midnerve glabrous or short-pubescent. Apex of lemma narrow, obtuse or notched, often short-awned. Base of lemma with a short, hairy callus. Palea broad, well developed. Basic chromosome number, $x = 10$.

Species about ten, mostly in Africa; one species in Texas and two in South America. *Trichoneura elegans* Swallen (Fig. 5-90), a rather robust annual, is restricted to southern Texas, where it is infrequent on sandy soils.

137. *Gymnopogon* Beauv.

Perennials with stiff, erect, many-noded culms. Base of plant often with short, knotty rhizomes. Leaves stiff and firm with rounded, overlapping sheaths and short, flat, stiff, usually spreading blades. Spikelets subsessile, rather widely spaced in two rows on slender, spreading primary inflorescence branches, these distributed on the upper portion of the culm. Spikelets 1- to 3-flowered, the rachilla prolonged behind the terminal fertile floret as a slender stipe bearing a rudimentary floret. Disarticulation above the glumes. Glumes narrow, nearly equal, 1-nerved, acuminate, the second exceeding the lemmas in length. Lemmas narrow, rounded on the back, 3-nerved, usually bearing a delicate awn from a minutely bifid apex. Basic chromosome number, $x = 10$.

Species about fifteen, in the American tropics and subtropics, four species in eastern and southeastern United States. *Gymnopogon ambiguus* (Michx.) B.S.P., a coarse perennial with a knotty, rhizomatous base, grows in sandy, mostly woodland soils from New Jersey and Florida to Kansas and Texas.

138. *Tripogon* Roth.

Low, tufted perennials (the American species) with filiform leaves, these mostly in a basal clump. Inflorescence a slender spike, the spikelets sessile

Fig. 5-90. *Trichoneura elegans*. Inflorescence and spikelet. (From Gould and Box, 1965.)

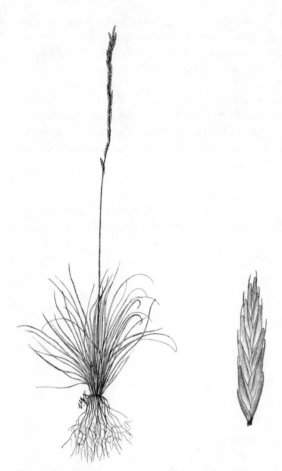

Fig. 5-91. *Tripogon spicatus*. Plant and spikelet.

or subsessile and solitary at the nodes of a straight or somewhat flexuous rachis. Spikelets several-flowered, disarticulating above the glumes and between the florets. Glumes unequal, acute or acuminate, 1-nerved, shorter than the lemmas. Lemmas 3-nerved, with a tuft of hair at the base, short-awned from between the lobes of a minutely notched apex. Basic chromosome number, $x = 10$.

Species about ten, one in the Americas and the others in Africa and the East Indies. *Tripogon spicatus* (Nees) Ekman (Fig. 5-91) is occasional on rocky outcrops in central Texas and has also been reported from Mexico, Cuba, and South America.

Tribe 13. Chlorideae

139. *Willkommia* Hack.

Low annuals or perennials, with flat or somewhat involute blades. Inflorescence with few to several short, closely flowered branches. Spikelets appressed and closely imbricated in two rows from the base to the apex of the branch rachis. Spikelets 1-flowered, disarticulating above the glumes. Glumes unequal, thin, the first short, rounded at the apex, narrow, and nerveless, and the second about as long as the lemma, 1-nerved, acute at the apex. Lemma 3-nerved, awnless, rounded dorsally, pubescent between the nerves and on the margins. Palea well developed, densely pubescent on the 2 nerves. Basic chromosome number, $x = 10$.

Fig. 5-92. *Willkommia texana.*
Plant and spikelet. (From Gould
and Box, 1965.)

A genus of four species, one in Texas and Argentina and three in South Africa. *Willkommia texana* Hitchc. (Fig. 5-92), a low, tufted, short-lived perennial, is occasional in hard, clayey soils bordering swales, ponds, and small lakes in central and southern Texas. It also occurs in Argentina.

140. *Schedonnardus* Steud.

Low, tufted perennial with slender, wiry, erect or decumbent culms. Leaves mostly in a basal clump, the blades short, flat, usually spirally twisted and

Fig. 5-93. *Schedonnardus paniculatus.* Inflorescence and spikelet. (From Gould and Box, 1965.)

wavy-margined, about 1 mm broad. Ligule membranous, 2 to 3 mm long, decurrent. Panicles half to three-fourths the entire length of the culm, with a stiff, curved, wiry central axis and slender, spreading, undivided primary branches. At maturity the panicle breaks off at the base and becomes a tumbleweed. Spikelets slender, 1-flowered, sessile, widely spaced and appressed on the branches and at the tip of the main axis. Disarticulation above the glumes. Glumes narrow, lanceolate or acuminate, 1-nerved, the second about as long as the lemma, the first shorter. Lemma narrow, rigid, 3-nerved, awnless or with a minute awn-tip. Palea similar to the lemma and about as long. Basic chromosome number, $x = 10$.

An American genus of a single species, *Schedonnardus paniculatus* (Nutt.) Trel., TUMBLEGRASS (Fig. 5-93). Tumblegrass is frequent on prairies and plains from Saskatchewan through the central United States to northern Mexico, and it also occurs in Argentina. Although widespread and locally abundant, it produces little herbage and is of no significance as a forage plant.

141. *Cynodon* L. Rich.

Low, mostly mat-forming stoloniferous and rhizomatous perennials. Culms much-branched, mostly with short internodes. Leaf blades flat, short, narrow, soft, and succulent. Ligule a fringe of hairs. Inflorescence with 2 to several slender, spicate branches, these digitately arranged at the culm apex. Spikelets sessile in two rows on a narrow, somewhat triangular branch rachis. Spikelets with a single floret, the rachilla prolonged behind the palea as a bristle and occasionally bearing a rudimentary lemma. Glumes slightly unequal, lanceolate, awnless, 1-nerved, the second nearly as long as the lemma. Lemma firm, laterally compressed, awnless, 3-nerved, usually puberulent on the midnerve. Palea narrow, 2-nerved, as long as the lemma. Basic chromosome number, $x = 9$.

Species about ten, mostly in Africa and Australia; one species, *Cynodon dactylon* (L.) Pers., COMMON BERMUDAGRASS (Fig. 5-94), distributed throughout the warmer parts of the world. This species, now widespread in the United States, is believed to have originated in tropical Africa. It is drought-resistant and alkali-tolerant and can stand moderate amounts of freezing temperatures. It is the most generally used and most satisfactory lawn grass of the southern states. Not only is common Bermudagrass of value for lawns, but it is also one of the outstanding forage grasses of the South, flourishing on relatively sterile soils and persisting under heavy grazing pressures. Hybrids of *C. dactylon* and other *Cynodon* species, such as the widely used "coastal Bermuda" of the Gulf Coast area of Texas, have been successful as improved pasture grasses. These hybrids are highly sterile and must be propagated by "sprigging." Reference: Harlan et al., 1970.

Fig. 5-94. *Cynodon dactylon*. Plant with both stolons and rhizomes, inflorescence, spikelet with glumes separated from the floret, and caryopsis. (From Gould, 1951.)

142. *Microchloa* R. Br.

Tufted annuals and perennials with slender culms and densely clumped, filiform leaves. Inflorescence a slender, curved, unilateral spike, with 1-flowered spikelets closely imbricated in two rows on one side of a narrow, flattened rachis. Disarticulation usually between the glumes. Glumes firm, lanceolate, subequal, 1-nerved, acute. Lemma thin and membranous, slightly shorter than the glumes, awnless, acute, with a midnerve and 2 short lateral nerves, the latter not always apparent. Palea similar to the lemma in texture but slightly shorter. Caryopsis oval, flattened, reddish brown. Basic chromosome number, $x = 10$.

A genus of four species, all native to Africa; one, *Microchloa kunthii* Desv., also widespread in tropical and subtropical regions of the world. Although *M. kunthii* is not infrequent in Mexico, it is known in the United States only from Carr Canyon of the Huachuca Mountains in southern Arizona.

143. *Chloris* Swartz

Annuals and perennials with flowering culms from erect or decumbent and stoloniferous bases. Lower portion of culms often flattened, and laterally compressed and sharply keeled leaf sheaths. Blades flat or folded. Inflorescence with few to several persistent, spicate branches, these mostly digitate or clustered at the culm apex but loosely distributed along the main axis in a few species. Spikelets with a single perfect floret and 1 to 3 staminate or rudimentary florets above. Disarticulation above the glumes. Glumes thin, lanceolate, 1-nerved (except in cleistogamous, subterranean spikelets), strongly unequal to nearly equal. Lemmas broadly to narrowly ovate or oblong, awnless or more frequently awned from a minutely bifid apex, 1- to 5-nerved, but usually with 3 strong nerves, glabrous or ciliate-pubescent or puberulent on the nerves. Palea well developed, strongly 2-nerved. Basic chromosome number, $x = 10$.

Fig. 5-95. *Chloris verticillata*. Plant and spikelet. (From Gould and Box, 1965.)

Species about seventy, distributed in the warmer regions of both hemi-spheres; 15 species native to the United States. At least 10 species have been reported as introduced or adventive in this country. Although none of the *Chloris* species are outstanding forage grasses, a number contribute to the available range forage of the Southwest. Included are *C. verticillata* Nutt., TUMBLE WINDMILLGRASS (Fig. 5-95), and *C. cucullata* Bisch., HOODED WINDMILLGRASS (Fig. 5-96). *Chloris gayana* Kunth, RHODESGRASS, an introduced perennial with stout stolons and tall culms, now is frequent in pastures, in abandoned fields, and on roadsides in south-ern Texas. *Chloris virgata* Swartz, FEATHER FINGERGRASS, with con-spicuously ciliate spikelets, is a weedy annual of wide distribution in the central and southern United States. Infrequent but of special interest is *C. chloridea* (Presl) Hitchc., BURYSEED CHLORIS, which develops cleisto-gamous spikelets on slender, underground branches (Fig. 5-97).

Following Clayton (1967), the genus *Trichloris* is merged with *Chloris*.

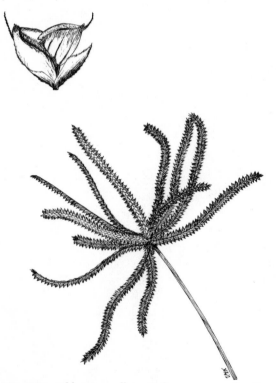

Fig. 5-96. *Chloris cucullata*. Inflorescence and spike-let. (From Gould and Box, 1965.)

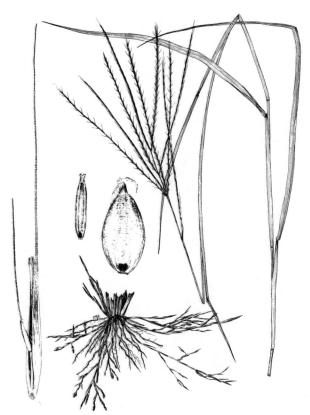

Fig. 5-97. *Chloris chloridea*. Base of plant showing cleistogenes on rhizomes, leaf culm with inflorescence, aerial spikelet (left) and caryopsis of aerial spikelet (left center), and large caryopsis of subterranean spikelet (center). (From Gould and Box, 1965.)

While *C. pluriflora* (Fourn.) Clayton, (*T. pluriflora* Fourn.) MULTIFLOWERED FALSE RHODESGRASS (Fig. 5-98), and *C. crinita* Lag. (*T. crinita* [Lag.] Parodi), FALSE RHODESGRASS, are readily distinguishable from the New World species of *Chloris* by their 3-awned lemmas, this distinction breaks down in Australia, where no sharp lines of generic separation can be maintained. References: Anderson, 1974; Vignal, 1979.

144. *Bouteloua* Lag.

Annuals and perennials of diverse habit, some cespitose, others with rhizomes or stolons. Leaves mostly basal, with flat or folded blades. Ligule usu-

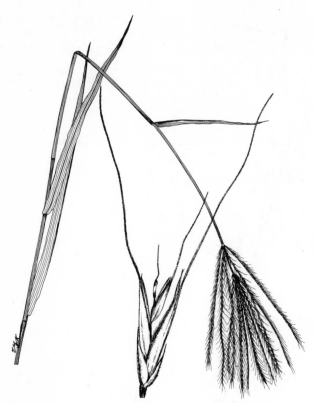

Fig. 5-98. *Chloris pluriflora*. Inflorescence and spikelet.
(From Gould and Box, 1965, as *Trichloris pluriflora*.)

ally a ring of hairs. Inflorescence of 1 to numerous short, spicate branches, these closely or distantly spaced along the main axis and bearing 1 to numerous sessile spikelets in two rows along the margins of an angular or flattened rachis. Disarticulation at the base of the branch rachis or above the glumes. Spikelets with 1 fertile floret and 1 to 3 reduced or rudimentary florets above. Glumes lanceolate, 1-nerved, unequal to nearly equal, awnless or short-awned. Lemmas membranous, 3-nerved, the mid-nerve often extending into an awn, the lateral nerves occasionally short-awned. Palea membranous, the 2 nerves occasionally awn-tipped. Basic chromosome number, $x = 10$.

Species about fifty, the majority in North America but several in Central and South America. Some seventeen species occur in the United States, mostly in the Southwest. The species of *Bouteloua* fall into two well-marked subgenera, *Bouteloua* and *Chondrosium*. In subgenus *Bouteloua*, disarticulation is at the base of the inflorescence branch, whereas in subgenus *Chondrosium* the spikelets disarticulate above the glumes, and the branch is persistent. Furthermore, in subgenus *Bouteloua* the inflorescence branches are several to numerous and usually with relatively few (1 to 15) large, non-pectinate spikelets, these with a single rudimentary floret. In subgenus *Chondrosium* the branches are 1 to several and typically with numerous (20 to 80) small, closely placed, pectinate spikelets, these usually with 2 or 3 rudiments.

Several species of *Bouteloua* are outstanding range forage grasses in the western United States. *Bouteloua curtipendula* (Michx.) Torr., SIDEOATS GRAMA (Fig. 5-99), and *B. gracilis* (Willd. ex H.B.K.) Lag. ex Griffiths,

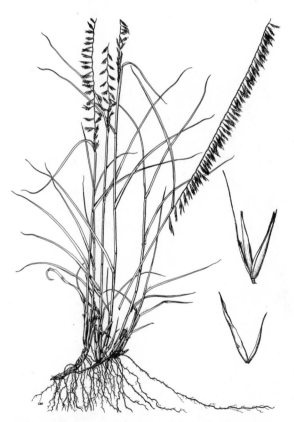

Fig. 5-99. *Bouteloua curtipendula* var. *curtipendula*. Plant and spikelet with glumes separated from the florets. (From Gould, 1951.)

BLUE GRAMA (Fig. 5-100), are perhaps the most widespread and valuable, the former at medium and low altitudes and the latter at medium and high altitudes. In the semiarid Southwest, *B. eriopoda* (Torr.) Torr., BLACK GRAMA, is an important range grass at medium and low altitudes. *Bou-*

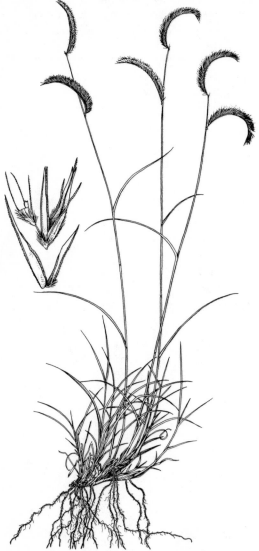

Fig. 5-100. *Bouteloua gracilis*. Plant and spikelet with glumes separated from the florets. (From Gould, 1951.)

teloua hirsuta Lag., HAIRY GRAMA, is widespread in the central and southwestern United States and northern Mexico but is a weak perennial of little forage significance. References: Gould and Kapadia, 1962, 1964; Kapadia and Gould, 1964a, 1964b; Mohamed and Gould, 1966; Gould, 1979; Reeder and Reeder, 1980.

Fig. 5-101. *Buchloë dactyloides*. (A) Pistillate plant with (B) inflorescence, (C) burr of female spikelets, and (D) pistillate floret. (E) Staminate plant and (F) male spikelet with glumes separated from the floret. (From Gould, 1951.)

145. *Buchloë* Engelm.

Low, mat-forming perennial, with extensive wiry stolons and flat, tufted leaves. Ligule a fringe of short hairs. Staminate and pistillate spikelets in separate inflorescences, usually on separate plants (dioecious) but not infrequently on the same plant (monoecious). Staminate spikelets 2-flowered, sessile, and closely crowded on 1 to 4 short, spicate inflorescence branches, these well exserted above the leafy portion of the plant. Pistillate spikelets 1-flowered, in deciduous, capitate, burrlike clusters of 2 to 4, these present in the leafy portion of the plant and usually partially included in expanded leaf sheaths. Rachis and lower two-thirds of second glume of pistillate spikelets thickened, indurate. Lemma of pistillate spikelet membranous, 3-nerved, usually glabrous and awnless. Basic chromosome number, $x = 10$.

Buchloë is the most widespread of five closely related, monotypic, dioecious North American genera of the tribe Chlorideae (Reeder and Reeder, 1963; Reeder et al., 1965). The single species *B. dactyloides* (Nutt.) Engelm., BUFFALOGRASS (Fig. 5-101), ranges from Montana to northern Mexico and is a dominant of the western "short-grass prairie." When growing in pure stands, it forms a soft, grayish-green turf. Although not considered a first-class forage grass, it is one of the most important range plants of the Great Plains area. Because of its low growth habit and general vigor, it persists under heavy grazing pressure and has considerable value as a soil binder.

146. *Cathestecum* Presl

Low, mostly stoloniferous annuals and perennials with tufted leaves and small, spicate, few-flowered inflorescence. Spikelets in subsessile clusters of 3, the terminal (central) one perfect, the 2 lower (lateral) ones usually staminate or sterile. Spikelet clusters few to several, scattered on the short culm rachis and deciduous as a whole. Terminal spikelet of each cluster with 1 perfect floret below and 1 or more reduced florets above. Glumes unequal, usually hairy, the first short, thin, nerveless or 1-nerved, and the second about as long as the lemma, 1-nerved. Lemmas typically 3-nerved, the nerves extending into awns at the broadly notched and 2-lobed apex. Palea well developed, the nerves usually extending into setae. Basic chromosome number, $x = 10$.

Species about six, mostly in Mexico; one in the southwestern United States and two ranging southward to Central America (Swallen, 1937). *Cathestecum erectum* Vasey & Hack., a low, tufted, stoloniferous perennial, is occasional in western Texas and southern Arizona. Pierce (1978) has segregated the monotypic genus *Griffithsochloa* (*G. multifida* [Griffiths] Pierce) from *Cathestecum*; however, this new genus does not occur within the United States.

147. *Aegopogon* Humb. & Bonpl.

Tufted annuals (or perennials?) with weak culms, short, flat blades and small, delicate inflorescences. Ligules membranous, the apex dentate or fimbriate. Spikelets 1-flowered, in pedunculate groups of 3, these alternately arranged on a slender rachis and disarticulating as a unit. Central spikelet of the cluster short-pediceled, fertile, the 2 lateral ones longer-pediceled, staminate or rudimentary and smaller. Glumes equal, membranous, 1-nerved, usually awned from a notched apex. Lemma of the perfect spikelet membranous,

Fig. 5-102. *Aegopogon tenellus*. Plant and spikelet cluster. (From Gould, 1951.)

longer than the glumes, 3-nerved, the nerves usually excurrent as short awns. Palea hyaline, the nerves extended as short awns. Caryopsis turgid, obovate, brownish, the surface slightly ridged and roughened. Basic chromosome number, $x = 10$.

A genus of three species, with distribution from the southwestern United States to northern South America. *Aegopogon tenellus* (D.C.) Trin. (Fig. 5-102) is occasional in the mountains of southeastern Arizona, growing in the partial shade of trees and boulders at medium to high altitudes. In Arizona this species grows as a delicate annual mostly 10 to 25 cm tall, but in the tropics it assumes a more coarse, robust habit. Hitchcock (1935, 1951) placed *Aegopogon* in the tribe Zoysieae, but it seems more appropriately grouped near *Cathestecum* in the Chlorideae. Reference: Beetle, 1948.

148. *Spartina* Schreb.

Perennials with slender or coarse, moderately tall or tall culms, these usually from a rhizomatous base. Leaves tough and firm, the blades long, flat or involute. Ligule a ring of long or short hairs. Inflorescence of few to numerous, racemosely arranged, short, usually appressed branches, bearing closely placed, sessile spikelets. Disarticulation below the glumes. Spikelets 1-flowered, laterally flattened. Glumes unequal, keeled, usually 1-nerved or the second with 3 closely placed nerves, acute or the second short-awned. Lemma firm, keeled, strongly 1- or 3-nerved and often with additional indistinct lateral nerves, tapering to a narrow but usually rounded, awnless tip. Palea as long as, or longer than, the lemma, with broad, membranous margins on either side of the closely placed nerves. Basic chromosome number, $x = 10$. Published counts of $2n = 28$, 42, and 56 appear to be in error.

A genus of sixteen species, one native to western Europe, the others American. Nine species are reported for the United States, all but two with distribution in coastal marshes. *Spartina pectinata* Link, PRAIRIE CORD-GRASS, is widespread in freshwater marshes and wet meadows throughout the United States except in the extreme Southeast and Southwest. *Spartina spartinae* (Trin.) Merr., GULF CORDGRASS (Fig. 5-103), is a coarse bunchgrass that grows in brackish marshes and saline coastal prairies from Florida to Texas, Mexico, and Central America. It is also reported from Paraguay and Argentina. Frequent along coastal flats of the East Coast and the Gulf Coast is *S. patens* (Ait.) Muhl., MARSHHAY CORDGRASS (Fig. 5-104).

Stebbins and Crampton (1961) followed Hubbard (1954) in segregating *Spartina* as a distinct tribe. This disposition was, at least in part, based on the erroneous belief that *Spartina* has no lodicules and that the basic chromosome number of the genus is $x = 7$ instead of $x = 10$, as has now been established. Reference: Mobberly, 1956.

Fig. 5-103. *Spartina spartinae*. Inflorescence and spikelet. (From Gould and Box, 1965.)

149. *Ctenium* Panzer

Perennials, some rhizomatous, mostly with tall, slender culms. Inflorescence (in ours) a short spike, this unilateral and usually curved, appearing as a branch that has assumed a terminal position. Spikelets several-flowered but with only 1 perfect floret, sessile in two rows on one side of the rachis, disarticulating above the glumes. First glume small, thin, 1-nerved. Second glume firm, 3- or 4-nerved, about as long as the lemmas, bearing from the middle of the back a stout, divergent awn. Lemmas thin, 3-nerved, pubescent in the lateral nerves, with a stout, divergent awn borne dorsally just below the tip. Lowermost 2 florets sterile, the third floret fertile, the upper 1 to 3 florets sterile and successively smaller. Palea of fertile floret about as long as the lemma. Basic chromosome number, $x = 10$.

A genus of twelve species in warm-temperate to tropical regions of the world, mostly in the Americas. *Ctenium floridanum* (Hitchc.) Hitchc., with creeping rhizomes and prominent glands on the second glume, is known

Fig. 5-104. *Spartina patens.* Inflorescence. (From Gould and Box, 1965.)

only from moist pine barrens of Florida. *Ctenium aromaticum* (Walt.) Wood, TOOTHACHE GRASS, without rhizomes or prominent glands on the glumes, is present in similar sites from Virginia to Florida and Louisiana.

150. *Hilaria* H.B.K.

Perennials, mostly rhizomatous or stoloniferous. Inflorescence a slender, dense, bilateral spike, the spikelets in clusters of 3 at each node of a zigzag rachis, the clusters deciduous as a whole. Spikelets of the cluster dissimilar, the 2 lateral ones 2-flowered, staminate, and the central one 1-flowered, perfect. Glumes firm, flat, usually asymmetrical, bearing an awn on one side from about the middle. Lemmas thin, 3-nerved, awned or awnless. Palea about as large as the lemma and similar in texture. Basic chromosome number, $x = 9$.

Species seven, mostly in the southern United States and northern Mexico, one ranging southward to Venezuela. Five species are present in the

United States, all in arid and semiarid regions of the Southwest. The stoloniferous *Hilaria belangeri* (Steud.) Nash, CURLY MESQUITE (Fig. 5-105), is distributed from southern Texas and Arizona to northern Mexico and furnishes considerable range forage. *Hilaria mutica* (Buckl.) Benth., TOBOSA, and *H. jamesii* (Torr.) Benth., GALLETA, are coarse, rhizomatous, sod-forming grasses of low palatability. *Hilaria rigida* (Thurb.) Benth., BIG GALLETA, is a large, robust desert species with culms in clumps from a more or less woody, rhizomatous base.

Tribe 14. Zoysieae

151. *Zoysia* Willd.

Low, sod-forming perennials, mostly with rhizomes and slender stolons. Inflorescence a slender, few-flowered spike, the spikelets solitary at the nodes of a slender, zigzag rachis. Disarticulation at the base of the 1-flowered, laterally compressed spikelets. First glume absent. Second glume firm, acute, mucronate or short awned. Lemma thin, membranous, awnless, narrow at the apex, shorter than the second glume and enclosed by it. Palea present or absent. Basic chromosome number, $x = 10$.

Species about five, native to southeastern Asia and New Zealand. Three

Fig. 5-105. *Hilaria belangeri*. Plant and two views of spikelet cluster. (From Gould, 1951.)

Fig. 5-106. *Tragus berteronianus*. Plant and
spikelet cluster. (From Gould, 1951.)

species, *Zoysia matrella* (L.) Merr., *Z. tenuifolia* Willd. ex Trin., and *Z. japonica* Steud., have been introduced into the southern and southeastern United States as lawn grasses and may be present as escapes in some areas.

152. *Tragus* Hall

Low annuals with weak stems; soft, flat blades; and slender, spikelike inflorescences of bristly, burrlike clusters of 2 to 5 spikelets. Ligule a ring of short, woolly hairs. Disarticulation at the base of each spikelet cluster, the inflorescence axis persistent. Spikelets 1-flowered. First glume small, thin, much reduced or wanting. Second glume of the lower 2 spikelets of a cluster large and firm, bearing three rows of stout, hooked spines. Lemmas of the lower spikelets thin and flat. Upper 1 to 3 spikelets sterile, the uppermost usually rudimentary. Basic chromosome number, $x = 10$.

A genus of three weedy annuals, these widely distributed in the tropics and subtropics of the world but none native to the United States. *Tragus berteronianus* Schult. (Fig. 5-106), with small, sessile spikelet burrs, ranges from Texas, New Mexico, and Arizona south through Mexico to South America. *Tragus racemosus* (L.) All., with larger, short-pediceled burrs, is adventive at a few locations from Maine to North Carolina and has also been reported from Texas and Arizona.

Tribe 15. Aeluropodeae

153. *Distichlis* Raf.

Low, dioecious (rarely monoecious) perennials with stout rhizomes and firm, glabrous culms and leaves. Culms with many nodes and internodes, usually completely covered to the inflorescence by the distichous and closely overlapping leaf sheaths. Ligule a short, fringed membrane. Spikelets large, several-flowered, laterally flattened, in short, contracted panicles or racemes. Disarticulation above the glumes and between the florets. Glumes unequal, acute, 3- to 5-nerved. Lemmas broad, indistinctly 9- to 11-nerved, awnless, laterally compressed, those of the pistillate spikelets coriaceous, thicker than those of the staminate spikelets. Palea large, strongly 2-keeled. Basic chromosome number, $x = 10$.

As presently interpreted, a genus of four species, with distribution in North and South America. *Distichlis spicata* (L.) Greene var. *spicata*, SEASHORE SALTGRASS (Fig. 5-107), is frequent on salty flats and along brackish ponds of the U.S. Atlantic and Pacific Coasts and the Gulf of Mexico. Although the herbage is somewhat coarse and tough, this is a fair to good forage grass in saline sites. INTERIOR SALTGRASS, *D. spicata* var. *stricta* (Torr.) Beetle, grows on alkaline and saline flats from Montana, Iowa, and Texas westward to the Pacific Coast and also provides considerable live-

Fig. 5-107. *Distichlis spicata*. Plant and inflorescence. (From Gould and Box, 1965.)

stock forage. The species listed by Hitchcock (1951) as *D. texana* (Vasey) Scribn. is now referred to the genus *Allolepis*. References: Beetle, 1943, 1955a.

154. *Allolepis* Soderstrom & Decker

Dioecious perennial with or without rhizomes and with stolons as much as 50 feet long. Culms decumbent at base, 30 to 60 cm tall. Blades flat, firm, glabrous or scabrous, with bicellular microhairs not sunken in the epidermis. Inflorescence a contracted panicle of large, 4- to 8-flowered, unisexual spikelets. Pistillate spikelets about twice as large as the staminate ones. Glumes unequal, acute. Lemmas of pistillate spikelets broad, coriaceous, laterally compressed, acute at the apex, with 3 strong nerves and indistinct intermediate nerves, the margins thin and erose. Paleas of the pistillate spikelets broad and bowed out below, narrow above, keeled, the 2 keels with narrow, erose or toothed wings. Caryopsis tightly enclosed by the base of the palea. Lemmas of the staminate spikelets thin, 3-nerved. Paleas of the staminate spikelets not bowed out and lacking winged keels. Basic chromosome number, $x = 10$.

A genus of a single species, *Allolepis texana* (Vasey) Soderstrom & Decker, this reported from sandy flats in Pecos, Crane, Presidio, Brewster, and El Paso counties, Texas, and northern Mexico. Reference: Soderstrom and Decker, 1965.

155. *Monanthochloë* Engelm.

Low, mat-forming, dioecious perennials with wiry, decumbent, much-branched culms and tufted leaves, these with blades mostly less than 1 cm long. Spikelets 3- to 5-flowered, only the lower fertile, borne in the axils of fascicled leaves terminating the short, erect branches. Glumes absent. Lemmas several-nerved, those of the pistillate spikelets like the leaf blades in texture. Palea narrow, 2-nerved, enfolding the caryopsis. Chromosome number unreported.

A genus of three species, one in the southern United States and Mexico and the other two in Argentina (Parodi, 1954). *Monanthochloë littoralis* Engelm., SHOREGRASS (Fig. 5-108), grows on saline, muddy or sandy

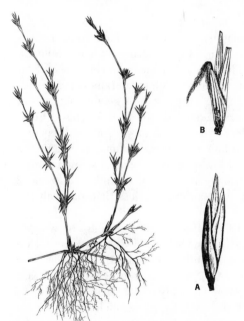

Fig. 5-108. *Monanthochloë littoralis*. Plant, (A) pistillate spikelet, and (B) staminate spikelet. (From Gould and Box, 1965.)

coastal flats in southern Florida, Louisiana, Texas, and southern California. It is also present on the coasts of Mexico and Cuba. This species is usually associated with *Distichlis spicata*, but in some saline sites it is the only grass vegetation.

156. *Swallenia* Soderstrom & Decker

Stout perennial with a woody, rhizomatous base and short, flat, stiff leaves. Spikelets large, 3- to 7-flowered, the florets crowded and persistent on the continuous rachis of a narrow panicle. Glumes subequal, awnless, about as long as the lemmas, 7- to 11-nerved. Lemmas broad, 5- to 7-nerved, awnless, densely hairy on margins below. Palea thin, broad, as long as the lemma or longer, villous on the margins with long hairs and somewhat hairy on the base and back. Caryopsis readily falling from the floret at maturity or even before. Basic chromosome number, $x = 10$.

A genus of a single species, *Swallenia alexandrae* Soderstrom & Decker, known only from the type locality in Eureka Valley, Inyo County, California. An extremely odd and interesting endemic of uncertain affinities, adapted for survival on dry, sandy sites.

Tribe 16. Unioleae

157. *Uniola* L.

Tall, coarse perennials, one rhizomatous, the other stoloniferous. Leaves long, the blades flat but involute on drying. Ligule a fringe of hairs. Inflorescence an open or a contracted panicle of 5- to many-flowered spikelets. Lower 2 to 6 florets of the spikelet sterile. Spikelets laterally compressed, disarticulating below the glumes, falling entire. Glumes subequal, acute, awnless or slightly mucronate, 3-nerved, keeled, the keel serrulate. Lemmas acute to narrowly obtuse, awnless or mucronate, 3- to 9-nerved, keeled, the keels serrulate. Palea shorter to longer than the lemma, 2-keeled, the keels winged and serrate to ciliate. Flowers perfect, with 3 stamens and 2 lodicules, these fleshy, cuneate. Ovary glabrous, with a single style and 2 plumose stigmas. Caryopsis linear, with an embryo less than one-half the length of the grain. Basic chromosome number, $x = 10$.

In the interpretation of Yates (1966a, 1966b, 1966c), a genus of two species, one, *Uniola paniculata* L., SEA OATS (Fig. 5-109), distributed on beaches and sand dunes from Alabama south to Florida, the Bahamas, the Gulf Coast of the United States and Mexico, and islands of the Caribbean, and the other, *U. pittieri* Hack., growing on sandy beaches along both coasts of Mexico and south to Ecuador. Of the six U.S. species included in *Uniola* by Hitchcock (1951), Yates retained only *U. paniculata* and grouped the other species in *Chasmanthium*. Among the numerous significant differ-

Fig. 5-109. *Uniola paniculata*. Plants on coastal dunes and inflorescence showing many-flowered spikelets.

ences noted between *Uniola* and *Chasmanthium* are those of chromosome number, leaf anatomy, type of epidermal bicellular microhairs, embryo type (P − PF in *Uniola*, P + PP in *Chasmanthium*), spikelet disarticulation, and number of stamens.

Tribe 17. Pappophoreae

158. *Pappophorum* Schreb.

Erect, cespitose perennials with slender, contracted, usually spikelike panicles of bristly spikelets. Ligule a ring of hairs. Blades long, narrow, flat or
folded. Spikelets 3- to 6-flowered, only the lower 1 to 3 fertile, disarticulating above the glumes, the florets falling together. Glumes subequal, thin
and membranous, 1-nerved. Lemmas firm, rounded on the back, indistinctly many-nerved, the nerves extending into 11 or more unequal, glabrous or scabrous awns. Palea about as long as the body of the lemma. Basic
chromosome number, $x = 10$.

A small genus of North and South American grasses, with two species in
the United States. *Pappophorum mucronulatum* Nees, WHIPLASH PAP-
PUSGRASS, with a tightly contracted, whitish panicle, is frequent in grasslands of southern Texas. It ranges from Texas to southeastern Arizona and
northern Mexico and also occurs in South America. *Pappophorum bicolor*
Fourn., PINK PAPPUSGRASS (Fig. 5-110), with a more loosely contracted,
pink- or purple-tinged panicle, has about the same range as *P. mucronulatum* in the United States and northern Mexico.

Fig. 5-110. *Pappophorum bicolor*.
Plant, spikelet, and floret. (From
Gould and Box, 1965.)

159. *Enneapogon* Desv. ex Beauv.

Low, tufted perennials, with narrow, often spikelike panicles. Ligule a ring of hairs. Spikelets several-flowered, the upper florets reduced. Disarticulation above the glumes and tardily between the florets. Glumes subequal, lanceolate, with 5 to numerous nerves. Lemmas broad, much shorter than the glumes, firm, rounded on the back, strongly 9-nerved and with 9 equal, plumose awns. Palea slightly longer than the body of the lemma. Basic chromosome number, $x = 10$.

A genus of about thirty-five species, these mostly in Australia, Africa, and Asia, with one species in North and South America. *Enneapogon desvauxii* Beauv., SPIKE PAPPUSGRASS, an inconspicuous, tufted, weak perennial adapted to dry, sterile soils, is distributed from Utah to Texas and Arizona and southward in Mexico to Oaxaca.

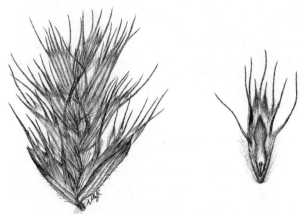

Fig. 5-111. *Cottea pappophoroides*. Spikelet and floret.

160. *Cottea* Kunth

Tufted perennial with culms 30 to 70 cm tall. Leaves usually pilose, the blades flat or folded. Ligule a ring of hairs. Inflorescence a narrow but open panicle with rather stout, short, erect-spreading branches. Spikelets large, with 6 to 10 florets, the upper ones reduced. Disarticulation above the glumes and between the florets. Glumes subequal, about as long as the lemmas, broadly lanceolate, with 7 to 13 fine nerves, the midnerve sometimes continued as a short awn. Lemmas broad, irregularly lobed and cleft, hairy below, with 9 to 13 strong nerves, these extended into awns of varying lengths. Palea broad, slightly longer than the body of the lemma. Basic chromosome number, $x = 10$.

A genus of one species, *Cottea pappophoroides* Kunth (Fig. 5-111), this ranging in North America from western Texas and southern Arizona to central Mexico and in South America from Ecuador to Argentina and Peru. *Cottea pappophoroides* occurs in scattered stands on dry, rocky slopes at medium altitudes.

Tribe 18. Orcuttieae

161. *Orcuttia* Vasey

Low, tufted annuals. Leaves sheathing at base and flattened into a blade above but without a definite differentiation of sheath and blade. Ligule absent. Inflorescence a spikelike raceme, the spikelets several-flowered. Disarticulation above the glumes and, when the spikelet is wet, between the florets (Reeder, 1965). Glumes about equal, shorter than the lemmas, irregularly 2- to 5-toothed and many-nerved. Lemmas firm, mostly 9- to 15-nerved, the apex 5- to 11-toothed. Palea broad, 2-nerved, as long as the lemma. Basic chromosome number probably $x = 10$, but counts of $2n = 24$, 26, and 32 are reported.

Species five, for the most part endemic to the Central Valley of California. One species, *Orcuttia californica* Vasey, ranges south into Lower California. Rare grasses of no economic significance, present along the margins of drying (vernal) pools and in moist, open depressions. References: Crampton, 1959; Reeder, 1965.

162. *Neostapfia* Davy

Tufted annual with spreading, leafy culms 7 to 30 cm tall. Leaves viscid-glandular on the nerves and margins, loosely sheathing below, broading into a flat blade above but without definite demarcation between sheath and blade. Ligule absent. Inflorescence a short, cylindrical, spikelike raceme or panicle, partially enclosed by the broad, sheathing upper leaves. Spikelets several-flowered, disarticulating above the glumes. Glumes and lemmas similar, large, thin, flat, awnless, strongly 9- to 11-nerved, with viscid glands on the nerves and margins. Palea thin, 2-nerved, expanded above to a broad, rounded apex. Basic chromosome number, $x = 10$.

A monotypic genus, closely related to *Orcuttia* and endemic to the same region as that genus. The single species, *Neostapfia colusana* (Davy) Davy, is known from Colusa, Stanislaus, Solano, and Merced counties, California, where it grows along the margins of vernal pools and "potholes" on hard alkali soil. Reference: Crampton, 1959.

Tribe 19. Aristideae

163. *Aristida* L.

Low to moderately tall annuals and perennials, these lacking rhizomes or stolons. Blades narrow, usually involute. Inflorescence an open or con-

tracted panicle of usually large, 1-flowered spikelets. Disarticulation above
the glumes. Glumes thin, lanceolate, with a strong central nerve and occasionally 2 lateral nerves. Lemma indurate, terete, 3-nerved, with a hard,
sharp-pointed callus at the base, tapering gradually or less frequently
abruptly to an awn column bearing usually 3 stiff awns, the lateral awns partially or totally reduced in a few species. In some African taxa the awns are
plumose and feathery. Caryopses long and slender, permanently enclosed in
the firm lemma. Basic chromosome number, $x = 11$.

Species about two hundred, in the warmer parts of the world. Chippendahl (1955) listed sixty-six species for South Africa, noting that some are
among the most important and widespread grasses in that country. *Aristida*
is represented in the United States by forty species (Hitchcock, 1951). Most
of the U.S. species grow in semiarid habitats of the Southwest, but many
occur in sandy, sterile sites in the East and Southeast. Although frequent on
western rangelands, the "threeawns," for the most part, have little forage
value, and from the range viewpoint few are considered desirable. Species

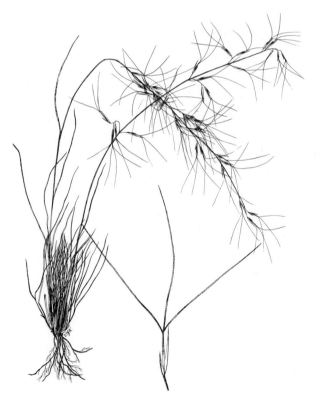

Fig. 5-112. *Aristida oligantha*. Plant and spikelet. (From
Gould and Box, 1965.)

such as *A. purpurea* Nutt., PURPLE THREEAWN; *A. longiseta* Steud., RED THREEAWN; and *A. oligantha* Michx., PRAIRIE THREEAWN (Fig. 5-112), often predominate on dry, overutilized, and disturbed range sites. The stout, usually geniculate, twisted awn and sharp-pointed calluses of the mature *Aristida* fruit are highly effective mechanisms for seed dispersal and seedling establishment.

As treated by Hitchcock (1951), the North American species of *Aristida* fall into three sections: *Arthratherum*, *Streptachne*, and *Chaetaria*. In grasses of the small section *Arthratherum*, the awn column disarticulates from the body of the lemma at a well-defined joint. Representative of this group are *A. desmantha* Trin. & Rup., CURLY THREEAWN, a rather tall annual of woodland habitats from Illinois to Nebraska and Texas, and *A. californica* Thurb., a low, tufted perennial of southwestern Arizona to southern California and northwestern Mexico.

Sections *Streptachne* and *Chaetaria* are separated on the basis of the

Fig. 5-113. *Aristida hamulosa*. Plant and spikelet. (From Gould, 1951.)

absence of, or great reduction in, the lateral awns of the former. This separation appears artificial, at least to a certain extent, as it places such similar species as *A. ternipes* Cav. and *A. hamulosa* Henr. in separate sections. *Aristida ternipes*, SPIDERGRASS, a bunchgrass with lateral awns greatly reduced or absent, is frequent in semiarid grasslands from Arizona south to South America. *Aristida hamulosa*, HOOK THREEAWN (Fig. 5-113), very similar in general aspect to *A. ternipes* but with 3 well-developed awns, ranges from Arizona and Texas to Central America.

Subfamily IV. Bambusoideae

Tribe 20. Bambuseae

164. *Arundinaria* Michx.

Small shrubby to large woody perennials with culms from stout, creeping rhizomes. Leaves of the main culms and larger branches with thin, papery sheaths and reduced or rudimentary blades. Leaves of the branch tips with firm sheaths; broad, flat blades; and a petiole-like constriction between the sheath and blade. Sheaths with a few stiff hairs or bristles on either side at the apex and a short, firm, erose to ciliate membrane across the collar. Ligule a firm, short, membranous rim. Spikelets produced at intervals of five to ten years in small panicles or racemes on branchlets of the leafy shoots and also on special flowering shoots arising from the rhizomes. Spikelets (in our species) large, many-flowered, mostly 3 to 7 cm long, borne on slender pedicels. Disarticulation above the glumes and between the florets. Glumes similar to the lemmas but shorter, the first sometimes minute or absent. Lemmas large, thin, many-nerved, gradually tapering to an acute, acuminate or short-awned apex. Palea large, 2-nerved, and 2-keeled. Stamens and lodicules 3. Stigmas usually 3, the style solitary. Caryopsis narrowly elliptic, terete, about 1 cm long. Basic chromosome number, $x = 12$.

A genus of about one hundred species, distributed principally in southeastern Asia and the adjacent islands, Japan to Madagascar (Swallen, 1955). *Arundinaria* is represented in the United States by *A. gigantea* (Walt.) Muhl., GIANT CANE (Fig. 5-114), and *A. tecta* (Walt.) Muhl., SWITCH-CANE. Giant cane is distributed from southern Ohio, Illinois, and Missouri south to Florida and Texas, whereas switchcane ranges from southern Maryland to Alabama and Mississippi (Hitchcock, 1951). These grasses form extensive colonies in low, moist woodlands and along streams and rivers. They are highly palatable to livestock, but the plants reportedly are easily destroyed by overgrazing and trampling. *Arundinaria gigantea* and *A. tecta* are closely related and possibly intergrading taxa. McClure (1963) has reported that the rhizomes of *A. tecta* have large, longitudinally continuous, peripheral air canals, whereas air canals have not been observed in rhizomes of typ-

Fig. 5-114. *Arundinaria gigantea*. Sterile shoot and inflorescences.

ical *A. gigantea.* Calderón and Soderstrom (1980) support the placement of this genus in the small tribe Arundinareae proposed by Nees (1835).

Tribe 21. Phareae

165. *Pharus* L.

Herbaceous perennials with erect or stoloniferous culm bases. Leaves with a petiole-like constriction to 1 cm long between the sheath and blade. Blades broad, tapering to both ends, with numerous fine nerves radiating out from the midnerve to the margins. Inflorescence a panicle of unisexual, 1-flowered spikelets, in pairs of 1 large, pistillate, and sessile and 1 small, staminate, and long-pediceled, the spikelet pairs appressed along the branches. Disarticulation below the spikelets. Glumes unequal or nearly equal, shorter than the lemmas. Lemma of the pistillate spikelet bearing uncinate hairs, at least near the beaked tip, becoming indurate and tightly inrolled over the palea at maturity. Stigmas 3, stamens 6. Basic chromosome number, $x = 12$.

A tropical genus of about eight species, represented in the United States by *Pharus parvifolius* Nash, the distribution of which is southern Florida and the West Indies to Brazil. *Pharus* was placed in the tribe Zizanieae by Hitchcock (1951) and in the tribe Phareae of subfamily Oryzoideae by Stebbins and Crampton (1961). Parodi (1961) grouped the tribe Phareae in subfamily Bambusoideae. Reference: Soderstrom and Calderón, 1979.

Subfamily V. Oryzoideae

Tribe 22. Oryzeae

166. *Oryza* L.

Annuals and a few perennials, with herbaceous culms; flat, lanceolate blades; and large, apparently 1-flowered spikelets borne in open or somewhat contracted panicles. Ligules membranous. Spikelets laterally flattened, with a thick, firm, 5-nerved, awnless or awned lemma and a large, 2-nerved palea of similar texture tightly enclosing the caryopsis. At the base of the lemma (in *Oryza sativa*) are 2 short, pointed bracts, these superficially resembling glumes but more correctly interpreted as being reduced lemmas of rudimentary florets. Disarticulation below the spikelet. Stamens usually 6, lodicules 2. Basic chromosome number, $x = 12$.

A genus of fifteen to twenty species, these widely distributed in moist tropical-subtropical habits of both the New and Old Worlds but none native to the United States. *Oryza sativa* L. (Fig. 5-115), RICE, is one of the world's major crop plants. In the United States rice is grown most extensively along the Gulf Coast in Louisiana and Texas and in the Central Valley of California.

Fig. 5-115. *Oryza sativa*. Inflorescence and spikelet.
(From Gould and Box, 1965.)

167. *Leersia* Swartz

Rhizomatous perennials with flat blades and open panicles. Ligule a short, firm membrane, often continued laterally as short sheath auricles. Spikelets 1-flowered, strongly compressed laterally, awnless, subsessile and crowded at the branch tips. Disarticulation below the spikelet. Glumes absent. Lemma firm or indurate, boat-shaped, 5-nerved, tightly enclosing the margins of a firm, narrow, usually 3-nerved palea. Stamen number varying from 1 to 6. Basic chromosome number, $x = 12$.

Species about ten, in marshy or mesic woodland habitats of temperate and tropical regions of both hemispheres. Five species are present in the United States, these mostly in the eastern states. *Leersia oryzoides* (L.) Swartz, RICE CUTGRASS, is widely distributed throughout the country.

Leersia hexandra Swartz (Fig. 5-116), a species of the tropics of both hemi-spheres, is present along coastal marshes and waterways from Virginia to Florida and Texas. Reference: Pyrah, 1969.

Fig. 5-116. *Leersia hexandra*. Plant and spikelet. (From Gould and Box, 1965.)

168. *Zizania* L.

Tall, reedlike marsh grasses with broad leaves and large panicles of 1-flowered, unisexual spikelets. Ligule membranous, often large. Staminate spikelets pendulous on the lower inflorescence branches. Pistillate spikelets erect on the short, stiffly erect upper panicle branches. Disarticulation below the spikelet. Glumes absent. Pistillate spikelets slender, about 2 cm long, with a firm, 3-nerved, long-awned lemma tightly clasping a narrow, 2-nerved palea. Staminate spikelets with a thin, 5-nerved, acuminate or short

Fig. 5-117. *Zizania aquatica*. Inflorescence, staminate spikelet (awnless), and pistillate spikelet (awned).

awn-tipped lemma, 3-nerved palea, and 6 stamens. Basic chromosome number, $x = 15$.

Species four, three native to the United States and one, *Zizania latifolia* (Griseb.) Turcz. ex Stapf, native to eastern Asia. *Zizania aquatica* L., ANNUAL WILDRICE (Fig. 5-117), is found along the Atlantic Coastal Plain to Louisiana and extends into New England and westward into Wisconsin; *Z. palustris*, NORTHERN WILDRICE, is reported from New England to the Great Lake states and central Canada; this species has long been used as a food source by Indians. *Zizania texana* Hitchc., TEXAS WILDRICE, is known only from the type locality, San Marcos, Texas, and has been reported as a doubtful species; however, Terrell et al., (1978) presented substantial evidence for the recognition of this taxon. References: Dore, 1969; Emery, 1967, 1977; Terrell and Wiser, 1975.

169. *Zizaniopsis* Doell & Aschers.

Large, robust, succulent perennials with stout rhizomes; broad, flat leaves; and large, open panicles. Ligule thin, membranous, to over 1 cm long. Spikelets large, 1-flowered, unisexual, the staminate and pistillate on the same branches, the pistillate above and the staminate below. Disarticulation below the spikelet. Glumes absent. Pistillate spikelet with a thin, 7-nerved, short-awned lemma and a large, 3-nerved palea. Caryopsis obovate, asymmetrical, beaked with a persistent style. Staminate spikelet with a thin, awnless, 5-nerved lemma and a 3-nerved palea of similar texture. Stamens 6. Basic chromosome number, $x = 12$.

Species three, one in the United States and two in South America. *Zizaniopsis miliacea* (Michx.) Doell & Ascherson, MARSHMILLET or SOUTHERN WILDRICE (Fig. 5-118), grows in shallow water along lakes and rivers from Maryland to Florida and eastern Texas. It is also present in dense stands along brackish coastal marshes. Marshmillet has value in supplying cover for wildlife, but tends to be objectionable in obstructing and filling up waterways and lake shores.

170. *Luziola* Juss.

Low, rather delicate perennials, some plants possibly annual, with weak, often submerged culms and small panicles of 1-flowered, unisexual spikelets. Ligule large, thin, membranous. Pistillate and staminate spikelets in the same inflorescence or more often in separate inflorescences on the same plant, the staminate usually terminal, the pistillate at the middle and upper culm nodes. Disarticulation below the spikelet. Glumes absent. Lemma and palea about equal, thin, several- to many-nerved. Caryopsis globose or oblong, smooth or striate, free from the lemma and palea. Stamens usually 6.

Based on Terrell and Robinson (1974), the monotypic genus *Hydrochloa* (*H. caroliniensis* Beauv.) is merged with *Luziola*. As currently inter-

Fig. 5-118. *Zizaniopsis miliacea.* Plant, (A) staminate spikelet, (B) pistillate spikelet, and (C) caryopsis. (From Gould and Box, 1965.)

preted, there are twelve species in tropical and subtropical regions of the New World (Swallen, 1965). The three species present in the southeastern United States are *Luziola peruviana* Gmel. in Louisiana and Florida, *L. bahiensis* (Steud.) Hitchc. in Alabama, and *L. caroliniensis* (Beauv.) Raspail from North Carolina to Texas.

Subfamily VI. Arundinoideae

Tribe 23. Arundineae

171. *Arundo* L.

Tall, stout, rhizomatous perennials with broad blades and large, plumose panicles. Ligule a short, lacerate, ciliate membrane. Spikelets with several florets, these successively smaller from the basal one upward. Rachilla glabrous, disarticulating above the glumes and between the florets. Glumes large, somewhat unequal, thin, 3-nerved, tapering to a slender point. Lemmas thin, 3-nerved, villous on the lower half, tapering to a point or short awn, the lateral nerves often extended as short teeth. Palea thin, 2-keeled. Chromosome records of $2n = 60$, 72, and 110 have been reported for *Arundo donax*.

A genus of about six species native to Asia. One species, *A. donax* L., GIANT REED (Fig. 5-119), has been introduced into the eastern and southern United States. In Texas, giant reed has been used extensively by the highway department for erosion control and beautification along culverts and bridges. It forms dense clumps or colonies, with culms 3 to 6 meters tall and panicles 20 to 40 cm or more long.

Fig. 5-119. *Arundo donax*. Portion of the inflorescence, and spikelet. (From Gould and Box, 1965.)

172. *Phragmites* Trin.

Tall, rhizomatous and stoloniferous perennial reed grasses with broad leaves and large, plumose panicles. Ligule a ring of short hairs. Spikelets several-flowered, disarticulating above the glumes and between the florets. Lower floret staminate or sterile, the terminal floret also reduced. Rachilla villous with long hairs. Glumes and lemmas thin, acute or acuminate, glabrous, the first glume 1-nerved, the second glume and the lemmas usually 3-nerved. Basic chromosome number, $x = 12$.

A genus of two or three species, one, *Phragmites australis* (Cav.) Trin.

ex Steudel., COMMON REED (Fig. 5-120), widespread in temperate and tropical regions of the world. This reed grass is present along lakes and waterways over most of the United States. It is similar in general aspect to *Arundo donax*. Reference: Clayton, 1968.

Fig. 5-120. *Phragmites australis*. Spikelet. (From Gould and Box, 1965, as *P. communis*.)

173. *Cortaderia* Stapf

Large, cespitose perennials with culms in dense clumps. Leaves densely clustered at the base of tall, slender, floriferous shoots. Leaf blades long and narrow, tough and fibrous, with rough, serrulate margins. Ligule absent (in our species). Plants dioecious, the pistillate spikelets plumose, in showy, silvery panicles 25 to 100 cm long, and the staminate spikelets glabrous, in large but nonshowy panicles. Disarticulation above the glumes and between the florets, the rachilla internodes jointed, the lower part glabrous, the upper bearded, forming a stipe for the floret. Spikelets 2- to 4-flowered, with narrow, papery, 1-nerved glumes and lemmas, the latter tapering to a narrow point or delicate awn. Lemmas of the pistillate spikelets with long, spreading hairs on the back and base. Chromosome number, $x = 12$.

A small genus of South American grasses, one species, *Cortaderia selloana* (Schult.) Ascherson & Graebn., PAMPASGRASS (Fig. 5-121), widely introduced in the warmer parts of the world. In the southern United States the female plants of pampasgrass, with their large, plumose panicles, commonly are grown as lawn ornamentals.

174. *Molinia* Schrank

Cespitose perennials with slender, leafy culms; flat blades; and open but narrow panicles. Ligule a fringe of soft hairs. Spikelets with 2 to 4 florets, these widely spaced on a relatively long rachilla. Rachilla bearing at its tip a rudimentary floret. Disarticulation above the glumes and between the florets.

Fig. 5-121. *Cortaderia selloana.* Flowering plant. (Photo by R. I. Lonard.)

Glumes shorter than the lemmas, 1-nerved. Lemmas thin, 3-nerved, acute. Palea bowed out below, equaling or slightly exceeding the lemma.

About five species, native to Europe and Asia. *Molinia caerulea* (L.) Moench. has been introduced at a few localities in the northeastern United States.

Tribe 24. Danthonieae

175. *Danthonia* Lam. & DC.

Low to moderately tall cespitose perennials with few-flowered panicles or racemes. Inflorescence commonly reduced to 1 spikelet in *Danthonia uni-*

spicata. Spikelets several-flowered, disarticulating above the glumes and between the florets. Glumes about equal, 1- to 5-nerved, much longer than the lemmas. Lemmas rounded and hairy on the back, indistinctly several-nerved, with a well-developed callus and a 2-toothed apex, the midnerve diverging as a stout, flat, twisted, geniculate awn at the base of the apical cleft. Palea broad, well developed. Basic chromosome number, $x = 6$.

Species over one hundred, throughout the temperate regions of the world but mostly in the Southern Hemisphere. Species of *Danthonia* are

Fig. 5-122. *Danthonia spicata.* Inflorescence and spikelet.

among the most valuable of the forage grasses in Australia and New Zealand and are also important on the ranges of South Africa. Seven species are listed for the United States, but none are of special forage significance. *Danthonia spicata* (L.) Beauv., POVERTY OATGRASS (Fig. 5-122), is present throughout the United States except in the far Southwest. Three of the other six species are in the eastern states, and three in the western part of the country. All the U.S. species produce cleistogamous spikelets in the axils of the lower sheaths (Hitchcock, 1951). References: de Wet, 1954, 1956.

176. *Sieglingia* Bernh.

Tufted perennial with short, narrow, flat, mostly basal leaves and a narrow, few-flowered panicle or raceme. Spikelets several-flowered, 8 to 12 mm long, disarticulating above the glumes and between the florets. Glumes long, subequal, the first 1- or 3-nerved, the second 3- to 5-nerved, much longer than the lemmas. Lemmas broad, firm, 7- to 9-nerved, notched at the apex and with the midnerve usually exserted as a short, flat mucro. Basic chromosome number, $x = 6$ or 12.

A genus of a single species, *Sieglingia decumbens* (L.) Bernh., this native to Europe and introduced in Washington and California and on the East Coast in Newfoundland and Nova Scotia.

177. *Schismus* Beauv.

Low, tufted annuals with narrow, short blades and small, usually contracted panicles. Spikelets several-flowered, the uppermost floret usually reduced. Rachilla disarticulating above the glumes and between the florets. Glumes subequal, large, lanceolate, several-nerved, slightly shorter than to about equaling the terminal floret. Lemmas several-nerved, broadly rounded and notched at the apex, the midnerve extended into a short mucro. Palea broad, 2-keeled, as long as the lemma or slightly shorter. Basic chromosome number, $x = 6$.

Species about five, native to Europe, Asia, and Africa. Adventive in southwestern United States are *Schismus barbatus* (L.) Thell. (Fig. 5-123), ranging from western Texas to southern California, and *S. arabicus* Nees, in Arizona, Nevada, and southern California. Growing as cool-season annuals in open, often disturbed sites, these tufted grasses have become locally abundant in California and Arizona.

Tribe 25. Centosteceae

178. *Chasmanthium* Link

Moderately tall perennials, some rhizomatous, with broad, flat blades. Ligule a fringe of hairs or a hyaline, ciliate membrane. Inflorescence open or

Fig. 5-123. *Schismus barbatus*. Plant and spikelet. (From Gould, 1951.)

Fig. 5-124. *Chasmanthium latifolium*. Plant, spikelet, and floret. (From Gould and Box, 1965.)

Fig. 5-125. *Chasmanthium sessiliflorum*. Inflorescence and spikelet.

contracted, the spikelets long- or short-pediceled. Spikelets 2- to many-flowered, the lower 1 to 4 sterile. Disarticulation above the glumes and between the florets. Glumes subequal, shorter than the lemmas, acute to acuminate, 3- to 7-nerved, laterally compressed and keeled, the keel serrulate. Lemmas 5- to 15-nerved, compressed-keeled, the keel serrate or ciliate. Palea about as large as the lemma, bowed out at the base, 2-keeled, the keels serrate-winged. Flowers perfect, with a single stamen and 2 fleshy, cuneate, lobed-truncate, 2- to 4-nerved lodicules. Ovary glabrous, with a single style and 2 plumose stigmas. Caryopsis ovate to elliptic, laterally compressed, with an embryo less than half the length of the grain. Basic chromosome number, $x = 12$.

As delimited by Yates (1966c), a North American genus of five woodland species, these mostly in the southeastern United States. Hitchcock (1951) included these grasses in *Uniola*, but, as has been discussed in the section on *Uniola*, they differ in many characteristics. The names of the species as transferred by Yates are *Chasmanthium latifolium* (Michx.) Yates (Fig. 5-124), *C. sessiliflorum* (Poir.) Yates (Fig. 5-125), *C. laxum* (L.) Yates, *C. ornithorhynchum* (Steud.) Yates, and *C. nitidum* (Baldw. ex Ell.) Yates.

LITERATURE CITED

Ali, M. A. 1967. The *Echinochloa crusgalli* complex in the United States. Ph.D. diss., Texas A&M Univ., College Station, Tex.

Anderson, D. E. 1961. Taxonomy and distribution of the genus *Phalaris*. *Iowa State J. Sci.*, 36:1–96.

———. 1974. Taxonomy of the genus *Chloris* (Gramineae). *Brigham Young Univ. Sci. Bull.*, 19:1–133.

Anton, A. M., and A. T. Hunziker. 1978. El genero *Munroa* (Poaceae): sinopsis morfologica y taxonomica. *Boletin de la Academia Nacional de Ciencias*, 52: 229–252.

Baker, W. H. 1964. Notes on the flora of Idaho. IV. *Leafl. West. Bot.*, 10:108–110.

Banks, D. J. 1966. Taxonomy of *Paspalum setaceum* (Gramineae). *Sida*, 2:269–284.

Baum, B. R. 1967. Kalm's specimens of North American grasses: Their evaluation for typification. *Can. J. Bot.*, 45:1845–1852.

Beetle, A. A. 1943. The North American variations of *Distichlis spicata*. *Bull. Torrey Bot. Club*, 70:638–650.

———. 1948. The genus *Aegopogon* Humb. & Bonpl. *Univ. Wyo. Publ.*, 13:17–23.

———. 1955a. The grass genus *Distichlis*. *Revista Agr. Agron.*, 22:86–94.

———. 1955b. *Wheatgrasses of Wyoming*. Wyo. Agr. Exp. Sta. Bull. 336.

Black, G. A. 1963. Grasses of the genus *Axonopus*. *Advanc. Front. Sci.*, 5:vi–186.

Blake, S. T. 1958. New criteria for distinguishing genera allied to *Panicum* (Gramineae). *Proc. R. Soc. Queensland*, 70:15–19.

———. 1969. Taxonomic and nomenclatural studies in the Gramineae. *Proc. R. Soc. Queensland*, 81:1–26.

Boivin, B., and D. Löve. 1960. *Poa agassizensis*: A new prairie bluegrass. *Le Naturaliste Can.*, 87:173–180.

Bowden, W. M. 1958. Natural and artificial × *Elymordeum* hybrids. *Can. J. Bot.*, 36:101–123.

———. 1962. Cytotaxonomy of the native and adventive species of *Hordeum*, *Eremopyrum*, *Secale*, *Sitanion*, and *Triticum* in Canada. *Can. J. Bot.*, 40: 1675–1711.

———. 1964. Cytotaxonomy of the species and interspecific hybrids of the genus *Elymus* in Canada and neighboring areas. *Can. J. Bot.*, 42:547–601.

———. 1965. Cytotaxonomy of the species and interspecific hybrids of the genus *Agropyron* in Canada and neighboring areas. *Can. J. Bot.*, 43:1421–1448.

Boyle, W. S. 1945. A cytotaxonomic study of the North American species of *Melica*. *Madroño*, 8:1–26.

———, and A. H. Holmgren. 1955. A cytogenetic study of natural and controlled hybrids between *Agropyron trachycaulum* and *Hordeum jubatum*. *Genetics*, 40:539–545.

Brown, R. H., and W. V. Brown. 1975. Photosynthesis characteristics of *Panicum milioides*, a species with reduced photorespiration. *Crop. Sci.*, 15:681–685.

Brown, W. V. 1949. A cytological study of cleistogamous *Stipa leucotricha*. *Madroño*, 10:97–107.

———. 1952. The relation of soil moisture to cleistogamy in *Stipa leucotricha*. *Bot. Gaz.*, 113:438–444.

———. 1977. The Kranz syndrome and its subtypes in grass systematics. *Mem. Torrey Bot. Club*, 23:1–97.

———, and G. A. Pratt. 1960. Hybridization and introgression in the grass genus *Elymus*. *Amer. J. Bot.*, 47:669–676.

———, and B. N. Smith. 1975. The genus *Dichanthelium* (Gramineae). *Bull. Torrey Bot. Club*, 102:10–13.

Calderón, C. E., and T. R. Soderstrom. 1980. The genera of Bambusoideae (Poaceae) of the American continent: Keys and comments. *Smithsonian Contr. Bot.*, 44:1–27.

Celarier, R. P.; J. M. J. de Wet; and W. L. Richardson. 1961. Species relationships in *Dichanthium*. I. Hybrids between *D. caricosum*, *D. aristatum*, and *D. annulatum*. *Phyton. Buenos Aires*, 16:63–67.

Chambers, K. L., and L. J. Dennis. 1963. New distributions of four grasses in Oregon, *Madroño*, 17:91–92.

Chippendahl, L. K. A. 1955. A guide to the identification of grasses in South Africa (part 1 of *The grasses and pastures of South Africa*). Cape Times Limited, Parow., C.P., pp. 7–527.

Church, G. L. 1949. A cytotaxonomic study of *Glyceria* and *Puccinellia*. *Amer. J. Bot.*, 36:155–156.

Clark, C. A., and F. W. Gould. 1975. Some epidermal characteristics of palea of *Dichanthelium*, *Panicum*, and *Echinochloa*. *Amer. J. Bot.*, 62:743–748.

———, and ———. 1978. *Dichanthelium* subgenus Turfosa. (Poaceae). *Brittonia*, 30:54–59.

Clausen, R. T. 1952. Suggestion for the assignment of *Torreyochloa* to *Puccinellia*. *Rhodora*, 54:42–44.

Clayton, W. D. 1967. Studies in the Gramineae: XIII. *Kew Bull.*, 21:99–110.

———. 1968. The correct name of the common reed., *Taxon*, 17:157–158.

Covas, G. 1949. Taxonomic observations on the North American species of *Hordeum. Madroño*, 10:1–121.

Crampton, B. 1959. The grass genera *Orcuttia* and *Neostapfia*: A study in habit and morphological specialization. *Madroño*, 15:97–110.

Cutler, H. C., and E. Anderson. 1941. A preliminary survey of the genus *Tripsacum. Ann. Mo. Bot. Gard.*, 28:249–269.

Davidse, G. 1978. A systematic study of the genus *Lasiacis* (Gramineae: Paniceae). *Ann. Missouri Bot. Gard.*, 65:1133–1254.

Decker, H. F. 1964. An anatomic-systematic study of the classical tribe Festuceae (Gramineae). *Amer. J. Bot.*, 51:453–463.

DeLisle, D. G. 1963. Taxonomy and distribution of the genus *Cenchrus. Iowa State J. Sci.*, 37:259–351.

de Wet, J. M. J. 1954. The genus *Danthonia* in grass phylogeny. *Amer. J. Bot.*, 41:204–211.

———. 1956. Leaf anatomy and phylogeny in the tribe Danthonieae. *Amer. J. Bot.*, 43:175–182.

———. 1978. Systematics and evolution of *Sorghum* section *Sorghum* (Gramineae). *Amer. J. Bot.*, 65:477–484.

Dewey, D. R. 1961. Hybrids between *Agropyron repens* and *Agropyron desertorum. J. Hered.*, 52:13–21.

———. 1963. Natural hybrids of *Agropyron trachycaulum* and *Agropyron scribneri. Bull. Torrey Bot. Club*, 90:111–122.

———. 1964a. Genome analysis of *Agropyron repens* × *Agropyron cristatum* synthetic hybrids. *Amer. J. Bot.*, 51:1062–1068.

———. 1964b. Synthetic hybrids of New World and Old World agropyrons. I. *Agropyron spicatum* × diploid *Agropyron cristatum. Amer. J. Bot.*, 51:763–769.

———. 1965a. Morphology, cytology, and fertility of synthetic hybrids of *Agropyron spicatum* × *Agropyron dasystachyumriparium. Bot. Gaz.*, 126:269–275.

———. 1965b. Synthetic hybrids of New World and Old World agropyrons. II. *Agropyron riparium* × *Agropyron repens. Amer. J. Bot.*, 52:1039–1045.

———. 1967a. Synthetic hybrids of New World and Old World agropyrons. III. *Agropyron repens* × tetraploid *Agropyron spicatum. Amer. J. Bot.*, 54:93–98.

———. 1967b. Synthetic hybrids of New World and Old World agropyrons. IV. Tetraploid *Agropyron spicatum* f. *inerme* × tetraploid *Agropyron desertorum. Amer. J. Bot.*, 54:403–409.

———, and A. H. Holmgren. 1962. Natural hybrids of *Elymus cinereus* × *Sitanion hystrix. Bull. Torrey Bot. Club*, 89:217–228.

Dore, W. G. 1969. *Wild-rice*. Can. Dept. Agric. Publ. 1393.

Emery, W. H. P. 1957a. A cyto-taxonomic study of *Setaria macrostachya* (Gramineae) and its relatives in the southwestern United States and Mexico. *Bull. Torrey Bot. Club*, 84:95–105.

———. 1957b. A study of reproduction in *Setaria macrostachya* and its relatives in the southwestern United States and Mexico. *Bull. Torrey Bot. Club*, 84:106–121.

———. 1967. The decline and threatened extinction of Texas wildrice (*Zizania texana* Hitchc.). *Southw. Naturalist*, 12:203–204.

————. 1977. Current status of Texas wildrice. *Southw. Naturalist*, 22:393–394.

Erdman, K. S. 1965. Taxonomy of the genus *Sphenopholis*. *Iowa State J. Sci.*, 39: 259–336.

Estes, J. R.; R. J. Tyrl; and J. N. Brunken (eds.). 1982. *Grasses and grasslands: Systematics and ecology*. University of Oklahoma Press, Norman.

Fairbrothers, D. E. 1952. A cytotaxonomic investigation within the genus *Echinochloa*. M.S. thesis, Cornell Univ., Ithaca, N.Y.

————. 1953. Relationships in the capillaria group of *Panicum* in Arizona and New Mexico. *Amer. J. Bot.*, 40:708–714.

Fernald, M. L. 1950. *Gray's Manual of Botany*, 8th ed. American Book Company, New York.

Gould, F. W. 1947. Nomenclatorial changes in *Elymus* with a key to the Californian species. *Madroño*, 9:120–128.

————. 1950. *Eriochloa* in Arizona. *Leafl. West. Bot.*, 6:50–51.

————. 1951. *Grasses of southwestern United States*. Univ. of Ariz. Biol. Sci. Bull. 7.

————. 1953. A cytotaxonomic study in the genus *Andropogon*. *Amer. J. Bot.*, 40:297–306.

————. 1957. New andropogons with a key to the native and naturalized species of section *Amphilophis* in the U.S. *Madroño*, 14:18–29.

————. 1959. The glume pit of *Andropogon barbinodis*. *Brittonia*, 11:182–187.

————. 1967. The grass genus *Andropogon* in the United States. *Brittonia*, 19: 68–73.

————. 1975. *The grasses of Texas*. Texas A&M University Press, College Station.

————. 1979. The genus *Bouteloua* (Poaceae). *Ann. Missouri Bot. Gard.*, 66: 348–416.

————. 1980. The Mexican species of *Dichanthelium* (Poaceae). *Brittonia*, 32: 353–364.

————; M. A. Ali; and D. E. Fairbrothers. 1972. A revision of *Echinochloa* in the United States. *Amer. Midl. Nat.*, 87:36–59.

————, and T. W. Box. 1965. *Grasses of the Texas Coastal Bend (Calhoun, Refugio, Aransas, San Patricio and northern Kleberg counties)*. Texas A&M University, College Station, Tex.

————, and C. A. Clark. 1978. *Dichanthelium* (Poaceae) in the United States and Canada. *Ann. Missouri Bot. Gard.*, 65:1088–1132.

————, and Z. J. Kapadia. 1962. Biosystematic studies in the *Bouteloua curtipendula* complex. I. The aneuploid, rhizomatous *B. curtipendula* of Texas. *Amer. J. Bot.*, 49:887–891.

————, and ————. 1964. Biosystematic studies in the *Bouteloua curtipendula* complex. II. Taxonomy. *Brittonia*, 16:182–208.

Hackel, E. 1887. Gramineae. In A. Engler and K. Prantl, *Die naturlichen Pflanzerfamilien*, vol. II, pp. 1–97. W. Englemann, Leipzig.

Hammel, B. E., and J. R. Reeder. 1979. The genus *Crypsis* (Gramineae) in the United States. *Syst. Bot.*, 4:267–280.

Harlan, J. R.; J. M. J. de Wet; W. W. Huffine; and J. R. Deakin. 1970. *A guide to the species of* Cynodon *(Gramineae)*. Ok. St. Univ. Agr. Exp. Stat. Bull. 8–673.

Henrard, J. T. 1950. *Monograph of the genus Digitaria*. Universitare Pers Leiden, Leiden.

Hitchcock, A. S. 1920. *The genera of grasses of the United States, with special reference to the economic species.* U.S. Dep. Agr. Bull. 772.

――――. 1935. *Manual of the grasses of the United States.* U.S. Dep. Agr. Misc. Publ. 200.

――――. 1936. *The genera of grasses of the United States, with special reference to the economic species,* rev. ed. U.S. Dep. Agr. Bull. 772.

――――. 1951. *Manual of the grasses of the United States,* 2d ed. (Revised by Agnes Chase.) U.S. Dep. Agr. Misc. Publ. 200.

――――, and A. Chase. 1910. The North American species of *Panicum. Cont. U.S. Nat. Herb.,* 15:1–396.

Hsu, C-C. 1965. The classification of *Panicum* (Gramineae) and its allies, with special reference to the characters of lodicules, style-bases and lemma. *J. Fc. Sci. Univ. Tokyo, Bot.,* 9:43–150.

Hubbard, C. E. 1954. *Grasses: A guide to their structure, identification, uses, and distribution in the British Isles.* Penguin Books, Bungay, Suffolk.

Hunziker, A. T., and A. M. Anton. 1979. A synoptical revision of *Blepharidachne. Brittonia,* 31:446–453.

Johnson, B. L. 1945. Natural hybrids between *Oryzopsis hymenoides* and several species of *Stipa. Amer. J. Bot.,* 32:599–608.

――――. 1960. Natural hybrids between *Oryzopsis* and *Stipa.* I. *Oryzopsis hymenoides* × *Stipa speciosa. Amer. J. Bot.,* 47:736–742.

――――. 1962a. Amphiploidy and introgression in *Stipa. Amer. J. Bot.,* 49:253–262.

――――. 1962b. Natural hybrids between *Oryzopsis* and *Stipa.* II. *Oryzopsis hymenoides* × *Stipa nevadensis. Amer. J. Bot.,* 540–546.

――――. 1963. Natural hybrids between *Oryzopsis* and *Stipa.* III. *Oryzopsis hymenoides* × *Stipa pinetorum. Amer. J. Bot.,* 50:228–234.

――――, and G. S. Rogler. 1943. A cytotaxonomic study of an intergeneric hybrid between *Oryzopsis hymenoides* and *Stipa viridula. Amer. J. Bot.,* 30:49–56.

Kapadia, Z. J., and F. W. Gould. 1964a. Biosystematic studies in the *Bouteloua curtipendula* complex. III. Pollen size as related to chromosome numbers. *Amer. J. Bot.,* 51:166–172.

――――, and ――――. 1964b. Biosystematic studies in the *Bouteloua curtipendula* complex. IV. Dynamics of variation in *B. curtipendula* var. *caespitosa. Bull. Torrey Bot. Club,* 91:465–478.

Kawano, S. 1963. Cytogeography and evolution of the *Deschampsia caespitosa* complex. *Can. J. Bot.,* 41:719–742.

Keck, D. D. 1965. *Poa.* In Munz, *A California flora.* University of California Press, Berkeley, Calif., pp. 1482–1490.

Koch, S. D. 1978. Notes on the genus *Eragrostis* (Gramineae) in the southeastern United States. *Rhodora,* 80:390–403.

――――. 1979. The relationship of three Mexican Aveneae and some new characters for distinguishing *Deschampsia* and *Trisetum* (Gramineae). *Taxon,* 28:225–235.

Komarov, V. L. (ed.). 1934. *Flora of the U.S.S.R., vol. 2, Gramineae.* (English translation, 1963, by Israel Program for Scientific Translations.) Leningrad.

Lonard, R. L., and F. W. Gould. 1974. The North American species of *Vulpia* (Gramineae). *Madroño,* 22:217–230.

Lorch, J. 1962. A revision of *Crypsis* Ait. (Gramineae.) *Bull. Res. Counc. Israel*, 11D:91–102.

McClure, F. A. 1963. A new feature in bamboo rhizome anatomy. *Rhodora*, 65: 134–136.

McNeill, J. 1979. *Diplachne* and *Leptochloa* (Poaceae) in North America. *Brittonia*, 31:399–404.

Mangelsdorf, P. C., and R. G. Reeves. 1939. *The origin of Indian corn and its relatives.* Tex. Agr. Exp. Sta. Bull. 574.

———, and ———. 1942. A proposed taxonomic change in the tribe Maydeae (family Gramineae). *Amer. J. Bot.*, 29:815–817.

———, and ———. 1959. The origin of corn. *Bot. Mus. Leafl. (Harvard Univ.)*, 18:329–440.

Mobberly, D. G. 1956. Taxonomy and distribution of the genus *Spartina. Iowa State Coll. J. Sci.*, 30:471–574.

Mohamed, A. H., and F. W. Gould. 1966. Biosystematic studies in the *Bouteloua curtipendula* complex. V. Megasporogenesis and embryo sac development. *Amer. J. Bot.*, 53:166–169.

Munz, P. A. 1958. California miscellany. *Aliso*, 4:87–100.

Nees von Esenbeck, C. G. D. 1835. Bambuseae Brasilienses: Recentsui, et alias in India Oriental: Provenientes adjecit. *Linnaea*, 9:461–494.

Nicora, E. G. 1962. Revalidacion del género de Gramineas *Neeragrostis* de la flora Norteamericana. *Rev. Argent. Agron.*, 29:1–11.

Parodi, L. R. 1940. Estudio crítico de las Gramineas austroamericanas del género *Agropyron. Rev. Mus. Plata (N.S.)*, 3 *(Secc. Bot.)*: 1–63.

———. 1954. Nota preliminar sobre el género *Monanthochloë* (Gramineae) en la Argentina. *Physis*, 20:1–3.

———. 1961. La taxonomia de las Gramineae Argentinas a la luz de las investigaciones mas recentes. *Recent Advances in Bot.*, 1:125–129.

———. 1969. Estudios sistematicos sobre las Gramineae Paniceae Argentinas y Uruguayas. *Darwiniana*, 15:65–111.

Pierce, G. J. 1978. *Griffithsochloa*, a new genus segregated from *Cathestecum* (Gramineae). *Bull. Torrey Bot. Club*, 105:134–138.

Pohl, R. W. 1959. Morphology and cytology of some hybrids between *Elymus canadensis* and *E. virginicus. Iowa Acad. Sci.*, 66:155–159.

———. 1962. *Agropyron* hybrids and the status of *Agropyron pseudorepens. Rhodora*, 64:143–147.

———. 1966. × *Elyhordeum iowense*: A new intergeneric hybrid in the Triticeae. *Brittonia*, 18:250–255.

Pyrah, G. L. 1969. Taxonomic and distributional studies in *Leersia* (Gramineae). *Iowa St. J. Sci.*, 44:215–270.

Reeder, J. R. 1953. Affinities of the grass genus *Beckmannia* Host. *Bull. Torrey Bot. Club*, 80:187–196.

———. 1965. The tribe Orcuttieae and the subtribes of the Pappophoreae (Gramineae). *Madroño*, 18:18–28.

———. 1976. Systematic position of *Redfieldia* (Gramineae). *Madroño*, 23:434–438.

————, and M. A. Ellington. 1960. *Calamovilfa*: A misplaced genus of Gramineae. *Brittonia*, 12:71–77.

————, and C. G. Reeder. 1963. Notes on Mexican grasses. II. *Cyclostachya*: A new dioecious genus. *Bull. Torrey Bot. Club*, 90:193–201.

————, and ————. 1980. Systematics of *Bouteloua breviseta* and *B. ramosa*. *Syst. Bot.*, 5:312–321.

————; ————; and J. Rzedowski. 1965. Notes on Mexican grasses. III. *Buchlomimus*: Another dioecious genus. *Brittonia*, 17:26–33.

Riggins, R. 1977. A biosystematic study of the *Sporobolus asper* complex (Gramineae). *Iowa St. J. Res.*, 51:287–321.

Rominger, J. M. 1962. *Taxonomy of* Setaria *(Gramineae) in North America*. III. Biol. Monogr. 29.

Shaw, R. B., and F. E. Smeins. 1979. Epidermal characteristics of the callus in the genus *Eriochloa* (Poaceae). *Amer. J. Bot.*, 66:907–913.

————, and ————. 1981. Anatomical and morphological characteristics of the *Eriochloa* (Poaceae) of North America. *Bot. Gaz.*, 142:534–544.

Shinners, L. H. 1954. Notes on north Texas grasses. *Rhodora*, 56:33–34.

————. 1956. Illegitimacy of Persoon's species of *Koeleria* (Gramineae). *Rhodora*, 58:93–96.

Soderstrom, T. R. 1967. Taxonomic study of subgenus *Podosemum* and section *Epicampes* of *Muhlenbergia* (Gramineae). *Contrib. U.S. Nat. Herb.*, 34:75–189.

————, and C. E. Calderón. 1979. A commentary on the bamboos (Poaceae: Bambusoideae). *Biotropica*, 11:161–172.

————, and H. F. Decker. 1965. *Allolepis*: A new segregate of *Distichlis* (Gramineae). *Madroño*, 18:33–39.

Spellenberg, R. 1975a. Synthetic hybridization and taxonomy of western North American *Dichanthelium*, group *Lanuginosa* (Poaceae). *Madroño*, 23:134–153.

————. 1975b. Autogamy and hybridization as evolutionary mechanisms in *Panicum*, subgenus *Dichanthelium* (Gramineae). *Brittonia*, 27:87–95.

Stebbins, G. L., and B. Crampton. 1961. A suggested revision of the grass genera of temperate North America. *Recent Advances in Bot.*, 1:133–145.

————, and R. Singh. 1950. Artificial and natural hybrids in the Gramineae, tribe Hordeae. IV. Two triploid hybrids of *Agropyron* and *Elymus*. *Amer. J. Bot.*, 37:388–392.

————; J. I. Valencia; and R. M. Valencia. 1946a. Artificial and natural hybrids in the Gramineae, tribe Hordeae. I. *Elymus*, *Sitanion*, and *Agropyron*. *Amer. J. Bot.*, 33:338–351.

————; ————; and ————. 1946b. Artificial and natural hybrids in the Gramineae, tribe Hordeae. II. *Agropyron*, *Elymus*, and *Hordeum*. *Amer. J. Bot.*, 33:579–586.

————, and D. Zohary. 1959. Cytogenetic and evolutionary studies in the genus *Dactylis*. I. Morphology, distribution, and interrelationships of the diploid species. *Univ. Calif. Publ. Bot.*, 31:1–40.

Swallen, J. R. 1937. The grass genus *Cathestecum*. *J. Wash. Acad. Sci.*, 27:495–501.

————. 1947. The awnless annual species of *Muhlenbergia*. *Contrib. U.S. Nat. Herb.*, 29:203–208.

————. 1950. Some introduced forage grasses of the genus *Andropogon* and related species. *Contrib. Tex. Res. Found.*, 1:15–19.

————. 1955. Flora of Guatemala. III. Grasses of Guatemala. *Fieldiana: Botany*, 24:1–390.

————. 1965. The grass genus *Luziola*. *Ann. Mo. Bot. Gard.*, 52:472–475.

————, and O. Tovar. 1965. The grass genus *Dissanthelium. Phytologia*, 11:361–376.

Tateoka, T. 1957. Notes on some grasses. IV. (7. Systematic position of the genus *Brachyleytrum.*) *J. Jap. Bot.*, 32:111–114.

————. 1959. Notes on some grasses. VII. Cytological evidence for the phylogenetic difference between *Lepturus* and *Monerma*. *Cytologia*, 23:447–451.

————. 1961. A biosystematic study of *Tridens* (Gramineae). *Amer. J. Bot.*, 48:565–573.

Terrell, E. E. 1967. Meadow fescue: *Festuca elatior* L. or *F. pratensis* Hudson? *Brittonia*, 19:129–132.

————; W. H. P. Emery; and H. E. Beaty. 1978. Observations on *Zizania texana* (Texas wildrice), an endangered species. *Bull. Torrey Bot. Club*, 105:50–57.

————, and H. Robinson. 1974. Luziolinae, a new subtribe of oryzoid grasses. *Bull. Torrey Bot. Club*, 101:235–245.

————, and W. J. Wiser. 1975. Protein and lysine contents in grains of three species of wildrice. (Zizania: Gramineae). *Bot. Gaz.*, 136:312–316.

Thieret, J. W. 1966. Synopsis of the genus *Calamovilfa* (Gramineae). *Castaneae*, 31:145–152.

U.S. Forest Service. 1937. *Range plant handbook*. Government Printing Office, Washington, D.C.

Vignal, C. 1979. Etude histologique des chloridaea: I, *Chloris*, Sw. *Andansonia*, 19:39–70.

Wagnon, H. K. 1952. A revision of the genus *Bromus*, section *Bromopsis*, of North America. *Brittonia*, 7:415–480.

Weber, W. A. 1952. *Phippsia algida* in the United States. *Rhodora*, 54:142–145.

————. 1966. *Additions to the flora of Colorado*. III. Univ. Colo. Stud., *Ser. Biol.*, no. 23, p. 2.

Wiegand, K. M. 1921. The genus *Echinochloa* in North America. *Rhodora*, 23:49–65.

Wilson, F. D. 1963. Revision of *Sitanion* (Triticeae, Gramineae). *Brittonia*, 15:303–323.

Witherspoon, J. T. 1977. New taxa and combinations in *Eragrostis* (Poaceae). *Ann. Missouri Bot. Gard.*, 64:324–329.

Yates, H. O. 1966a. Morphology and cytology of *Uniola* (Gramineae). *Southw. Naturalist*, 11:145–189.

————. 1966b. Revision of grasses traditionally referred to *Uniola*. I. *Uniola* and *Leptochloöpsis. Southw. Naturalist*, 11:372–394.

————. 1966c. Revision of grasses traditionally referred to *Uniola*. II. *Chasmanthium. Southw. Naturalist*, 11:415–455.

6
Grassland Associations in North America

Throughout central North America, grasses occur as the climax dominants. The grasslands of this area extend from central Canada south into central Mexico and from the deciduous forest on the east to the woodlands of the Rocky Mountains on the west and the warm desert of the Southwest. Separated from the main body are the Palouse Prairie of the Northwest and the Pacific Prairie of California. Collectively these regions of grass domination are spoken of as the *grassland formation of North America.*

In considering this great grassland formation, seven associations may be recognized: the True Prairie, the Coastal Prairie, the Mixed Prairie, the Fescue Grassland, the Palouse Prairie, the Pacific Prairie, and the Desert Plains Grassland (Fig. 6-1). Each association is characterized by a particular set of dominants, with the exception of the Fescue Grassland, which has a single dominant. Within each association, local areas may be dominated by different grasses, and in some sites the association dominants are conspicuous by their absence. These local differences are largely the result of variations in edaphic, topographic, and climatic factors. Some of the changes, however, are not due only to physical features of the environment but also to land-use practices and intensity of grazing. For example, heavy grazing in the past has resulted in the elimination or extreme reduction of some of the original dominant grasses from the western Mixed Prairie, which is now frequently referred to as the *short-grass prairie.*

The grassland formation of North America is the most extensive plant formation of this hemisphere and appears to have been developed under many different conditions. It occurs from sea level in the Coastal Prairie to an elevation of approximately 7,000 feet in the southern Rocky Mountains. Along the eastern edge of the formation, annual precipitation varies from 35–40 inches in the south to 25–30 inches in the north. About 75 percent of the precipitation occurs during the growing season, with a distinct wet season followed by a marked dry period. In moving west across the main body of the formation, the annual precipitation decreases to approximately 10 inches near the foothills of the Rockies.

Three distinct precipitation patterns occur within the grassland formation. Typical continental climate, with the major portion of precipitation coming during the summer, is characteristic of the True Prairie, the Coastal Prairie, the Mixed Prairie, and the Fescue Grassland. A summer-winter precipitation pattern brings limited amounts of moisture to the Desert Plains Grassland. The California and Palouse Prairies have developed in a Mediterranean-type climate, with winter precipitation followed by dry summers.

Temperature variations are equal in magnitude to those of precipitation. In the southern part of the central grasslands and in the Desert Plains

Contributed by J. D. Dodd.

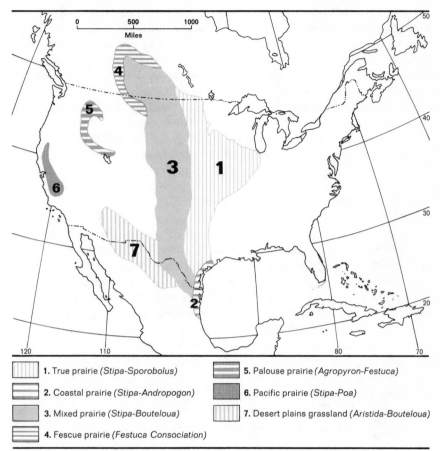

1. True prairie *(Stipa-Sporobolus)*   5. Palouse prairie *(Agropyron-Festuca)*

2. Coastal prairie *(Stipa-Andropogon)*   6. Pacific prairie *(Stipa-Poa)*

3. Mixed prairie *(Stipa-Bouteloua)*   7. Desert plains grassland *(Aristida-Bouteloua)*

4. Fescue prairie *(Festuca Consociation)*

Fig. 6-1. Grassland associations in North America.

Grassland, maximum temperatures of 120° F are not uncommon. In the northern True Prairie, the Mixed Prairie, and the Fescue Grassland, temperatures as low as −50° F have been recorded during the winter season.

Throughout the grassland formation, the grass dominants have relatively long periods of dormancy. In the north, the growing season varies from three to four months, with dormancy caused by low temperatures. In the southern portion of the formation, favorable temperatures for growth may occur throughout the year, but the herbaceous plants usually become dormant during the long, hot, dry summers. Thus, all perennial grasses of the formation have a resting period at some time during the year, as a result of low temperatures, low precipitation, or a combination of these factors.

As is true of grasses of all temperate and cooler subtropical areas, the dominants of the North American formation are of an herbaceous life-form. More or less arbitrarily, they have been placed into three groups based on culm height. Species varying from 6 to 24 inches in height at maturity are classified as *short grasses*, while those from 24 to 48 inches are *mid-grasses*. Grasses of the more mesic associations that exceed 48 inches in height are referred to as *tall grasses*. The original associations were made up of various combinations of tall grasses, mid-grasses, and short grasses. However, mismanagement of some range forage areas has resulted in the striking decrease in, or elimination of, the tall grasses and mid-grasses. This change in species composition constitutes a considerable loss in biomass production in these native habitats.

TRUE PRAIRIE (*STIPA-SPOROBOLUS*)

The True Prairie association constitutes the eastern edge of the grassland formation. It is bordered on the east by the deciduous forest and on the west by the Mixed Prairie. This once majestic prairie extends from southern Manitoba to the northern half of Texas (Allred and Mitchell, 1954; Clements and Shelford, 1939). Although the eastern edge is delimited by the deciduous forest border extending through Minnesota, Wisconsin, Illinois, and Missouri, areas of True Prairie may be found dispersed within the forest. Transeau (1935) described such a "prairie opening" located in Ohio. A broad ecotone with the Mixed Prairie extends through central Oklahoma, Kansas, and Nebraska to form the west boundary of the True Prairie. In the north it contacts the Aspen Grove region, the Fescue Grassland, and the Boreal Forest, while in the southern regions it joins the Coastal Prairie.

Climatically the True Prairie is less humid than the forest but more humid than most other associations. Precipitation varies from about 40 inches in the southeast to slightly more than 20 inches in the north. Along the western boundary the variation is from approximately 20 inches in the north to 30 inches in the south. Although adequate moisture is usually present for lush growth, late summer and fall droughts are common (Weaver, 1954).

Stipa spartea and *Sporobolus heterolepis* are the characteristic dominants of the upland communities of the True Prairie. Of equal importance in many areas is *Schizachyrium scoparium*. Other associated species of considerable importance are *Bouteloua curtipendula* and *Koeleria pyramidata*.

Three communities characterize the True Prairie uplands. *Schizachyrium scoparium* contributes approximately 55 percent of the vegetation in one community type. Locally, however, it may comprise as much as 90 percent of the grass cover. *Andropogen gerardii*, *Sporobolus heterolepsis*, *Stipa spartea*, and *Poa pratensis* are the commonly associated grasses in the central portion of the True Prairie. All are climax species except *P. pratensis*, an introduced grass that assumes importance in areas of disturbance.

From Kansas northward, a community dominated by *Stipa spartea* assumes great importance. *Stipa* is abundant in the more xeric habitats, where it comprises 50 to 80 percent of the plant cover. Important associates are *Schizachyrium scoparium*, *Andropogon gerardii*, *Koeleria pyramidata*, and *Bouteloua curtipendula*.

Sporobolus heterolepis is widespread throughout the True Prairie uplands and is the major species of the third recognized community. It occurs as a dominant on local areas of the driest upland, where it may form 80 percent or more of the species composition. The *S. heterolepis* community is smaller in area than the other two, but intermingles freely with both. Associated species are similar to the other two communities, including the dominants and major associates.

A number of grass and sedge species of limited importance occur in the upland communities. *Dichanthelium oligosanthes* var. *scribnerianum* and *D.* var. *wilcoxianum* are short-grass understory species that may be present along with the sedges *Carex pennsylvanica*, *C. meadii*, *C. festucaceae*, and *C. scoparia*. Other short grasses of importance in the True Prairie are *Eragrostis spectabilis*, *Bouteloua gracilis*, *B. hirsuta*, and *Buchloë dactyloides*. Tall grasses or mid-grasses of secondary importance are *Sporobolus asper*, prominent during drought cycles, and *Agropyron smithii*. The latter has increased since the drought of the mid-1930s (Weaver, 1954).

Lowland areas of the True Prairie association are typified by three major plant communities. On the drier lowland area, *Andropogon gerardii* dominates, usually occurring in dense stands. Limited amounts of other tall grasses, including *Sorghastrum nutans*, *Panicum virgatum*, *Stipa spartea*, and *Schizachyrium scoparium*, may be present. Tester and Marshall (1961) found that *A. gerardii*, *S. scoparium*, *S. nutans*, and *S. spartea* were of about equal importance on the Waubun Prairie of western Minnesota. The introduced, shade-tolerant *Poa pratensis* occurs interspersed between the bunches of tall grass.

Spartina pectinata is a dominant in wet lowlands throughout the True Prairie. The *S. pectinata* community represents the last stage in succession from wet land or open water to climax prairie. Dense stands of *Spartina* reduce the light intensity to 1 percent or less of full sunlight (Weaver, 1954). Other grasses are consequently excluded from this community type.

An intermediate community, between the mesic *Spartina* and drier *Andropogon* communities, is dominated by *Panicum virgatum* and *Elymus canadensis*. *Panicum virgatum* is more important from Nebraska south and southeast, while *E. canadensis* increases in importance westward and northward.

Other grasses of limited importance in lowland areas of the True Prairie are mid-grasses and tall grasses such as *Tripsacum dactyloides*, *Agrostis stolonifera* L. (*A. alba* L.), *Calamagrostis inexpansa*, and *Phalaris arun-*

dinacea. Distichlis spicata var. *stricta*, a salt-tolerant short grass, occurs on the margins of saline areas, while *Agropyron smithii* is characteristic of low-lands with a clay pan.

Since the True Prairie, with its fertile soils and usually gentle, undulating topography, lies within the region of adequate precipitation for agricultural crops, most of the original prairie has been destroyed. The areas that have not been cultivated are mainly rocky or poorly drained sites that have been fenced off and heavily grazed by livestock. As a result, the former mid-grass and tall-grass dominants have been replaced by short grasses such as *Poa pratensis*.

COASTAL PRAIRIE (*STIPA-ANDROPOGON*)

As indicated by the name, this association is located in the region around the long curve of the Gulf of Mexico. It reaches greatest prominence on the level, low-lying plains immediately along the Gulf Coast. On the north and northwest it grades into the True Prairie, and on the west and south into the Mixed Prairie. It extends from the edge of the deciduous forest subclimax in eastern Texas for about two hundred miles into northeastern Mexico.

Since the Coastal Prairie is adjacent to the Gulf, the annual rainfall of the association is relatively high. Precipitation in the Coastal Bend area, centered around Corpus Christi, ranges from about 26 to 34 inches (Gould and Box, 1965). Associated with the abundant precipitation is a high relative humidity. These factors favoring a mesic condition, however, are partially offset by extreme temperatures, high evaporation losses, and a long growing season.

The original cover of the Coastal Prairie considered of tall bunchgrasses and mid-bunchgrasses with *Bothriochloa saccharoides* var. *longipaniculata* and *Stipa leucotricha* as the major dominants. An abundance of *S. leuco-tricha* distinguishes this association from the True Prairie. The mild climate and long growing season favor the development of subtropical grasses. In addition to the two dominants noted, grasses of major importance are *Schi-zachyrium scoparium* var. *littoralis*, *Andropogon gerardii*, and *Heteropogon contortus*. *Sorghastrum nutans* and species of *Trachypogon*, *Bouteloua*, *Elyonurus*, *Coelorachis*, *Paspalum*, and *Panicum* are locally abundant. *Bou-teloua curtipendula*, *Sporobolus indicus*, and *S. purpurascens* were of limited importance in the former cover. Under favorable growing conditions, little differentiation can be made between mid-grasses and tall grasses since both obtain considerable stature.

With settlement and the cessation of grass fires, the area of the Coastal Prairie was greatly reduced by cultivation and cropping. The remaining area, for the most part, has been heavily pastured, resulting in extensive changes in the vegetation composition. At present, much of the former tall-grass and mid-grass vegetation has been replaced by native short grasses, most prominent of which are *Buchloë dactyloides, Hilaria belangeri, Bouteloua trifida, B. rigidiseta, B. hirsuta, Cenchrus incertus,* and *Panicum obtusum* (Weaver and Clements, 1938). *Cynodon dactylon,* an introduced species, is present throughout the association, in some areas almost to the complete exclusion of other grasses. Along with the invasion and establishment of short grasses, native shrubs and shrubby trees such as species of *Prosopis, Celtis, Berberis, Acacia, Zizyphus, Opuntia,* and *Zanthoxylum* have increased markedly.

Thus, the original vegetation of the Coastal Prairie association, like that of the True Prairie association, has been destroyed to a considerable extent, not only by cultivation but also by heavy grazing and the cessation of fire. Areas still supporting grassland vegetation are relatively low in forage productivity because of the influx of short grasses and woody species.

MIXED PRAIRIE (*STIPA-BOUTELOUA*)

The Mixed Prairie is the largest association of the grassland formation, extending from central Canada to south-central Texas and from western Nebraska and Kansas to the foothills of the Rocky Mountains. More specifically, in the north it includes the southern parts of Saskatchewan, and Alberta and the western two-thirds of North Dakota, South Dakota, and Nebraska. It extends northward through Wyoming into the eastern and northern half of Montana, and further south it occupies approximately the western half of Kansas and eastern third of Colorado. It covers the western third of Oklahoma and portions of western Texas. At the higher elevations and under good soil moisture conditions, it extends across eastern and northern New Mexico and northern Arizona (Weaver and Albertson, 1956).

As would be expected with an area covering such a great range of latitude, longitude, and altitude, many different climatic conditions are encountered in the Mixed Prairie. However, similarities do exist. The major portion of the precipitation is received during the spring and early summer, with periodic summer drought conditions. The annual precipitation varies from a high of about 27 inches in the south and southeast to a low of about 10 inches in the west and northwest. High evaporation losses and longer growing seasons (over 200 days) in the south have resulted in grasslands similar to the northern region, where the growing season is 120 days or less and evaporation losses are lower. Throughout the Mixed Prairie region, mean annual temperatures, as well as the length of the frost-free season, decrease as the latitude and altitude increase. There is little variation in summer temperatures, but a wide variation in winter temperatures.

Species of *Stipa* and *Bouteloua* are present throughout the association, and these genera are considered to be characteristic of the Mixed Prairie.

Coupland (1961), in a discussion of the Canadian Mixed Prairie, delimited five community types that are of frequent occurrence.

In the more mesic eastern portion of this prairie type, *Stipa spartea* var. *curtiseta* and *Agropyron dasystachyum* are the dominants, and *S. comata* and *A. smithii* are the principal associated grasses. These four mid-grass species produce approximately 75 percent of the forage. Of the minor grasses, *Bouteloua gracilis* and *Koeleria pyramidata* are most frequent. *Carex stenophylla, C. filifolia,* and other sedges are often an important element of the prairie.

In the more arid regions, *Bouteloua gracilis* occurs as the dominant, with *Stipa spartea* and *Agropyron dasystachyum* the major associated species. Other grasses of importance are *Koeleria pyramidata, Calamagrostis montanensis, Poa sandbergii,* and *Muhlenbergia cuspidata. Stipa comata* and *B. gracilis* are dominant in a limited number of areas with sandy soil.

Other communities of lesser importance occur under either pre-climax or post-climax conditions. *Stipa comata* and *Bouteloua gracilis* are dominants on xeric sandy loam, while *B. gracilis, Agropyron smithii,* and *A. dasystachyum* are the major species on soils with relatively impermeable subsoil. *Agropyron dasystachyum* and *Koeleria pyramidata* are the dominants on soils of high moisture-holding capacity and are thus the best soils for cultivation.

On two ungrazed mesas with precipitous slopes in western North Dakota, Quinnild and Cosby (1958) found that *Agropyron smithii* and *A. dasystachyum* were the most abundant grasses. Associated with these species were the mid-grasses *Stipa comata* and *S. viridula* and the short grasses *Bouteloua gracilis* and *Koeleria pyramidata.*

A number of ecological studies have been made of Mixed Prairie areas in west-central Kansas. Albertson (1937) gave extensive descriptions of the vegetation of this region, and these have been supplemented more recently by Weaver and Albertson (1956). The characteristic community in western Kansas is the normal upland type, with the short grasses *Bouteloua gracilis* and *Buchloë dactyloides* the major dominants. Typically these two species contribute up to 80 percent of the vegetation. An overstory of taller grasses is usually present, giving the grassland a Mixed Prairie appearance. Frequent in the overstory layer are *Bouteloua curtipendula, Aristida longiseta, A. purpurea, Schizachyrium scoparium,* and *Sitanion hystrix.* Associated short grasses are *Munroa squarrosa* and *Bouteloua hirsuta.* Also frequent are the sedges *Carex praegracilis, C. pennsylvanica,* and *C. heliophila.*

In the mesic lowlands, a post-climax community occurs that is dominated by *Andropogon gerardii.* As this species normally contributes 50 percent or more of the mesic lowland cover, the community has the general appearance of True Prairie. Associated mid-grasses and tall grasses are *Sorghastrum nutans, Panicum virgatum, Elymus canadensis, E. virginicus,* and species of *Agropyron*, *Bouteloua*, and *Sporobolus*. Short grasses are absent, apparently because of the lack of adequate light at the soil surface.

Schizachyrium scoparium dominates a community that topographically is located between the upland Mixed Prairie and the lowland post-climax prairie. Associated species are the dominants and major associated species from the mesic lowland and the xeric Mixed Prairie.

Allred (1956), in a description of the Mixed Prairie of Texas, listed *Bouteloua curtipendula*, *Bothriochloa* (*Andropogon*) *saccharoides*, *Sporobolus cryptandrus*, and *Aristida purpurea* as the major mid-grasses, and *Bouteloua gracilis*, *Buchloë dactyloides*, and *Bouteloua hirsuta* as the prominent short grasses. Allred noted that a number of northern Mixed Prairie species drop out in the Texas region and are replaced by species better adapted to the subtropical climate. For example, *Agropyron smithii* is replaced by *Setaria leucopila*, *Aristida purpurea* by *A. roemeriana* and *Stipa comata*, and *S. spartea* var. *curtiseta* by *S. leucotricha* and *S. neomexicana*. Other species common to the Mixed Prairie of western and southern Texas are *Hilaria mutica*, *H. belangeri*, *Bothriochloa* (*Andropogon*) *barbinodis*, and *Digitaria* (*Trichachne*) *californica*.

Limited, but important, sandhill areas occur throughout the Mixed Prairie. The largest and most significant are the sandhills of western Nebraska. Here succession is initiated in a "blowout," with the pioneer grass *Redfieldia flexuosa* the first to be established. Following stabilization, a number of grasses including *Oryzopsis hymenoides*, *Eragrostis trichodes*, *Stipa comata*, *Calamovilfa longifolia*, and *Muhlenbergia pungens* assume importance (Weaver and Albertson, 1956). *Muhlenbergia pungens* regularly dominates upland areas of relatively stable sand, and Tolstead (1942) listed *Bouteloua hirsuta* as a common associate. The bunchgrass community is characteristic of the stabilized sandhill region. *Schizachyrium scoparium*, *Andropogon hallii*, *C. longifolia*, and *S. comata* are the most important post-climax dominants.

With soil disturbance, overgrazing, and drought, a complete change occurs in the vegetation of the Mixed Prairie. As has been noted, the taller grasses drop out, and only the short grasses persist. This change has given rise to the use of the name "short-grass plains" or, more properly, "short-grass disclimax." Thus, the former "storied vegetation" becomes a single layer, usually with *Bouteloua gracilis* and *Buchloë dactyloides* as dominants in the central and northern regions, and *Hilaria belangeri* or *H. mutica* in the south and southwest.

The invasion of trees and shrubs following over-use of the prairie is of considerable importance. Most subject to invasion has been the southern Mixed Prairie, where *Prosopis*, *Acacia*, *Condalia*, *Opuntia*, and *Larrea* have changed the prairie to a brushland type in many localities.

FESCUE PRAIRIE (*FESTUCA* CONSOCIATION)

The Fescue Grassland association extends northward along the eastern foothills of the Rocky Mountains from about central Montana to central Alberta. It contacts the Mixed Prairie at the lower elevations and the Rocky Mountain forest complex at the higher elevations. Eastward the Fescue Grassland extends across Alberta to central Saskatchewan. Here in the black soils of the aspen grove region it contacts the Boreal Forest on the north and the Mixed Prairie on the south. It also intermingles with the Mixed Prairie and True Prairie in the southwest corner of Manitoba. An "island" of this association occurs in Mixed Prairie on the bench and upper slopes of the Cypress Hills of southwestern Saskatchewan and southeastern Alberta. Cosby (1965) reported scattered stands in northern North Dakota.

The principal distribution of this assocation is in regions of greater moisture efficiency than in the Mixed Prairie. The greater amount of available moisture is due to lower temperatures, lower evaporation rates, and slightly higher precipitation. A short growing season is also associated with the lower temperatures. Approximately 70 percent of the annual precipitation is received from May to September in the form of rain.

Festuca scabrella is the sole dominant of the Fescue Grassland association (Coupland and Brayshaw, 1953). Numerous authors have reported on the composition of this grassland type (Cosby, 1965; Coupland, 1961; Coupland and Brayshaw, 1953; Johnston, 1961; Moss, 1944; Moss and Campbell, 1947; Stickney, 1961). In the Cypress Hills, *F. scabrella* contributes over 55 percent of the vegetation. Species of *Carex* are important associates, as are the mid-grasses *Agropyron subsecundum*, *Danthonia intermedia*, *Helictotrichon hookeri*, and *Stipa spartea* var. *curtiseta*. In Saskatchewan, the main body of the Fescue Grassland is forty to sixty miles wide. On the southern edge of this strip, *S. spartea* var. *curtiseta* and *Koeleria pyramidata* are important, along with *Bouteloua gracilis* and *Muhlenbergia richardsonis*. Species of *Carex* are also well represented in the lower vegetative layer. Mid-grasses present are *A. trachycaulum*, *A. subsecundum*, and *H. hookeri*. Northward, there is an increase in *F. scabrella*, with a corresponding decrease in the associated species (Coupland and Brayshaw, 1953).

According to Moss (1944), in the *Festuca scabrella*–dominated grassland of Alberta, this species is associated primarily with *F. idahoensis*, *Danthonia intermedia*, and, in a few localities, *D. parryi*. Less frequent here are *Koeleria pyramidata*, *Agropyron dasystachyum*, *A. trachycaulum*, *Stipa spartea* var. *curtiseta*, and *Helictotrichon hookeri*.

Although little information is available on the response of the *Festuca scabrella* association to grazing, Johnston (1961) reported that in Alberta, light use of the range resulted in *Danthonia parryi* assuming the position of a codominant. No other minor species showed significant changes in abundance. Moss and Campbell (1947) report that with heavy grazing or mowing, the *F. scabrella* bunches become smaller, and *F. idahoensis, Stipa spartea* var. *curtiseta, Agropyron dasystachyum*, or other commonly associated grasses increase in abundance. In disturbed situations, *F. scabrella* may be reduced to a grass of minor importance or may be completely excluded from the community.

PALOUSE PRAIRIE (*AGROPYRON-FESTUCA*)

Agropyron spicatum and *Festuca idahoensis* are the major dominants of a relatively arid grassland, the Palouse Prairie, located in the northwestern United States. According to Weaver and Clements (1938), this association reaches its best development in the Palouse wheat-producing area of eastern Washington and Oregon, southern Idaho, and northern Utah.

In the Palouse Prairie, precipitation varies from 8 inches in the west to approximately 25 inches in the eastern portion. The seasonal pattern is similar to that of the Pacific Prairie, with the major portion of the precipitation coming between November and February. Much of the moisture is received in the form of snow, resulting in deep water percolation and relatively high water efficiency by the vegetation. Excellent grass production is thus obtained with respect to the total amount of annual precipitation (Clements and Shelford, 1939). During the growing period June, July, and August, however, rainfall seldom is sufficient to be of benefit to the vegetation (Stoddart, 1941). Coupled with this low precipitation during the growing season are low relative humidities and high temperatures. Because of these conditions, the growth of grasses in the Palouse Prairie takes place in the spring, with summer dormancy due to lack of soil moisture and winter dormancy due to low temperatures.

The original bunchgrass type of vegetation of the Palouse Prairie resembled that of the Pacific Prairie. Formerly it probably connected with the Pacific Prairie in the valleys of northern California, but this is not evident now, possibly because of heavy pasturing of livestock. Other plant associations which are in contact are the Cold Desert association on the south, the Fescue Grassland on the north, and the Montane Forest associations at higher elevations.

Agropyron spicatum was of widespread distribution and importance in the original climax cover. *Festuca idahoensis* was abundant in most regions, particularly at the higher elevations. *Elymus cinereus* was frequent in the moist valleys. Important associated species were *Poa sandbergii, Stipa comata, S. occidentalis, Koeleria pyramidata, A. pauciflorum*, and *A. smithii*.

Stoddart (1941), working in northern Utah, described the Palouse Prairie vegetation of that region. Originally, in the drier areas *Artemisia tridentata* provided a shrub cover, with *Agropyron spicatum* as the major understory grass. In many areas, *A. spicatum* comprised up to 95 percent of the herbaceous cover. *Poa sandbergii* and *Koeleria pyramidata* were secondary grass species, and with an increase in soil moisture *Elymus cinereus* also became prominent.

Subclimax communities dominated by *Oryzopsis hymenoides, Stipa comata, Poa sandbergii, Aristida longiseta,* and *Sporobolus cryptandrus* occurred on sandy and gravelly soils. *Elymus cinereus* and *Agropyron smithii* were frequent in communities of moist bottomlands and slopes of heavy clay soil.

Because of cultivation and heavy grazing practices, the boundaries, or ecotones, of the Palouse Prairie are difficult to locate. Much of the uncultivated portion of this association is now dominated by *Artemisia tridentata,* and a close resemblance to the Cold Desert or Sagebrush association exists. The understory grass species are *Bromus tectorum, B. japonicus,* and other annuals (Clements and Shelford, 1939; Stoddart, 1941; Weaver and Clements, 1938). With the invasion and establishment of these annual grasses, the association now has the general appearance of the Pacific Prairie association, except for the abundance of *A. tridentata.*

PACIFIC PRAIRIE (*STIPA-POA*)

This association might well be called the California Prairie, a name frequently used, since it is entirely confined to California and Lower California. Basically this association is composed of two grassland communities, differentiated by both climate and dominant species (Burcham, 1957). The larger community is located in the Central Valley of California, the associated warm valleys of the Coast Ranges, and the coastal areas of central and southern California and northern Mexico. Precipitation comes mostly during the winter months, and the summer period is hot and dry. In the Central Valley of California, precipitation ranges from 10 to 20 inches per year. Coupled with this low to moderate amount of annual precipitation are high summer temperatures and a long growing season (from about 277 to 305 days per year). This climate, often referred to as "Mediterranean," is not conducive to summer growth, and most plants are dormant during the hot, dry period.

Burcham (1957) pointed out that the original dominants were perennial bunchgrasses, with *Stipa pulchra* and *S. cernua* the most important grasses in the Central Valley. Frequent associates in this area were *Elymus glaucus, Poa scabrella,* and *Muhlenbergia rigens.* All the major species had a cespitose habit, except *E. triticoides,* a sod former. In the more southern regions, *S. lepida* and *S. coronata* shared the dominance. *Koeleria pyramidata, Melica imperfecta,* and several species of *Aristida* were common associates. On low, poorly drained, salines areas, *Distichlis spicata* var. *stricta* and *Sporobolus airoides* occurred as dominants.

The second community of the Pacific Prairie association is the North Coastal Prairie, a discontinuous grassland of medium and high altitudes. This has cooler temperatures and higher rainfall than the Central Valley. At the high altitudes, the average annual precipitation is as much as 50 inches.

The North Coastal Prairie differs significantly in species composition from the Central Valley community. *Danthonia californica, Deschampsia caespitosa, Festuca occidentalis, F. idahoensis,* and *Calamagrostis nutkaensis* were the original perennial grass dominants (Burcham, 1957).

Although all the grassland formations have been altered considerably since settlement, the changes in the Pacific Prairie association are greater than those of any of the others. The original perennial, mid-grass dominants have been replaced for the most part by annual grasses, including *Avena fatua, A. barbata, Bromus mollis, B. diandrus, B. rubens, Hordeum maritimum, H. murinum, H. pusillum, Vulpia myuros,* and *V. megalura.* Coupled with this change has been a corresponding decrease in productivity. The annual grasses now dominate the area so completely that present grazing management systems are based on their production capacities. Little or no attempt is being made to restore the former perennial grass cover of the Pacific Prairie by land-management practices.

DESERT PLAINS GRASSLAND (*ARISTIDA-BOUTELOUA*)

The Desert Plains Grassland association occupies much of southwestern United States and north-central Mexico. Whitfield and Beutner (1938) report it to include grasslands from southwestern Texas westward through central and southern New Mexico and Arizona below about 6,000 feet elevation. On the west, it meets the Warm Desert formation, while the Mixed Prairie, Woodland, and Montane associations form an ecotone on the north. In eastern New Mexico and Texas the Desert Plains Grassland is bounded by the Mixed Prairie. Although the original eastern boundary is now difficult to establish, Whitfield and Beutner (1938) concluded that the Desert Plains Grassland does not extend far enough eastward in Texas to contact either the Coastal or the True Prairie. However, present vegetation does indicate the possibility of such a connection.

Climatically, the Desert Plains Grassland is the driest of all grassland associations, with the average annual precipitation varying from 11 to 17 inches. Normally, precipitation is received during two periods of the year, summer and winter, with the greatest amounts in July or August and December or January. The summer precipitation comes as local showers of short duration and high intensity. The winter precipitation is more widespread, of longer duration, and of less intensity. Snow seldom occurs below 3,000 feet but becomes a factor at the higher elevations. Maximum temperatures are high throughout the association, particularly below 3,000 feet, where daytime temperatures often exceed 100° F. Temperatures reach 90° F even at the higher elevations. The combination of high temperatures and low relative humidity results in high rates of water loss through evaporation and transpiration. It has been determined that annual water losses in excess of 77 inches occur over most of the association (Whitfield and Beutner, 1938).

Two community types characterize the Desert Plains Grassland. *Bouteloua eriopoda* and *Hilaria mutica* are the major dominants of a community occupying the lower elevations. Prominent associated species are *B. rothrockii*, *Aristida divaricata*, and *A. purpurea*, all of which are short grasses. *Muhlenbergia porteri* and *Sporobolus cryptandrus* are mid-grasses of importance. The grasses of this community occur in a scattered cover of *Larrea tridentata*, creosotebush, and other desert shrubs (Whitfield and Beutner, 1938). Whitfield and Anderson (1938) reported that few of the former dominant grasses still occur in this community type. Intensive grazing practices have resulted in a shift from perennial grasses to annual grasses and perennial woody plants. At present, *B. aristidoides*, *B. barbata*, and annual species of *Aristida* constitute the main herbaceous species. *Hilaria mutica* may occur to the almost complete exclusion of other plants on the flood plains. With heavy grazing, however, the sod-forming short grass *Scleropogon brevifolius* invades these areas (Whitfield and Anderson, 1938). In alkaline areas, the typical variety of *Sporobolus airoides* and *S. airoides* var. *wrightii* dominate.

From about 3,500 to 5,000 feet elevation, *Hilaria belangeri* and *Bouteloua gracilis* are the major grasses, associated with other short grasses, such as *B. hirsuta*, *B. eriopoda*, *H. mutica*, and species of *Aristida*. Midgrasses are *Stipa neomexicana*, *B. curtipendula*, and *Muhlenbergia porteri*. Semidesert shrubs and low trees, such as *Prosopis glandulosa* and species of *Acacia*, are frequent.

A post-climax community, dominated by mid-grasses, was originally prominent at the higher elevations in areas of greater soil moisture. Climax vegetation of such areas was primarily of perennial grasses including *Bouteloua curtipendula*, *Bothriochloa saccharoides*, *Heteropogon contortus*, *Muhlenbergia emersleyii*, and *Koeleria pyramidata* (Whitfield and Beutner, 1938). Disturbance has resulted in the invasion, establishment, and dominance by annual species of *Bouteloua* and *Aristida*. These disclimax species are frequently referred to as "six-week grasses" (Clements and Shelford, 1939). As in the case of the lowland community, woody shrubs have generally increased and now dominate much of the community.

LITERATURE CITED

Albertson, F. W. 1937. Ecology of mixed prairie in west central Kansas. *Ecol. Monogr.*, 7:481–547.

Allred, B. W. 1956. Mixed prairie in Texas. In Weaver and Albertson, *Grasslands of the Great Plains*, pp. 267–283.

———, and H. C. Mitchell. 1954. *Major plant types of Arkansas, Louisiana, Oklahoma and Texas*. U.S. Dep. Agr. Soil Conserv. Serv.

Burcham, L. T. 1957. *California range land*. Divis. Forest. Dep. Natur. Resources. Sacramento, Calif.

Clements, F. E., and V. E. Shelford. 1939. *Bio-ecology*. John Wiley & Sons, Inc., New York.

Cosby, H. E. 1965. Fescue grassland in North Dakota. *J. Range Manag.*, 18:284–285.

Coupland, R. T. 1961. A reconsideration of grassland classification in the Northern Great Plains of North America. *J. Ecol.*, 49:135–167.

———, and T. C. Brayshaw. 1953. The fescue grassland in Saskatchewan. *Ecology*, 34:386–405.

Gould, F. W., and T. W. Box. 1965. *Grasses of the Texas Coastal Bend (Calhoun, Refugio, Aransas, San Patricio and northern Kleberg counties)*. Texas A&M University Press, College Station, Texas.

Johnston, A. 1961. Comparison of lightly grazed and ungrazed range in the fescue grassland of southwestern Alberta. *Can. J. Plant Sci.*, 41:615–622.

Moss, E. H. 1944. The prairie and associated vegetation of southwestern Alberta. *Can. J. Res. C.*, 22:11–31.

———, and J. A. Campbell. 1947. The fescue grassland of Alberta. *Can. J. Res. C.*, 25:209–227.

Quinnild, C. L., and H. E. Cosby. 1958. Relicts of climax vegetation on two mesas in western North Dakota. *Ecology*, 39:29–32.

Stickney, P. F. 1961. Range of rough fescue (*Festuca scabrella* Torr.) in Montana. *Proc. Mont. Acad. Sci.*, 20:12–17.

Stoddart, L. A. 1941. The Palouse Grassland Association in northern Utah. *Ecology*, 22:158–163.

Tester, J. R., and W. H. Marshall. 1961. *A study of certain plant and animal interrelations on a native prairie in northwestern Minnesota*. Minn. Mus. Natur. Histor. Occas. Paper no. 8.

Tolstead, W. L. 1942. Vegetation of the northern part of Cherry County, Nebraska. *Ecol. Monogr.*, 12:255–292.

Transeau, E. N. 1935. The prairie peninsula. *Ecology*, 16:423–437.

Weaver, J. E. 1954. *North American Prairie*. Johnson Publishing Co., Lincoln, Nebr.

———, and F. W. Albertson. 1956. *Grasslands of the Great Plains*. Johnson Publishing Co., Lincoln, Nebr.

———, and F. E. Clements. 1938. *Plant ecology*. McGraw-Hill Book Company, New York.

Whitfield, C. J., and H. L. Anderson. 1938. Secondary succession of the desert plains grassland. *Ecology*, 19:171–180.

———, and E. L. Beutner. 1938. Natural vegetation in the desert plains grassland. *Ecology*, 19:26–37.

APPENDIX A

The Herbarium and the Preparation and Handling of Grass Specimens

The place of the herbarium and the value of herbarium specimens in many fields of natural science are well known. Some of the multiple uses of dried, pressed, permanently preserved, and properly labeled plant specimens are as follows:

1. *The primary basis of taxonomic studies.* Not only do herbarium specimens provide series of plant materials for studies of grass morphology, but they also furnish material for the observation of the microscopic structure of stems, roots, leaves, and flowers.
2. *Voucher specimens.* Permanent specimen mounts validate records of plant structure and form and of taxonomic identification. They are of special importance as voucher records in investigations of chromosome numbers and polyploid series. Type specimens, upon which the names of species and varieties are based, are the most valuable of the voucher specimens.
3. *Records of distribution.* Plant distributions as reported in floras and checklists are, for the most part, based on herbarium records.
4. *Plant identification in classroom studies.* The "teaching herbarium" is an essential tool in many courses of plant study, including taxonomy, ecology, range management, and crop science. Agrostology is especially dependent upon mounted specimens for demonstration of plant characters and for the representation of species, genera, and other taxa.
5. *Identification of plant materials.* Much demand is placed on the herbarium in respect to identification of range, crop, ornamental, and weedy plants. In few regions of the United States are the taxonomic publications adequate to provide the sole basis for specimen identification.
6. *Records of host plants.* The value of host plant records in studies of such organisms as rusts, smuts, fungi, lichens, and algae, as well as animal organisms such as aphids, beetles, and scale insects, has not generally been recognized, but is becoming increasingly evident.

GRASS SPECIMEN COLLECTION

Equipment necessary for field collection of grass specimens includes a digging implement, a plant press, and a field notebook. Various tools, ranging from an ordinary table knife to weed digger, trowel, geological pick, and small shovel, are used in digging out grass root systems. Experience has shown that the geological pick is most satisfactory when working in rocky soil. Although grass roots have little diagnostic value, the base of the shoot system is highly important in determining the annual or perennial nature of the plant, the development of rhizomes, the pattern of basal branching, and the character of basal leaves.

The plant press should be of standard size and light enough for convenient handling (Fig. A-1). As the conventional herbarium mounting sheet is 11½ by 16½ inches, grass specimens need to be folded to a size of no more than 12 to 14 inches in length. Press frames, however, should be 12 by 17 or 12 by 18 inches. Plant specimens are dried in a single fold of newspaper (Fig. A-2) and pressed between absorbent blotters or corrugated cardboard. Specimens are kept in the newspaper folder until they are mounted on herbarium sheets to become part of the permanent herbarium file. Premature mounting on inferior paper or with improper adhesives destroys much of the value of the specimen as a permanent record.

Plants of a single species collected at the same time in a given area are entered in the field book under one number. If special designation is needed for specific plants of the collection, then letters should be used with the numbers. For example, special plants of the collection 309 would be designated 309a, 309b, 309c, etc. Such designations are desirable in respect to

Fig. A-1. The standard plant press, complete with press frames, straps, cardboard corrugates, felt blotters, and newspaper folders for the specimens.

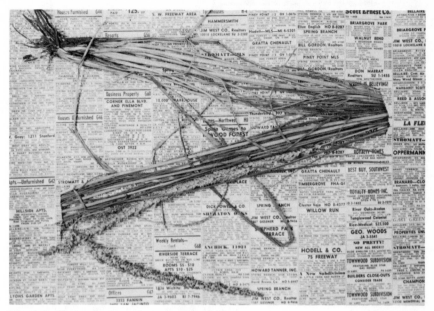

Fig. A-2. Grass specimen in newspaper folder.

voucher material for chromosome counts or similar data pertaining to individual plants. To spread the folded grass plant in the newspaper properly, it is often desirable to make slight tears or cuts in the paper as shown in Fig. A-2. Plants should not be removed from the newspaper during the drying process. Daily changing of the dryers and replacement with warm, dry blotters is necessary for the retention of color and the prevention of mold. Ordinarily, three to eight days of drying in the plant press is adequate for grasses. With artificial heat, the time necessary for drying is reduced to one or two days. In using artificial heat, however, care must be taken to ensure proper circulation of air through the plant press. Cardboard or aluminum corrugates should be used in conjunction with the blotters in this process.

The field book and field data are essential to the preparation of a good herbarium specimen record. It is recommended that all data be kept in the field book and only numbers be entered on the newspaper collection folder. The field book should be a pocket-sized, permanently bound, hard-cover notebook. Figure A-3 shows an efficient system of entering data in the field book. At each collection site, the date and the exact distance and direction from a town or other map location are noted. If all collections are from the same habitat, this is noted. Plant association, soil type and disturbance, minor habitat differences, and differences in populations of a single species are valuable data to be entered in the field book. Field-book entries should

13 Oct. 1961
8 mi. so. Archer City
Archer Co.
rolling grassy hills; brown loamy soil
9762 Tridens albescens (Vasey) Woot. & Standl.
- low, poorly drained area
9763 Panicum filipes Scribn. ?
- graded roadbank, loose soil.
9764 Aristida purpurea Nutt.
- frequent in heavily grazed pasture
2.5 mi. no. Springtown
Wise Co.
low, open hills.
9765 Andropogon ternarius Michx.
on "poor" soil — associated with Aristida and Bothriochloa saccl.

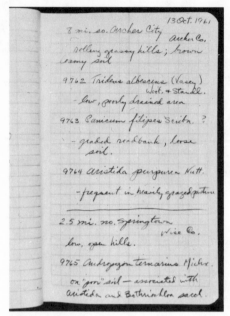

Fig. A-3. Page from field book showing suggested method of entering data.

be in pencil to prevent loss of data as a result of possible moisture damage to the book.

The plant label is an important part of the final plant specimen. The minimum information that must appear on the label includes the scientific

TRACY HERBARIUM, TEXAS A&M UNIVERSITY

PLANTS OF TEXAS

Aristida purpurea Nutt.

With Chloris cucullata, Leptochloa dubia and other perennial grasses in heavily grazed pasture, the grasses mainly in protection of cacti and shrubs. Four miles northeast of Freer.

Duval County

Charles E. Duncan 4729 8 August 1966

Fig. A-4. Suggested form for a herbarium sheet label.

plant name, the exact locality, the state and county of collection, the collector's name and collection number, and the date. More complete labels include habit data such as plant height, color, and growth characteristics that are not observable from the specimen and also habitat data such as soil type, plant associates, topography, and elevation. Professional herbarium labels are of all-rag bond paper and have printed headings. For student collections, temporary labels typed on 3- by 5-inch file cards are adequate. An example of a herbarium sheet label is shown in Fig. A-4.

APPENDIX B

The Naming of Grasses: Rules of Nomenclature

The *specific name* of any plant consists of two parts: the *generic name* and the *specific epithet*. It consists of Latin words or words that have been put into Latin form. The generic name is always capitalized, while the specific name is written with a small initial letter. For precise citation, the name of the author who first published the specific name is given, usually in abbreviated form, following the specific name; thus *Poa annua* L. was first proposed by Linnaeus in 1753. The specific name is normally written in italics, whereas that of the publishing author is not.

The names of plants are regulated by a series of formal rules set forth in the *International Code of Botanical Nomenclature*, the latest edition of which appeared in 1972. Successive editions of the *Code* are the product of deliberations at international botanical congresses. Through international agreement, each species of plant has only one valid name, and this name is understood by the scientific community everywhere in the world.

Some of the pertinent provisions of the *Code* are as follows: The correct name for a vascular plant is the oldest validly published name starting with Linnaeus' *Species Plantarum*, published in 1753, and the fifth edition of his *Genera Plantarum*, published in 1754. This principle of priority is modified, however, by the recognition of *nomina conservanda* (names that are to be conserved). Under this provision, generic names of long-standing and common use may be conserved by vote of an international congress in preference to older names that have remained obscure. Such conserved names are listed in an appendix following the *Code*, together with their rejected counterparts.

There is, however, no comparable provision for the conservation of specific names, and here the oldest valid name must be used in all cases. Subsequent names become synonyms and may not be used for any other species. Similarly, generic names that apply to a genus which has an earlier valid name are synonyms unless they are officially placed on the list of *nomina conservanda* and the older name is suppressed. When a species is shifted from one genus to another, the earliest valid epithet must be transferred to the new genus. Thus, *Panicum dactylon* L. became *Cynodon dactylon* (L.) Pers., and *Festuca kingii* S. Wats. became *Hesperochloa kingii* (S. Wats.) Rydb. and then *Leucopoa kingii* (S. Wats.) Weber. Note that the name of the original author of the epithet is placed in parentheses following the new

Contributed by Peter Raven.

name and is followed by the name of the combining author. An epithet is not transferred, however, if by doing so a name already in existence in the new genus would be duplicated. In this case, the next oldest specific epithet, if one exists, must be utilized, or a new one created. This can be illustrated by the following example. When the name *Triticum pauciflorum* Schwein., published in 1825, was recognized as the oldest valid name for a species of *Agropyron* in 1932, the species could not be called *Agropyron pauciflorum* because of the prior existence of the name *A. pauciflorum* Schur (1859), which applies to another species of the genus. Thus, the species for which *T. pauciflorum* is the oldest valid name is currently known as *A. trachycaulum* (Link) Malte (1932), based on *T. trachycaulum* Link (1833), the next validly published name that applies to this species.

In similar fashion, the correct name for an order, family, subfamily, or tribe of plants is the oldest validly published name in that rank. Names of these ranks are not written in italics. The correct name for a subspecies or variety is written in italics and must be or be based on the oldest validly published name at that level. There must always be an indication of the rank of such names, and this is achieved by the insertion of the abbreviations "subsp." or "var." between the specific and subspecific or varietal epithets. When a subspecies or variety is raised to specific rank or vice versa, it is customary but not obligatory to use the earliest valid epithet together with the proper parenthetical author citation. Examples are the change from *Panicum scribnerianum* Nash to *P. oligosanthes* Schult. var. *scribnerianum* (Nash) Fernald then to *Dichanthelium oligosanthes* (Shult.) Gould var. *scribnerianum* (Nash) Gould and the change from *Aristida dichotoma* Michx. var. *curtissii* A. Gray to *A. curtissii* (A. Gray) Nash. As an example of the opposite sort, *Sporobolus cryptandrus* (Torr.) A. Gray var. *robustus* Vasey (1890) is the oldest validly published name for the species now generally known as *S. giganteus* Nash (1898).

The application of names at all levels is fixed by the designation of *types*. Thus a species, subspecies, or variety is based on a *type specimen*, and this specimen determines the entity to which the name is permanently attached. A genus, on the other hand, is based on a *type species*, a family on a *type genus*, and so forth.

One additional aspect of the *Code's* provisions should be mentioned at this point. When a species is subdivided into subspecies or varieties, the one which includes the type specimen is designated by a repetition of the specific epithet with no author citation, as *Bouteloua curtipendula* (Michx.) Torr. var. *curtipendula*. The names of other subspecies or varieties must be accompanied by author citations, as *Bouteloua curtipendula* (Michx.) Torr. var. *caespitosa* Gould & Kapadia.

For further information concerning the construction and application of plant names, the reader is referred to the current edition of the *International Code of Botanical Nomenclature* and to the chapter on plant nomenclature in Lawrence's text on plant taxonomy (1951).

LITERATURE CITED

Lawrence, G. H. M. 1951. *Taxonomy of vascular plants.* The Macmillan Company, New York.

Linnaeus, C. 1753. *Species plantarum.* Holmiae, impensis Laurentii Salvii. 2 vols.

————. 1754. *Genera plantarum,* 5th ed. Holmiae, impensis Laurentii Salvii.

Link, H. F. 1833. *Hortus regius botanicus berolinensis,* vol. 2. Reimer, Berlin.

Malte, M. O. 1932. *The so-called* Agropyron caninum *(L.) Beauv. of North America.* Ann. Rep. 1930 Nat. Mus. Can. Bull. 68, pp. 27–57.

Nash, G. V. 1898. New or noteworthy American grasses. *Bull. Torrey Bot. Club,* 25:83–89.

Schur, P. J. F. 1859. Bericht über eine botanische Rundreise durch Siebenbürgen, . . . *Verhandl. Mittheil. Vereins Naturwiss. Hermannstadt,* 10:71–77. (See p. 77 for *Agropyrum pauciflorum.*)

von Schweinitz, L. D. 1825. In *Keat. Narr. Exped. St. Peters River,* vol. 2. London, p. 383.

Vasey, G. 1890. Gramineae. pp. 52–61. In J. M. Coulter. Upon a collection of plants made by Mr. G. C. Nealley, in the region of the Rio Grande, in Texas, from Brazos Santiago to El Paso County. *Contrib. U.S. Nat. Herb.,* 1:29–61.

GLOSSARY

Abaxial. Located on the side away from the axis.

Acentric chromosome. A chromosome lacking a centromere.

Accessory chromosomes. See *B-chromosomes*.

Achene. A small, dry, hard, indehiscent, one-locular, one-seeded fruit. See *caryopsis*.

Acicular. Needlelike.

Acuminate. Gradually tapering to a point.

Acute. Sharp-pointed, not abruptly or long-extended, but making an angle of less than 90°.

Adaxial. Located on the side toward the axis.

Adventitious embryony. Apomictic reproduction in which the embryo arises in the ovule but outside the embryo sac, either in the nucellus or in the integument; a type of agamospermy.

Adventitious root. A root that arises from any organ other than the primary root or branches thereof.

Adventive. Introduced by chance or accidental seedlings and imperfectly naturalized.

Agamospermy. Apomictic reproduction involving seed production without fusion of gametes.

Agrostology. The branch of systematic botany that deals with grasses.

Allopolyploid. Pertaining to polyploid organisms in which the chromosome sets or genomes are different, usually derived from different species.

Alveolate. Honeycombed, with angular depressions separated by thin partitions.

Anaphase. The stage in mitosis or meiosis in which the chromosomes move from the equatorial plate toward the poles; the stage between metaphase and telophase.

Anatomy. The science, or branch of morphology, that deals with plant (or animal) structure. Plant anatomy is usually interpreted as being concerned with internal structure.

Androecium. The stamens of the flower referred to collectively.

Aneuploid. Having a chromosome number that is not an exact multiple of the basic haploid number.

Annual. Of one season's or year's duration, from seed to maturity and death.

Anther. The pollen-bearing part of a stamen.

Anthesis. The period during which the flower is open and functional.

Antipodal cells. The cells of the embryo sac positioned at the opposite end from the egg cell and synergids; in a majority of angiosperms, there are three antipodal cells.

Antrorse. Directed forward or toward the apex; the opposite of retrorse.

Apical meristem. Terminal growing point.

Apiculate. Terminating abruptly in a small, short point.

Apomixis. Reproduction which involves structures commonly concerned in sexual reproduction but in which there is no actual fusion of male and female gametes.

Apospory. A form of apomixis in which the embryo sac is formed from a cell of the inner integument by a series of mitotic divisions.

Arborescent. Treelike in size and habit.

Archesporial cell. A cell that gives rise to spore mother cells.

Aristate. Awned; with a stiff bristle at the tip or terminating nerves at the back or margins of organs such as the glumes, lemmas, and paleas.

Arm-cell. Cells of the leaf mesophyll with incomplete septa or partitions extending from the walls inward; these cells characteristic of the subfamilies Bambusoideae and Oryzoideae.

Articulation. A joint or node. See *disarticulation.*

Asperity. A minute projection or "prickle-hair" of an epidermal cell.

Asperous. Rough or harsh to the touch.

Asymmetrical. Without symmetry.

Auricle. An ear-shaped appendage; a structure which occurs in pairs laterally at the base of the leaf blade in some grasses and laterally at the sheath apex in others.

Autopolyploid. A plant with three or more basic sets of chromosomes that are essentially the same; autopolyploids generally arise by doubling the chromosome number of diploids with only one type of genome or by the functioning of an unreduced gamete.

Awn. A bristle or stiff, hairlike projection; in the grass spikelet, usually the prolongation of the midnerve or lateral nerves of the glumes, lemma, or palea.

Axils. The upper angle formed between two structures such as the leaf or spikelet and stem axis.

Axillary. In an axil.

Axis (of the culm, inflorescence, etc.). The central stem or branch upon which the parts or organs are arranged.

Basic chromosome number. The lowest, actual or theoretical, haploid chromosome number, designated by x as $x = 7$, 9, or 10; usually applied to a group of species.

Beaked. Ending in a firm, prolonged, slender tip.

Bearded. Bearing stiff, usually long hairs.

B-chromosomes. Short chromosomes, with usually terminal or subterminal centromeres, that do not pair with any of the normal longer ones, the A-chromosomes. Also termed *supernumerary* or *accessory chromosomes.*

Bidentate. Having two teeth.

Bifid. Deeply two-cleft.

Bilobed. Having two lobes.

Biomass. The total quantity at a given time of living organisms of one or more species units per unit of space (species biomass) or of all the species of a community (community biomass).

Bivalent. A pair of chromosome homologues associated in synapsis.

Blade. The expanded portion of a flattened structure such as a leaf or flower petal. The blade of the grass leaf is the usually flattened, expanded portion above the sheath.

Bract. A modified leaf subtending a flower or belonging to an inflorescence; the term is sometimes used in reference to the scales of a vegetative bud or other shoot structure.

Bracteal. Of or pertaining to bracts.

Bracteate. Having bracts.

Bracteole. Small bract.

Bristle. A stiff hair or hairlike projection.

Bulb. An underground or partially underground bud with swollen, fleshy leaves, like an onion.

Bulliform cells. Usually large, thin-walled, highly vacuolated, colorless epidermal cells present in the intercostal zones of the leaf blade. These are most commonly present at the base of furrows on the adaxial surface, but may also be present on the abaxial surface.

C_3. The standard photosynthetic pathway, whereby the entire process takes place in the chloroplast present in the chlorenchyma cells of the mesophyll.

C_4. A spatially segregated photosynthetic pathway whereby CO_2 is assimilated in the mesophyll of the leaf and converted into starch within the chloroplasts found in the parenchyma bundle sheath.

Callus. The hard, usually pointed base of the spikelet (as in *Heteropogon*, *Andropogon*, and related genera) or of the floret (as in *Aristida* and *Stipa*), just above the point of disarticulation. In the former case, the callus is a portion of the rachis; in the latter, it is a portion of the rachilla. In *Eriochloa* the callus is the thickened node and remnant first glume; in *Chrysopogon* it is part of the pedicel.

Capillary. As applied to hairs, very slender and fine.

Capitate. Head-shaped; collected into a head or dense cluster.

Capsule. A dry, more or less dehiscent fruit composed of more than one carpel.

Carpel. A unit of the pistil; a simple pistil is formed from a single carpel and a compound one from two or more carpels.

Cartilaginous. Firm and tough but flexible; like cartilage.

Caryopsis. A dry, hard, indehiscent, one-seeded fruit with the thin pericarp adnate to the seed coat; the characteristic grass fruit. This differs from the achene only in the fusion of the pericarp and seed coat.

Centromere. Spindle-fiber attachment region of a chromosome.

Cespitose (or **caespitose**). In tufts or dense clumps.

Chalaza. The portion of an ovule below the region where the integuments are united to the nucellus.

Chartaceous. With the texture of writing paper.

Chasmogamy. Pollination and fertilization of open flowers or florets; the opposite of *cleistogamy*.

Chlorenchyma. Parenchyma cells containing chloroplasts; generally used in reference to the photosynthetic tissue of the leaf mesophyll.

Chromomere. The smallest identifiable particle in the chromosome thread between the leptotene and pachytene stages of cell division.

Chromonemata. The filamentous constituents of a chromosome.

Cilitate. Fringed with hairs.

Clavate. Club-shaped, thickened or enlarged at the apex from a slender base.

Cleistogamy. Pollination and fertilization within closed flowers or florets; the opposite of *chasmogamy*.

Climax. That stage in ecological succession or evolution in a biotic community which is stable and self-perpetuating.

Clone. The aggregate of individual plants produced asexually from a single sexually reproducing plant.

Coleoptile. The sheath protecting the embryonic shoot, attached at the base of the plumule on the scutellar side of the shoot axis; interpreted by many as the first vegetative leaf of the shoot.

Coleorhiza. The sheath of the monocotyledonous embryo that protects the primary root or radicle.

Collar. The outer side of a grass leaf at the junction of the blade and sheath.

Consociation. A climax plant community with a single dominant.

Continuous rachis. A rachis that does not disarticulate at the nodes at maturity.

Coriaceous. Leathery in texture.

Cordate. Heart-shaped, with a broad, notched base and a pointed tip.

Corm. The enlarged, fleshy base of a stem.

Corpus. The central core of cells just below the tunica layers of the growing stem apex.

Cortex. The tissue of a stem between the epidermis and vascular tissue in stems having a solid central core of vascular tissue; in grasses, sometimes applied to the several layers of nonpithy parenchyma tissue just inside the epidermis.

Costal. Pertaining to a rib or costa; the region over the nerves of the grass leaf.

Culm. The stem of a grass.

Cuneate. Wedge-shaped; narrowly triangular and broadest at the tip.

Cyme. A usually broad and flattish determinate inflorescence, with the central or uppermost flower maturing first; grass inflorescences are cymose in that they are determinate, with the terminal spikelet maturing first.

Cytokinesis. The cytoplasmic changes involved in mitosis and meiosis, as distinguished from the nuclear changes.

Cytology. The science concerned with the structure, function, and multiplication of cells.

Dark reaction. Part of photosynthesis where chemical energy from the light reaction and carbon from CO_2 are formed into starch.

Decurrent. Extending downward from the point of insertion.

Depauperate. Impoverished, stunted.

Dermal appendage. Appendages of the epidermis.

Dermatogen. The primordial epidermis of the apical meristem of the shoot; the cell layer that becomes the epidermis.

Diakinesis. The late prophase stage of meiosis in which the chromosomes shorten and thicken.

Dicentric chromosome. A chromosome with two centromeres.

Dioecious. Unisexual, with staminate and pistillate flowers on separate plants.

Diploid. With respect to polyploid chromosome series, having twice the basic (x) number of chromosomes. With respect to gametophyte-sporophyte generations, the zygote or fertilized egg is referred to as *diploid*.

Diplospory. Apomictic reproduction in which the archesporial cell functions directly as a spore without going into meiosis.

Disarticulation. Separation at the joints or nodes at maturity.

Distichous. Distinctly two-ranked, in two rows.

Dorsal. The back side or surface; the surface turned away from the central stalk or axis; abaxial.

Double fertilization. The characteristic angiosperm fertilization process wherein one male gamete fuses with the egg nucleus to form the zygote, and the other male gamete fuses with the usually two polar nuclei to form the primary endosperm nucleus.

Ecology. The branch of biology that deals with the relations between organisms and their environment.

Ecotone. Mixed plant communities formed by the overlapping of adjoining communities in the transitional areas.

Elliptic. In the form of a flattened circle, more than twice as long as broad.

Embryo sac. The mature female gametophyte (megagametophyte) of higher plants. The embryo sac usually consists of eight haploid nuclei, one of which, the egg, is the female gamete.

Endemic. Indigenous or native in a restricted locality.

Endosperm. Nutritive tissue arising in the embryo sac of most angiosperms following the fertilization of the two fused polar nuclei (primary endosperm nuclei) by one of the two male gametes. In most organisms the cells of the endosperm have a $3n$ chromosome number.

Entire. Undivided, the margin continuous, without teeth or lobes.

Epiblast. A small, nonvasculated flap or flange of tissue on the side of the grass embryo axis opposite the scutellum. The epiblast is characteristically present on pooid and chloridoid embryos and absent on panicoid embryos.

Erose. Irregular and uneven, as if gnawed or worn away.

Euploid. With a chromosome number that is an exact multiple of the basic (x) complement.

Exine. The outer layer (outer wall) of the two-layered wall of the microspore or pollen grain.

Extravaginal. Growth of the shoot is extravaginal when the tip breaks through the enclosing sheath; the opposite of *intravaginal*.

Fascicle. A cluster or close bunch, usually used in reference to culms, leaves, or branches of the inflorescence.

Fertile floret. A floret capable of producing fruit; a fertile floret may be perfect or pistillate.

Flabellate. Fan-shaped, broadly wedge-shaped.

Floret. As applies to grasses, the lemma and palea with the enclosed flower. The floret may be perfect, pistillate, staminate, or sterile.

Fruit. As applied to grasses, the ripened ovary; the typical grass fruit is a caryopsis.

Funiculus. The stalk by means of which the ovule is attached to the placenta.

Fusoid-cells. Large, colorless cells present in the leaf blade mesophyll of most bamboos. These cells are fusoid (spindle-shaped) only when viewed from one side.

Gametophyte. In grasses, as in all other flowering plants, two types of gametophytes are produced: the *megagametophyte*, or female gametophyte, and the *microgametophyte*, or male gametophyte. The megagametophyte consists of the embryo sac with usually eight nuclei, one of which is the female gamete, or egg. The microgametophyte consists of the pollen grain. All gametophyte nuclei have the $1n$ number of chromosomes, as opposed to the $2n$ condition in all normal sporophyte nuclei.

Gametophytic apomixis. Apomictic reproduction in which reduction division is avoided and a diploid $(2n)$ megagametophyte is produced. With a diploid gametophyte, fertilization must be avoided if the apomictic plant is to be fertile and stable.

Gene. A genetic determiner, a unit of DNA, located at a fixed position in a chromosome.

Genetics. The branch of biology that deals with heredity and variation among related organisms.

Geniculate. Abruptly bent, as at the elbow or knee joint.

Genome. The basic haploid set of chromosomes.

Genotype. The genetic constitution of an individual.

Gibbous. Swollen on one side; with a pouchlike swelling.

Glabrous. Without hairs.

Gland. A protuberance, depression, or appendage that secretes or appears to secrete a fluid.

Glandular. Glandlike or bearing glands.

Glaucous. Covered or whitened with a waxy bloom, as a cabbage leaf or a plum.

Globose. Spherical or rounded; globelike.

Glomerate. In densely contracted, headlike clusters.

Glume. The pair of bracts usually present at the base of the spikelet.

Grain. In respect to grass, the threshed or unhusked fruit, usually a caryopsis; used to refer to the mature ovary alone or the ovary enclosed in persistent bracts (palea, lemma, glumes).

Gross morphology. The external structure and form.

Guard cell. One of two adjacent, paired cells in the epidermis which by their partial separation form a stomatal pore.

Gynoecium. The female portion of a flower as a whole; the pistil of the grass flower.

Haploid. With respect to polyploid chromosome series, the basic (x) number of chromosomes. With respect to sporophyte-gametophyte generations, the gametic ($1n$) chromosome number; half the number of chromosomes of the somatic (diploid) cells.

Herb. A plant without persistent, woody stems, at least above the ground.

Herbaceous. Without persistent woody stems; dying to the ground each year.

Herbarium. A collection of dried and pressed plant specimens, usually mounted for permanent preservation; the room, building, or institution in which such a collection is kept or to which it belongs.

Heterochromatic. In cytological preparations, some chromosomes and chromosome parts stain darkly in interphase nuclei or in the prophase of division, whereas others stain only faintly or not at all. The precociously staining chromosomes or segments are referred to as *heterochromatic*, and the remainder as *euchromatic*.

Hilum. Usually defined as the scar at the point of attachment of an ovule or seed; in grass terminology, the scar at the point of attachment of the ovary.

Hirsute. Provided with rather coarse and stiff hairs, these long, straight, and erect or ascending.

Hispid. Provided with erect, rigid, bristly hairs.

Histology. The branch of biology that deals with the microscopic structure of tissues.

Homologues. The two members of each pair of essentially identical chromosomes in sexually reproducing organisms; homologous chromosomes.

Hyaline. Transparent or translucent.

Hypodermis. The inner layer of the two-layered *tunica* that covers the irregularly arranged cells of the growing shoot tip; the outer layer is the *dematogen*.

Imbricate. Overlapping, as the shingles of a roof.

Imperfect. Unisexual flowers or florets; with either male or female reproductive structures but not both.

Indurate. Hard.

Inflorescence. The flowering portion of a shoot; in grass, the spikelets and the axis or branch system that supports them, the inflorescence being delimited at the base by the uppermost leafy node of the shoot.

Intercalary meristem. An actively growing primary tissue region somewhat removed from the apical meristem; intercalary meristems are present at the base of the internodes in young grass shoots.

Intercostal. The area between the vascular bundles or nerves (of a leaf).

Internode. The portion of the stem or other structure between two nodes.

Intine. The inner layer (inner wall) of the two-layered wall of the microspore or pollen grain.

Intravaginal. Growth of a shoot is intravaginal when the tip does not break through the enclosing sheath but emerges at the top; the opposite of *extravaginal*, and the more common condition in grasses.

Inversion. A condition in which a segment of the chromosome is inverted a complete 180°.

Involucre. A circle or cluster of bracts or reduced branchlets that surround a flower or floret, or a group of flowers or florets.

Involute. Rolled inward from the edges.

Joint. The node of a grass culm; an articulation.

Keel. A prominent dorsal ridge, like the keel of a boat. Glumes and lemmas of laterally compressed spikelets are often sharply keeled; the palea of some florets is two-keeled.

Kranz syndrome. All of the anatomical and physiological characteristics common to C_4 plants.

Lacerate. Irregularly cleft or torn.

Lamina. The blade or expanded part of a leaf.

Lanceolate. Lance-shaped; relatively narrow, tapering to both ends from a point below the middle.

Lemma. The lowermost of the two bracts enclosing the flower in the grass floret.

Leptomorph rhizome. Long, slender bamboo rhizome in which every node bears a shoot bud and a verticil of roots.

Light reaction. Part of photosynthesis where light energy is converted to chemical energy (ADP to ATP and NADP to NADPH) through the process of photophosphorylation.

Lignified. Woody; converted into lignin.

Ligule. A membranous or hairy appendage on the adaxial surface of the grass leaf at the junction of sheath and blade.

Linear. Long and narrow and with parallel margins.

Locule. The cavity of an ovary or anther.

Lodicule. Small, scalelike processes, usually two or three in number, at the base of the stamens in grass flowers. Lodicules are generally interpreted as rudimentary perianth segments.

Lumen. The central cavity of a cell.

Macrohair. As applied to grasses, unicellular, usually thick-walled hairs of the leaf epidermis; variable in size, shape, and wall thickness; grading into prickle-hairs when short and pointed, and also grading into the usually smaller microhairs.

Mechanical cells. Thick-walled cells that function in support; sclerenchyma cells.

Megagametophyte. The female gametophyte. See *gametophyte* and *megaspore*.

Megaspore. The 1*n* cell that gives rise to the female gametophyte. In angiosperms, four megaspores are produced from an archesporeal cell or megasporocyte by meiosis. Three megaspores usually disintegrate, and the fourth divides and redivides to become the embryo sac or female gametophyte.

Meiosis. The two nuclear divisions resulting in a reduction of the chromosome number from diploid to haploid in spore formation.

Membranous. With the character of a membrane; thin, soft, and pliable.

Meristem. Embryonic or undifferentiated tissue, the cells of which are capable of active division.

Mesic. A moist habitat; in respect to moisture, habitats are classified as *hydrophytic* (wet), *mesic* (moist), or *xeric* (dry).

Mesocotyl. The internode between the scutellar node and the coleoptile in the embryo and seedling of a grass.

Mesophyll. The photosynthetic parenchyma between the epidermal layers of a leaf and forming the internal ground tissue.

Mestome sheath. A term applied by some to the inner sheath of the double sheath surrounding vascular bundles of grass leaves. Cells of the inner sheath are generally small and thick-walled, whereas those of the outer sheath are large and thin-walled.

Metaphase. The stage in nuclear division when the chromosomes are arranged in a single plane about midway between the two poles of the spindle; the stage of nuclear division between prophase and anaphase.

Metaxylem. Xylem that appears after the protoxylem; in plants having no secondary growth, metaxylem constitutes the only water-conducting tissue of the mature plant.

Microgametophyte. The male gametophyte. See *gametophyte*.

Microhairs. Microscopic hairs of the leaf epidermis, commonly two-celled (*bicellular microhairs*) but infrequently one-celled, as in some species of *Sporobolus*, or three- or more-celled, as in some bamboos. Bicellular microhairs vary greatly with respect to the size and shape of the two cells. Generally the basal cell is relatively thick-walled, and the apical cell is thin-walled and delicate.

Micropyle. The minute opening in the integument or integuments of an ovule through which the pollen tube ordinarily enters.

Microsporangium. A sporangium with microspores; in angiosperms, the pollen sac of the anther.

Microspore. The immature pollen grain of seed plants at the one-nucleate stage.

Mitosis. Cell division in which nuclear division, involving longitudinal splitting of the chromosomes, precedes cytoplasmic fission; sometimes restricted to the process of nuclear division alone.

Monocotyledonous. Having a single cotyledon.

Monoecious. Flowers unisexual, with male and female flowers borne on the same plant.

Monoploid. Organism with only the haploid number of chromosomes, each chromosome being represented only once in the nucleus of the somatic cell.

Monosomic. A plant in which the somatic cells possess $2n - 1$ chromosomes, one less than the normal diploid $(2n)$ complement.

Monotypic. Having a single type or representative, as a genus with only one species.

Morphology. The science that deals with the form and structure of plants (or animals). Broadly interpreted, morphology includes anatomy, histology, and the nonphysiological aspects of cytology and embryology. In plant taxonomy, morphology is concerned mainly with external form.

Mucro. A short, small, abrupt tip of an organ, as the projection of a nerve of the leaf.

Mucronate. With a mucro.

Multivalent. The association of more than two chromosomes at the first (reduction) division of meiosis, as contrasted with the usual bivalent association of the paired homologues.

Mutation. Any change in the hereditary material that effects the nature of the individual phenotype.

Nerve. A simple vein or slender rib of a leaf or bract.

Neuter. Without functional stamens or pistils.

Node. The joint of a stem, the region of attachment of the leaves.

Non-Kranz. Having all of the anatomical and physiological characteristics common to C_3 plants.

Nucleolus. A comparatively large and usually spherical body containing RNA and proteins present in the nucleus of most cells; sometimes more than one nucleolus is present.

Nucleus. A rounded protoplasmic body present in the cytoplasm of nearly all cells; to a large extent, the elements of the nucleus control cell metabolism, growth, reproduction, and inheritance.

Nullasomic. A plant developing from a $2n - 2$ zygote; such plants are usually weak and highly sterile.

Oblong. Two to three times longer than broad and with nearly parallel sides.

Obovate. Inverted ovate.

Ovate. Egg-shaped, with the broadest end toward the base.

Ovule. The megasporangium of a seed plant; an immature seed, consisting of a central mass of tissue, the nucellus, enclosed in one or two integuments and containing a single megaspore, the embryo sac.

Pachymorph rhizome. Short, thick bamboo rhizome with lateral buds giving rise only to rhizomes and with culm axes arising only from the apex of the usually pear-shaped rhizome segment.

Pachynema. The chromatin threads in the pachytene stage of meiosis.

Pachytene stage. The stage in the first (reduction) division of meiosis where the chromatin threads that have come together as bivalents in synapsis become shorter and thicker; the stage between leptotene and diplotene.

Palea. The uppermost of the two bracts enclosing the grass flower in the floret; the palea is usually two-nerved and two-keeled.

Palisade parenchyma. The portion of the mesophyll that is composed of elongated cells lying directly below and at right angles to the upper (adaxial) epidermis of the leaf; this infrequently developed in grasses.

Panicle. As applied to grasses, all inflorescences where the spikelets are not sessile or individually pediceled on the main axis.

Papillate, papillose. Bearing minute nipple-shaped projections (papillae).

Parenchyma. Tissue composed of living cells, these usually more or less isodiametric, thin-walled, and not greatly differentiated; also the cells themselves.

Parthenogenesis. Apomictic reproduction in which the embryo develops by mitotic division of the egg cell without fertilization.

Pectinate. With narrow, closely set and divergent segments or units, like the teeth of a comb.

Pedicel. The stalk of a single flower; in grasses, applied to the stalk of a single spikelet.

Peduncle. A primary flower stalk, usually applied to the stalk of a flower cluster (spikelet cluster in grasses).

Pedunculate. With a peduncle.

Pendulous. Suspended or hanging.

Perennial. Lasting year after year.

Perfect. A flower (or spikelet) with both functional male and female reproductive structures.

Perianth. The floral envelope, undifferentiated or of calyx and corolla.

Pericarp. The fruit wall developed from the ovary wall.

Petiole. A leaf stalk.

Phenotype. The aggregate of visible characteristics of an organism.

Phloem. The principal food-conducting tissue of the vascular plant, composed basically of sieve tube elements, companion cells, parenchyma, fibers, and sclereids.

Photophosphorylation. The process by which light energy is converted to chemical energy during the light reaction of photosynthesis. Two types are cyclic and noncyclic.

Phylogeny. The evolutionary history of an organism or a group of organisms.

Phytomere. The basic unit of structure of the grass shoot, an internode together with the leaf and portion of the node at the upper end and a bud and portion of the node at the lower end.

Pilose. With soft, straight hairs.

Pistil. The female (seed-bearing) structures of the flower, ordinarily consisting of the ovary and one or more stigmas and styles.

Pistillate. Having a pistil but not stamens.

Plano-convex. Flat on one side and rounded on the other.

Plicate. Folded into pleats, usually lengthwise, as in a fan.

Plumose. Feathery, having fine, elongate hairs on either side.

PMC. Pollen mother cell, which by meiosis, involving two nuclear divisions, gives rise to four microspores.

Pollen grain. Microgametophyte at the stage when it is shed from the anther; usually two- or three-nucleate.

Polyploid. A plant with three or more basic sets of chromosomes; any ploidy level above the diploid.

Prickle-hairs. Short macrohairs, also termed *asperities*.

Primary shoot. The shoot developing from the plumule of the embryo.

Promeristem. The youngest cells of an apical meristem.

Prophase. The stage of nuclear division preceding metaphase.

Prophyll (prophyllum). The first leaf of a lateral shoot or vegetative culm branch. The prophyll is a sheath, usually with two strong lateral nerves and numerous fine intermediate nerves; a blade is never developed.

Protoxylem. The xylem tissue which appears at the beginning of vascular differentiation and occupies a characteristic position in the vascular system of a given plant organ. Because the protoxylem usually matures before the organ completes its elongation, these cells are stretched and frequently completely destroyed.

Pseudogamy. Parthenogenetic reproduction in which fertilization of the polar nuclei by a male gamete is necessary for development of the endosperm and the embryo; the embryo develops from the egg cell without fertilization.

Pseudospikelet. The highly reduced inflorescence of some bamboos that consists of a shortened axis bearing a single spikelet at the apex and few to several bracts below. Secondary, tertiary, and higher-order pseudospikelets may arise in the axils of the lower bracts of pre-existing pseudospikelets.

Puberulent. Minutely pubescent.

Pubescent. Downy or hairy.

Pulvinus. A swelling at the base of a leaf or of a branch of the inflorescence.

Pungent. Terminating in a rigid, sharp point.

Pustulate. With irregular, blisterlike swellings or pustules.

Quadrivalent. The association of four homologous chromosomes at meiosis.

Raceme. As applied to grasses, an inflorescence in which all the spikelets are borne on pedicels inserted directly on the main (undivided) inflorescence axis or some are sessile and some pediceled on the main axis.

Rachilla. The axis of a grass spikelet.

Rachis. The axis of a spike or raceme or of a spicate inflorescence branch; also applied to the axis of a compound leaf.

Radicle. The primary root of the grass embryo.

Reduced floret. A staminate or neuter floret; if highly reduced, then termed a *rudimentary floret.*

Reniform. Kidney-shaped.

Rhizome. An underground stem, usually with scale leaves and adventitious roots borne at regularly spaced nodes.

Rudiment. An imperfectly developed organ or part, as the rudimentary floret or florets of some spikelets.

Rugose. Wrinkled.

Satellite. A segment of a chromosome, separated from the rest of the chromosome by a long secondary constriction.

Scabrous. Rough to the touch, usually because of the presence of minute prickle-hairs in the epidermis.

Scarious. Thin, dry, and membranous, not green.

Sclerenchyma. A tissue or complex of thick-walled cells, often lignified, whose principal function is mechanical support.

Scutellar node. The "node" of the embryo axis at which the scutellum is attached; the scutellar node is marked by the divergence at that point of vascular traces to the scutellum.

Scutellum. Haustorial tissue of the embryo located between the endosperm and the plumule-radicle axis of the embryo; considered by some to be the grass cotyledon.

Seed. A ripened ovule.

Seminal root. Literally translated, "seed root"; the primary root and all other roots that arise from embryonic tissue below the scutellar node.

Serrate. Saw-toothed, with sharp teeth pointing forward.

Serrulate. Finely serrate.

Sessile. Attached directly at the base, without a stalk.

Seta. A bristle or a rigid, sharp-pointed, bristlelike organ.

Setaceous. Bristly or bristlelike.

Sheath. The tubular basal portion of a leaf that encloses the stem, as in grasses and sedges.

Silica-cells. Short epidermal cells of the grass leaf and culm that contain silica-bodies of various shapes (round, elongate, notched, cross-shaped, saddle-shaped, etc.). Each silica-cell contains one silica-body.

Sinuous (sinuate). With a strong, wavy margin.

Somatic cells. Body cells, with $2n$ chromosomes, as opposed to reproductive cells, gametes, with n chromosomes.

Spathe. A large bract enclosing an inflorescence.

Spicate. Spikelike.

Spicule. Short, stout, pointed projection of the leaf epidermis; spicules often grade into prickle-hairs.

Spike. An inflorescence with flowers or spikelets sessile on an elongated, unbranched rachis.

Spikelet. The basic unit of the grass inflorescence, usually consisting of a short axis, the rachilla, bearing two "empty" bracts, the glumes, at the basal nodes and one or more florets above. Each floret consists usually of two bracts, the lemma (lower) and the palea (upper), which enclose a flower. The flower usually includes two lodicules (vestigial perianth segments), three stamens, and a pistil.

Sporophyte. The $2n$, spore-producing generation in the plant life cycle.

Stamen. The male organ of the flower, consisting of a pollen-bearing anther on a filiform filament. Collectively, the stamens of a flower are referred to as the *androecium*.

Staminate. Having stamens but not pistils.

Stigma. The part of the ovary or style that receives the pollen for effective fertilization.

Stipe. A stalk.

Stipitate. Having a stalk or stipe, as an elevated gland.

Stolon. A modified horizontal stem that loops or runs along the surface of the ground and serves to spread the plant by rooting at the nodes.

Stoma. Singular of *stomata*.

Stomata. Apertures or pores in the epidermis bounded by two guard cells; stomata function in the interchange of gases between the atmosphere and the intercellular spaces of the stem and leaf parenchyma.

Style. The contracted portion of the pistil between the ovary and the stigma.

Sub-. Latin prefix meaning "almost," "somewhat," "of inferior rank," "beneath."

Subsidiary cells (accessory cells). Two or more cells adjacent to the stomatal guard cells that appear to be associated functionally with them.

Subulate. Awl-shaped.

Succulent. Fleshy or juicy.

Sucker. A vegetative shoot of subterranean origin.

Supernumerary chromosomes. See *B-chromosomes*.

Synapsis. The conjugation of pairs of homologous chromosomes early in meiosis.

Synergids. The two nuclei associated with the egg at the micropylar end of the embryo sac.

Syngamy. Cell union, as that of gametes in fertilization.

Tapetum. The innermost layer of the microsporangium (pollen sac of anther) which apparently functions in the nutrition of the developing pollen mother cells.

Taxon. Any taxonomic unit, as species, genus, or tribe.

Telophase. The final stage of nuclear division; in telophase the chromosomes at each end of the spindle reorganize as a metabolic nucleus.

Terete. Cylindrical, round in cross section.

Terminal meristem (or **apical meristem**). The terminal growing point.

Tetraploid. Having four of the basic (x) sets of chromosomes.

Tetrasomic. In tetrasomics, all chromosomes of the somatic cells are in pairs, but one pair is duplicated. The chromosome number is thus $2n + 2$.

Tiller. A subterranean or ground-level lateral shoot, usually erect, as contrasted with horizontally spreading stolons and rhizomes.

Tracheid. Thick-walled cells which, together with vessels, form the water and aqueous solution conducting elements of the xylem. Tracheid cells are not fused together in long tubes, as are vessels.

Translocation. Chromosome changes in which end segments of nonhomologous chromosomes have exchanged positions.

Translucent. Allowing the passage of light rays, but not transparent.

Trisomic. Plant with $2n + 1$ chromosomes in the somatic cells.

Truncate. Terminating abruptly as if cut off transversely.

Tunica. The two layers of cells that cover the irregularly arranged cells of the corpus of the apical shoot meristem. A one-layered tunica has been reported for a number of grasses, but the accuracy of this observation has been questioned.

Turgid. Swollen, tightly drawn by pressure from within.

Type specimen. The specimen upon which the name of a plant species, subspecies, or variety is based and to which it is permanently attached.

Uncinate. Hooked at the tip.

Unilateral. One-sided, developed or hanging on one side.

Unisexual. With either male or female sex structures but not both.

Univalent. In meiosis the chromosomes normally pair to form bivalents; unpaired chromosomes are termed *univalents*.

Utricle. A small, bladdery, one-seeded fruit.

Vascular bundle. A strand composed primarily of vascular tissue traversing some part of the plant; the vascular bundles of a grass stem and the larger bundles of the leaf consist largely of xylem and phloem elements and associated parenchyma and mechanical cells partially or completely enclosed by one or two bundle sheaths.

Versatile anther. The anther attached to the filament at or near the middle, as opposed to the *basifixed* anther, which is attached at the base.

Verticil. A whorl; with three or more members or parts attached at the same node of the supporting axis.

Vestigial. Rudimentary and almost completely reduced; with only a vestige remaining.

Villous. Bearing long and soft, not matted hairs; shaggy.

Viscid. Sticky, glutinous.

Vivipary. Apomictic reproduction in which the progeny are produced by means other than seed. In grasses the term *vivipary* generally is used in reference to reproduction involving spikelet structures; reproduction by stolons, rhizomes, etc., is referred to as *vegetative reproduction*.

Xeric. Characterized by, or pertaining to, conditions of scanty moisture supply.

Xylem. The principal water-conducting tissue in vascular plants, characterized by the presence of tracheary elements.

Zygomorphic. Bilaterally symmetrical; capable of division in only one plane of symmetry.

Zygote. The cell formed by the union of two gametes; the fertilized egg.

INDEX

Page references in **boldface** describe the taxon.